Circular Economy and Sustainable Management

This book summarizes both conventional and emerging waste management approaches for fly ash, particularly with a focus on sustainability dimension – including its applicability, scope, methods, and challenges of bio-coal production. It covers the sustainability aspects of fly ash management and recent developments in methods, processes, and scope of bio-coal production as an alternative to conventional coal along with its implications for industrial ecosystems. It focuses on climate change, sustainability, and circular economy aspects of fly ash management approaches.

- Comprehensively covers all the dimensions related to coal and bio-coal fly ash and circular economy associated with it
- Discusses cradle-to-grave technology for bio-coal
- Emphasizes the efficient, economically viable, and environmentally sustainable use of coal and bio-coal fly ash
- Explores renewable fuel solutions and circular economy associated with it
- Encourages the use of bio-coal for boosting sustainability and circular economy

This book is aimed at researchers and graduate students in environmental and civil engineering, as well as those working in clean technologies.

Circular Economy and Sustainable Management
Coal and Bio-Coal Fly Ash

Edited by Siddhartha Pandey, Nitin Kumar Singh,
Tuhin Sengupta and Sanchita Roy Chowdhury

CRC Press
Taylor & Francis Group
Boca Raton London New York

CRC Press is an imprint of the
Taylor & Francis Group, an **informa** business

Designed cover image: Shutterstock

First edition published 2026
by CRC Press
2385 NW Executive Center Drive, Suite 320, Boca Raton FL 33431

and by CRC Press
4 Park Square, Milton Park, Abingdon, Oxon, OX14 4RN

CRC Press is an imprint of Taylor & Francis Group, LLC

ISBN: 9781032796864 (hbk)
ISBN: 9781032806822 (pbk)
ISBN: 9781003498124 (ebk)

DOI: 10.1201/9781003498124

Typeset in Times
by codeMantra

Contents

Preface

The idea for this edited book, *Circular Economy and Sustainable Management: Coal and Bio-Coal Fly Ash*, was conceived while reflecting on the pressing challenges faced by various industrial development corporations in India, including Maharashtra Industrial Development Corporation (MIDC), in managing fly ash. Observing the issues related to the handling of fly ash, the environmental and health impacts on local communities, and the increasing demand for alternative raw materials, it became clear that a sustainable, circular economy approach was urgently needed. Further, the global focus on sustainability and the circular economy has intensified the search for innovative ways to manage industrial by-products such as fly ash – a residue of coal and bio-coal combustion. The ongoing challenges related to the environmental impacts and disposal of fly ash have sparked increased research efforts into its potential as a valuable resource. This book addresses various aspects of managing fly ash in a sustainable and circular way, turning waste into valuable resources.

The chapters in this book explore the generation of fly ash from industrial operations, emphasizing both coal and bio-coal fly ash. The analysis of their mechanical and chemical properties forms the foundation for understanding the opportunities in transforming fly ash into functional materials. As industries move toward greener technologies, utilizing fly ash for environmental remediation and as a sustainable alternative in building materials is becoming increasingly relevant. Notable contributions in this volume include insights into the valorization of fly ash for advanced applications such as zeolites and ceramic materials, highlighting how this waste can be converted into high-value products. Life cycle assessments (LCA) of fly ash and cost analyses further underscore the economic viability of sustainable management practices. Sustainable approaches to fly ash management are discussed, with particular emphasis on vermicomposting and its benefits to soil quality. Moreover, fly ash's low-cost and efficient use as an adsorbent in removing industrial dyes from wastewater exemplifies the environmental remediation potential of this material.

Additionally, this book delves into the usage of fly ash in bioenergy production, exploring the intersection of bio-coal fly ash with global bioenergy trends. Discussions on the ecosystem services related to fly ash, the global economy of bio-coal, and policy reviews aimed at fostering international collaboration round out the book's holistic approach to the topic.

The final chapters focus on circular economy strategies, assessing fly ash management through life cycle assessment tools and exploring the economic and environmental impacts of ash utilization. This book aims to serve as a comprehensive guide for researchers, policymakers, and industry stakeholders interested in unlocking the potential of fly ash in a circular economy framework.

Through its wide-ranging discussions, this book highlights both the challenges and the untapped opportunities associated with fly ash management, demonstrating how this industrial by-product can contribute to sustainability, resource efficiency, and a greener future.

Editors' Biography

Siddhartha Pandey holds a Ph.D. in Environmental Engineering from the Indian Institute of Technology Roorkee and is a researcher at the Indian Institute of Management, Ranchi, in the field of Operations Management. His research interests include the areas of environmental sustainability, eco-tourism, circular economy, water and wastewater treatment, environmental impact assessment, and sustainable development. He has published over 25 SCI articles (h-index 10) and have contributed to nine book chapters and he has edited one international book in CRC press (Taylor & Francis).

Nitin Kumar Singh holds a Ph.D. in Environmental Engineering from the Indian Institute of Technology, Roorkee (India) and is an Associate Professor in the Chemical Engineering Department at Marwadi University, Gujarat (India). He has spent more than 12 years in the field of wastewater treatment, waste management, green synthesis and processes, cost-benefit analysis, biological air pollution monitoring, and sustainability assessment of environmental systems. As of now, he has published 37 SCI articles (h-index: 21; i-index: 36), six SCOPUS indexed publications, 19 book chapters, 12 papers in international conferences, and edited one international book in CRC press, and one book for other reputed publishers. He has presented his research work at various international platforms (Malaysia, Israel, Taiwan, United Kingdom, Nepal).

Tuhin Sengupta earned his Ph.D. in Operations Management and Quantitative Techniques from the Indian Institute of Management, Indore and is an Assistant Professor in the Indian Institute of Management, Ranchi. His research experience lies in the areas of supply chain management, disruptive technologies, supply chain resilience, and sustainable development goals. He has over 40 quality publications (h-index 15). He has five years of industry and academic experience (three years in academics (post-PhD) and two years in industry (pre-PhD)).

Sanchita Roy Chowdhury completed her PhD in the field of economics from the Indian Institute of Management, Indore. She is a research fellow at the Anglia Ruskin University, Cambridge. Sanchita's research interests include international economics, trade, supply chain management, agri-food business, and circular economy.

Contributors

Amare Ayalew Abebe
Department of Chemistry,
Debre Berhan University, Debre Birhan, Ethiopia

Brahim Achiou
Laboratory of Membranes, Materials and Environment,
Faculty of Sciences and Technologies of Mohammedia,
Hassan II University of Casablanca, Morocco

Akash Anand
Civil Engineering Department,
KIT's College of Engineering (Autonomous), Kolhapur, Maharashtra, India

Redouane Beniazza
Institute of Science, Technology & Innovation (IST&I),
Mohammed VI Polytechnic University, Ben Guerir, Morocco

Dounia Beqqour
Laboratory of Membranes, Materials and Environment,
Faculty of Sciences and Technologies of Mohammedia,
Hassan II University of Casablanca, Casablanca, Morocco

Jamal Ait Brahim
Institute of Science, Technology & Innovation (IST&I),
Mohammed VI Polytechnic University, Ben Guerir, Morocco

Majda Breida
Laboratory of Membranes, Materials and Environment,
Faculty of Sciences and Technologies of Mohammedia,
Hassan II University of Casablanca, Casablanca, Morocco

Amit Chauhan
Department of Polymer and Process Engineering,
Indian Institute of Technology, Roorkee, Uttarakhand, India

Sanjana Das
Department of Chemistry,
Indian Institute of Technology Guwahati, Guwahati, Assam, India

Surajit Das
Department of Chemistry,
Tata Institute of Fundamental Research Hyderabad, Hyderabad, Telangana, India

Deepika
Department of Botany,
Mahila Vidyalaya Degree College, Lucknow, Uttar Pradesh, India

Simran Kaur Dhillon
Department of Civil and Environmental Engineering,
University of Alberta, Edmonton, Canada

Nishat Fatima
Department of Zoology, Deen Dayal Upadhyaya Gorakhpur University,
Gorakhpur, Uttar Pradesh, India

Rahul Gautam
Department of Materials Science and Environmental Engineering,
Tampere University, Hervanta Campus, Finland

Maanit Ghodasara
Chemical Engineering Department,
Faculty of Diploma Studies, Marwadi University, Rajkot, Gujarat, India

Arnab Ghosh
University Core Research Center for Disaster-Free Safe Ocean City Construction,
Dong-A University, Busan, Republic of Korea

N. Giri
Soundarya Institute of Management and Science,
Bangalore, Karnataka, India

S. Shreemathi Giri
Soundarya Institute of Management and Science,
Bangalore, Karnataka, India

Dikshita Goswami
Department of Chemistry,
Indian Institute of Technology Guwahati, Guwahati, Assam, India

Marttin Paulraj Gundupalli
Department of Agricultural, Food and Nutritional Science, University of Alberta,
Edmonton, Alberta, Canada

Amit K. Jaiswal
School of Food Science and Environmental Health,
Technological University Dublin-City Campus, Dublin, Ireland

Charu Jhamaria
Department of Environmental Science,
IIS (Deemed to be University) Jaipur, Jaipur, Rajasthan, India

Gaurav Kabra
Jindal Global Business School,
OP Jindal Global University, Sonipat, Haryana, India

Navneet Kumar
Department of Electronics Engineering,
Hanyang University, Seoul, South Korea

Tarun Kumar
Amnex Infotechnologies Pvt. Ltd.,
Mahalanobis National Crop Forecast Centre (MNCFC), Pusa, New Delhi, India

Angela Martins
Departamento de Engenharia Química, Instituto Superior de Engenharia de Lisboa,
Lisboa, Portugal

Suranjana V. Mayani
Department of Chemistry,
Marwadi University, Rajkot, Gujarat, India

Kedar Mehta
Institute of New Energy Systems,
TH Ingolstadt, Germany

Shivam M. Mod
Department of Chemistry,
Marwadi University, Rajkot, Gujarat, India

Nelson Nunes
Centro de Química Estrutural, Faculdade de Ciências,
Institute of Molecular Sciences, Universidade de Lisboa,
Campo Grande, Lisboa, Portugal

Mouhamed Ouammou
Laboratory of Membranes, Materials and Environment,
Faculty of Sciences and Technologies of Mohammedia,
Hassan II University of Casablanca, Casablanca, Morocco

Pooja M. Pancholi
Department of Chemistry,
Marwadi University, Rajkot, Gujarat, India

Siddhartha Pandey
Department of Operations Management,
Indian Institute of Management Ranchi, Ranchi, India

Jinsub Park
Department of Electronics Engineering,
Hanyang University, Seoul, South Korea

Alpeshkumar V. Patel
Department of Chemistry,
Government Science College, Valsad, Gujarat, India

Rajkumar Patel
Energy & Environmental Science and Engineering (EESE),
Integrated Science and Engineering Division (ISED),
Underwood International College, Yonsei University, Incheon, South Korea

Naresh V. Patil
Civil Engineering Department,
Jain College of Engineering, Belgaum, Karnataka, India

G. Pranavanathan
Aerospace Department,
Indian Institute of Science, Bangalore, Karnataka, India

Priyanka Ray
Department of Chemistry,
Indian Institute of Technology Guwahati, Guwahati, Assam, India

Debopriyo Roy
Center for Language Research, School of Computer Science & Engineering
University of Aizu, Aizuwakamatsu, Japan

Akash Saharan
Jindal Global Business School,
O. P. Jindal Global University, Sonipat, Haryana, India

Satya Prakash Sahu
Technical Secretariat,
Mahanadi Coalfields Limited, Sambalpur, Odisha, India

Tuhin Sengupta
Department of Operations Management,
Indian Institute of Management Ranchi, Ranchi, Jharkhand, India

Harish Sharma
Department of Paper Technology,
Indian Institute of Technology, Roorkee, Uttarakhand, India

Shalini Shashni
Energy Research Centre,
Panjab University, Chandigarh, Punjab, India

Alain Tèebwaoga Sina
Institute of Science, Technology & Innovation (IST&I),
Mohammed VI Polytechnic University, Ben Guerir, Morocco

Anjali Singh
Department of Zoology, Deen Dayal Upadhyaya Gorakhpur University,
Gorakhpur, Uttar Pradesh, India

Keshav Singh
Department of Zoology, Deen Dayal Upadhyaya Gorakhpur University,
Gorakhpur, Uttar Pradesh, India

Nitin Kumar Singh
Department of Chemical Engineering,
Marwadi University, Rajkot, Gujarat, India

Pankaj Kumar Singh
Department of Zoology, Deen Dayal Upadhyaya Gorakhpur University,
Gorakhpur, Uttar Pradesh, India

Sandeep Singh
Jindal Global Business School,
OP Jindal Global University, Sonipat, Haryana, India

Dharmesh Sur
Chemical Engineering Department,
Faculty of Technology, Marwadi University, Rajkot, Gujarat, India

Namita K. Talapatra
Department of Polymer and Process Engineering,
Indian Institute of Technology, Roorkee, Uttarakhand, India

Rucha N. Thaker
Department of Chemistry,
Marwadi University, Rajkot, Gujarat, India

Abhinay Thakur
Division of Research and Development,
Lovely Professional University, Phagwara, Punjab, India

Arti Thanki
Pre-Health Sciences Pathway to Certificates and Diplomas,
Fleming College, Peterborough, Canada

Divya Tirva
Chemical Engineering Department,
Faculty of Technology, Marwadi University, Rajkot, Gujarat, India

Swati Verma
Centre of Energy and Environment,
VSB TU Ostrava, Czech Republic

Manish Yadav
Central Mine Planning and Design Institute,
Bhubaneswar, Odisha, India

Saad Alami Younssi
Laboratory of Membranes, Materials and Environment,
Faculty of Sciences and Technologies of Mohammedia,
Hassan II University of Casablanca, Casablanca, Morocco

Bagia Maryam Ibrahim Yunus
Department of Chemistry,
School of Science and Technology, Vanita Vishram Women's University,
Surat, Gujarat, India

1 Introduction to Various Fly Ash (Coal & Bio-Coal) Generated from Industrial Operations

Siddhartha Pandey, Tuhin Sengupta, Nitin Kumar Singh, Gaurav Kabra, Navneet Kumar and Jinsub Park

1.1 INTRODUCTION

Fly ash, a byproduct obtained by the combustion of coal and bio-coal, is an industrial waste material that has gained significant importance in various applications due to its unique properties. Fly ash is primarily composed of fine particles that are carried out of the boiler with the flue gases during the combustion process. This material is captured by pollution control equipment, such as electrostatic precipitators or bag filters before the flue gases are released into the atmosphere. The importance of fly ash lies not only in its abundance but also in its potential as a resource for various industrial processes, particularly in the construction sector (Alrefaei et al., 2021). The composition and properties of fly ash can vary significantly depending on the type of coal or bio-coal used, the combustion conditions, and the efficiency of the pollution control equipment. Typically, fly ash from coal combustion contains silica (SiO_2), alumina (Al_2O_3), and iron oxides (Fe_2O_3), while fly ash from bio-coal combustion may also contain a higher proportion of unburned carbon and other organic compounds (Wang et al., 2020). Coal-fired power plants have traditionally been the largest source of fly ash, producing millions of tons of this material annually. The combustion of coal generates two main types of ash: bottom ash, which is collected at the bottom of the furnace, and fly ash, which is lighter and carried with the flue gases (Yao et al., 2015). The increasing use of bio-coal, a renewable and more sustainable alternative to traditional coal, has led to the production of bio-coal fly ash (BFA), which has similar characteristics to coal fly ash but may differ in terms of chemical composition and potential applications (Wang et al., 2020). The production of fly ash is influenced by several factors, including the type of coal or bio-coal used, the combustion temperature, and the design of the boiler and pollution control equipment (Chindaprasirt and Rattanasak, 2010). Pulverized coal combustion (PCC), which is the most common method used in power plants, produces fly ash with a relatively

DOI: 10.1201/9781003498124-1

high proportion of fine particles. In contrast, fluidized bed combustion, often used for burning low-grade coal or bio-coal, generates fly ash with a higher content of unburned carbon and larger particle sizes (Wang et al., 2020).

Fly ash has become a valuable resource in various industrial processes due to its pozzolanic properties, which allow it to react with lime in the presence of water to form compounds with cementitious properties. This makes fly ash an ideal material for use in the construction industry, particularly in the production of concrete and cement (Fernández-Jiménez and Palomo, 2003). The use of fly ash in concrete not only improves the strength and durability of the material but also reduces the need for Portland cement, leading to significant cost savings and a reduction in carbon emissions (Provis and Bernal, 2015). The use of fly ash in cement and concrete production is one of the most well-established applications, with several standards and guidelines in place to ensure the quality and performance of fly ash-based products. Fly ash can be used as a partial replacement for Portland cement, with replacement levels typically ranging from 15% to 30%, depending on the specific requirements of the project. The inclusion of fly ash in concrete mixtures can improve workability, reduce water demand, and enhance the long-term strength and durability of the concrete (Yao et al., 2015). In addition to its use in cement and concrete production, fly ash is also used in other industrial applications, such as road construction, soil stabilization, and the manufacture of building materials (Provis and Bernal, 2014). For example, fly ash can be used as a filler material in asphalt mixtures, improving the strength and stability of the pavement. It can also be used in the production of bricks, tiles, and other building products, offering a sustainable alternative to traditional materials (Chindaprasirt and Rattanasak, 2010). The chemical composition of fly ash, particularly its high silica and alumina content, makes it suitable for use in the production of geopolymers, a class of materials that are gaining popularity as a sustainable alternative to conventional cement. Geopolymers are formed by the reaction of aluminosilicate materials, such as fly ash, with alkaline activators, resulting in a hardened material with excellent mechanical and chemical properties. The use of fly ash in geopolymer production not only reduces the demand for natural resources but also helps mitigate the environmental impact of fly ash disposal (Provis and Bernal, 2015).

1.2 ENVIRONMENTAL MANAGEMENT

The management of fly ash has become a critical issue due to the large volumes of material generated and the potential environmental risks associated with its disposal. Fly ash contains trace amounts of heavy metals, such as arsenic, lead, and mercury, which can leach into the soil and groundwater if not properly managed. The improper disposal of fly ash can also lead to air pollution, as fine particles can become airborne and contribute to respiratory problems in nearby communities (Wang et al., 2020). To address these environmental concerns, several regulatory frameworks and guidelines have been established to govern the handling, storage, and disposal of fly ash. In many countries, fly ash is classified as a non-hazardous waste, allowing it to be reused in various applications, provided it meets certain quality standards. However, the classification of fly ash can vary depending on its chemical composition and the presence of hazardous substances (Chindaprasirt and Rattanasak, 2010). The reuse of fly ash in industrial applications offers a sustainable solution to the challenges of fly ash management. By incorporating fly ash into construction materials, road bases, and other products, the need for landfill disposal is reduced, and the environmental impact of fly ash is minimized (Wang et al., 2020). Moreover, the use of fly ash in these applications can lead to significant cost savings and contribute to the conservation of natural elements. In recent years, there has been a growing interest in the development of advanced technologies for the treatment and utilization of fly ash. These technologies aim to enhance the quality of fly ash, making it suitable for use in high-value applications, such as the production of advanced materials and the extraction of valuable minerals. For example, the beneficiation of fly ash, which involves the removal of unburned carbon and other impurities, can improve its performance in cement and concrete applications. Additionally, research on the use of fly ash in carbon capture and storage (CCS) technologies is underway, where it can be used to sequester carbon dioxide and reduce greenhouse gas emissions (Yao et al., 2015).

1.3 SOURCES OF FLY ASH

1.3.1 Coal Combustion

Fly ash is a byproduct of the combustion process in coal-based thermal power plants or industrial boilers. As coal is burned to generate heat, non-combustible mineral impurities in the coal, such as silica, alumina, and iron oxides, are left behind as ash. The combustion process generates two types of ash: bottom ash, which is too heavy to be carried by the flue gases and falls to the bottom of the furnace, and fly ash, which is carried by the flue gases and captured by electrostatic precipitators or bag filters before the gases are released into the atmosphere (Yao et al., 2021). The generation of fly ash is heavily influenced by the type of coal used and the combustion technology employed. In traditional PCC systems, coal is ground into a fine powder and blown into the furnace, where it burns at high temperatures, typically ranging from 1,300°C to 1,700°C. This high-temperature combustion results in the formation of spherical fly ash particles that are rich in glassy aluminosilicate phases. The ash is then carried out of thc boiler with the flue gases and captured using pollution control devices. One of the key factors that determine the characteristics of fly ash is the mineral content of the coal. During combustion, the organic matter in the coal is oxidized, leaving behind mineral residues that form fly ash. The exact composition of the fly ash depends on the mineralogy of the coal, the combustion temperature, and the efficiency of the pollution control equipment (Saikia et al., 2021). Fly ash typically consists of fine, powdery particles that are composed primarily of silica, alumina, and iron oxides, with smaller amounts of calcium, magnesium, and sulfur compounds.

1.3.2 Bio-Coal Combustion

Bio-coal, also known as bio-coal or torrefied biomass, is a renewable energy source that is produced by the thermal processing of biomass materials such as wood, agricultural residues, and organic waste. The torrefaction process involves heating the biomass in an oxygen-deprived environment at temperatures typically ranging from 200°C to 300°C. This process removes moisture and volatile compounds from the biomass, resulting in a solid fuel that is more energy-dense and has properties similar to traditional coal (Adeoye et al., 2023). Bio-coal briquettes are a sustainable energy source made from agricultural waste. These briquettes are created by compressing biomass such as cotton, maize, groundnut shells, etc. into dense blocks that can be burned as fuel. Unlike traditional coal, bio-coal is renewable and produces lower carbon emissions. It is an eco-friendly alternative that contributes to reducing the environmental impact of coal combustion. When bio-coal briquettes are burned, they generate fly ash, which can be used in various applications (Naik et al., 2023). This fly ash, known as bio-coal fly ash (BFA), has been studied for its potential in construction materials. Research has shown that BFA can mimic the properties of coal-based fly ash and can be used as a binder in geopolymer mortar. When mixed with materials like ground granulated blast furnace slag (GGBS) and activated by an alkaline solution, BFA can produce construction materials with high compressive strength. These materials, including bricks and paver blocks, offer a sustainable alternative to traditional cement-based products, reducing raw material consumption, carbon emissions, and environmental degradation. Bio-coal briquettes, therefore, represent a key component in the transition toward greener energy and construction practices. The combustion of bio-coal in thermal power plants generates fly ash in a manner similar to traditional coal. However, the composition and properties of BFA can differ significantly from those of coal fly ash due to the differences in the feedstock materials and the torrefaction process. Understanding these differences is essential for optimizing the use of BFA in industrial applications and for addressing potential environmental concerns.

1.4 DIFFERENCES IN FLY ASH GENERATED FROM BIO-COAL COMPARED TO TRADITIONAL COAL

The fly ash generated from the combustion of bio-coal differs from that produced by traditional coal in several key aspects, including chemical composition, particle size distribution, and environmental impact. These differences are primarily due to the nature of the biomass feedstock used to produce bio-coal and the conditions under which it is torrefied and combusted.

1.4.1 Chemical Composition

BFA generally has a higher content of unburned carbon compared to coal fly ash, which can affect its pozzolanic reactivity and suitability for use in cement and concrete applications (Li et al., 2023). The presence of higher levels of organic compounds in bio-coal feedstock contributes to the increased carbon content in the fly ash. Additionally, BFA may contain higher levels of potassium, phosphorus, and other elements derived from biomass, which can influence its behavior in industrial processes. Unlike coal fly ash, which is rich in silica and alumina, BFA ash may have a more varied composition depending on the type of biomass used. For example, fly ash from the combustion of woody biomass may have a higher silica content, while fly ash from agricultural residues may contain more calcium and potassium. These compositional differences can impact the performance of BFA in applications such as cement production, where consistency in chemical composition is important for maintaining product quality (Yao et al., 2015).

1.4.2 Particle Size Distribution

The particle size distribution of BFA can also differ from that of coal fly ash. BFA particles are generally larger and less spherical than those of coal fly ash, which can affect the workability and strength of concrete when used as a supplementary cementitious material (SCM) (Rawat and Kumar, 2022). The larger particle size of BFA may require adjustments to the mix design in concrete production to achieve the desired properties. The physical properties of BFA, such as surface area and particle shape, can also influence its performance in other industrial applications, such as soil stabilization and the production of building materials. For instance, the irregular shape and larger size of BFA particles may result in lower packing density and higher water demand in concrete mixtures, which could impact the overall strength and durability of the final product.

1.4.3 Environmental Impact

The environmental impact of BFA is generally considered to be lower than that of coal fly ash due to the renewable nature of the feedstock and the lower levels of toxic elements. However, the higher content of unburned carbon and organic compounds in BFA can pose challenges to its disposal and utilization. For example, the presence of high levels of unburned carbon can make BFA less suitable for use in certain applications, such as the production of high-performance concrete, where low carbon content is desired.

Additionally, BFA may have a higher nutrient content, such as potassium and phosphorus, which could make it suitable for use as a soil amendment or fertilizer. However, the variability in composition and the potential presence of heavy metals or other toxic elements necessitate thorough characterization and regulatory assessment before widespread agricultural application.

1.5 INDUSTRIAL APPLICATIONS OF FLY ASH

There have been multiple reports which illustrate industrial applications of fly ash. Figure 1.1 presents several pathways for utilizing fly ash, grouped into different industrial, agricultural, and environmental applications.

One of the most practical uses of fly ash is for mine filling. Fly ash can be used to fill abandoned or exhausted mines, providing structural support to prevent land subsidence and water pollution. Its fine texture and pozzolanic properties make it an ideal filler material, stabilizing voids in underground mines (Cao et al., 2022). Additionally, fly ash can help improve the geotechnical properties of the soil in mine areas, thus reclaiming the land for future use. The use of fly ash in mine filling helps

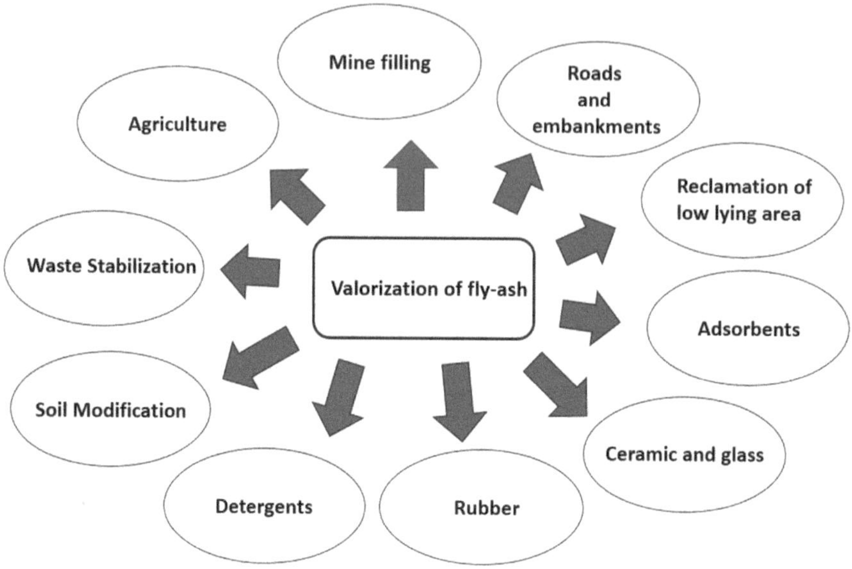

FIGURE 1.1 Valorization of fly ash in various industrial processes/practices to support circular economy.

to reduce the environmental hazards of abandoned mines, such as acid mine drainage and air pollution from fugitive dust. Fly ash is widely utilized in the construction of roads and embankments due to its excellent compaction and strength properties. When mixed with lime and water, fly ash forms a cementitious compound that can be used as a stabilizer in road base layers. This application enhances the durability and strength of roads, helping to reduce construction costs and material consumption. Fly ash is also used in the construction of embankments for highways, railways, and other infrastructure projects, providing a lightweight yet strong material that reduces the demand for natural aggregates and minimizes environmental degradation (Sahay and Bansal, 2022). Low-lying areas, especially in flood-prone regions, can benefit from the application of fly ash. Its lightweight and fine texture allow it to be used as a filling material to raise the ground level in areas susceptible to flooding. This helps in land reclamation and the prevention of waterlogging. Fly ash not only provides a cost-effective solution for filling low-lying areas but also contributes to mitigating land disposal issues by diverting fly ash from landfills. In addition, it improves the soil's physical properties, enhancing its suitability for future development or agricultural use.

Fly ash can also be processed into adsorbents for environmental applications. Due to its porous nature and chemical composition, fly ash can be modified to capture and remove pollutants from air, water, and wastewater. In water treatment, fly ash-based adsorbents are used to remove heavy metals, dyes, and organic contaminants. Similarly, in air purification, fly ash is employed to capture particulate matter (PM) and gaseous pollutants, contributing to pollution control efforts. The conversion of fly ash into adsorbents adds value to this waste material while addressing environmental challenges related to pollution. The ceramic and glass industries use fly ash as a raw material in the production of tiles, bricks, and other ceramic products. Fly ash contains silica and alumina, essential components in ceramic manufacturing, making it a suitable substitute for traditional clay. Moreover, the use of fly ash in glass production reduces the demand for virgin materials like sand and limestone (Zhang et al., 2022). The fly ash's fine texture and chemical stability ensure that ceramic and glass products maintain their strength and durability while promoting sustainability through resource conservation. Fly ash finds application in the rubber industry as a reinforcing filler. In the manufacturing of rubber products like tires and conveyor belts, fly ash is incorporated to enhance mechanical properties, such as tensile strength, abrasion resistance, and heat resistance. Using fly ash as a filler in rubber production reduces the consumption of other natural fillers like carbon black, which are more expensive and environmentally taxing to produce (Hoe-Woon et al., 2024). The incorporation of fly ash contributes to lowering production costs and the overall environmental footprint of the rubber industry. Another innovative use of fly ash is in the production of detergents. Fly ash can act as a filler and abrasive agent in detergent formulations, improving the cleaning properties of the detergent. It helps in the scrubbing action required for removing tough stains and serves as a bulking agent, reducing the need for more expensive raw materials. This application not only enhances the efficiency of detergents but also provides an outlet for large quantities of fly ash, further promoting waste reduction and resource optimization.

In agriculture, fly ash can be used to improve the structure and fertility of soils. Its addition to soil helps to increase water retention, reduce bulk density, and enhance aeration, which is beneficial for plant growth. Fly ash is also rich in essential nutrients like potassium, phosphorus, and trace elements, which contribute to soil fertility (Varshney et al., 2022). It has been effectively used to amend acidic soils by neutralizing their pH and improving their capacity to support crop production. Fly ash-based soil modification helps to enhance agricultural productivity while providing a sustainable use for this waste material (Hu et al., 2021). Fly ash is also employed in the stabilization of hazardous and non-hazardous waste. Its pozzolanic properties enable it to bind with other materials, encapsulating harmful substances and reducing their mobility. This is particularly useful in the treatment of industrial waste, sewage sludge, and contaminated soils. Fly ash is mixed with these wastes to form a stable, inert mass that prevents the leaching of contaminants into the environment. Waste stabilization with fly ash is a cost-effective and environmentally responsible way to manage hazardous waste and minimize its impact on ecosystems. Beyond soil modification, fly ash has broader applications in agriculture. It is used as a soil conditioner and fertilizer to improve soil structure and nutrient availability. Fly ash can enhance the cation exchange capacity of soils, allowing them to retain more nutrients and release them slowly to plants. Additionally, its use in composting can improve the decomposition process, producing nutrient-rich organic fertilizers. Fly ash also acts as a pest repellent when applied to crops, reducing the need for chemical pesticides. The use of fly ash in agriculture provides a sustainable alternative to chemical fertilizers, contributing to organic farming practices and environmental conservation.

1.6 FLY ASH VALORIZATION AND CIRCULAR ECONOMY

The uses of fly ash outlined in Figure 1.1 align closely with the principles of a circular economy, which aims to minimize waste and maximize resource efficiency by closing the loop on material usage through recycling, reuse, and regeneration. Here's how these applications contribute to the circular economy.

1.6.1 Waste Minimization and Resource Efficiency

Fly ash is typically considered a waste product generated from coal combustion in power plants. However, instead of disposing of it in landfills (a linear economy approach), valorizing fly ash allows it to be repurposed into valuable materials. This reduces the amount of waste that needs to be managed and helps conserve virgin resources like sand, lime, and clay. In the circular economy, the goal is to use materials for as long as possible. By converting fly ash into products like construction materials, adsorbents, ceramics, and soil conditioners, industries can extend the useful life of this material, contributing to the efficient use of available resources (Huang et al., 2022).

1.6.2 Reduction of Virgin Material Consumption

Using fly ash in roads and embankments, mine filling, and as a binder in geopolymer concrete reduces the need for virgin materials like cement, lime, and natural aggregates. Cement production, for example, is a significant source of greenhouse gas emissions, so substituting it with fly ash not only reduces the demand for raw cement but also helps mitigate environmental impacts like CO_2 emissions. In the ceramic and glass industries, fly ash serves as an alternative to natural raw materials, reducing the need to extract virgin sand, clay, and minerals. This substitution preserves natural resources and decreases the environmental degradation caused by mining (da Costa et al., 2022).

1.6.3 Closing the Loop Through Recycling and Reuse

Fly ash can be used in mine filling, helping to rehabilitate and reclaim land that has been mined, a key principle of the circular economy. This turns an environmental problem of unused fly ash and degraded land into a solution, restoring land for potential future use while safely disposing of industrial waste. The use of fly ash for waste stabilization ensures that hazardous waste is immobilized and safely contained, reducing the risk of environmental contamination. By incorporating fly ash into this process, industrial waste is repurposed in a way that prevents further pollution, thereby contributing to sustainable waste management.

1.6.4 Supporting Sustainable Agriculture

Fly ash can act as a soil conditioner or additive, improving soil fertility and structure. In agriculture, it enhances water retention, aeration, and nutrient availability, reducing the need for chemical fertilizers. This promotes more sustainable agricultural practices by utilizing fly ash as a renewable input rather than relying solely on synthetic fertilizers, which are resource-intensive to produce. By improving soil quality, fly ash helps regenerate soil ecosystems, a core principle of the circular economy. Fly ash is also used in composting to improve decomposition, which enhances organic matter recycling. This supports the circular economy by returning nutrients to the soil in a natural cycle.

1.6.5 Reducing Carbon Footprint

One of the major circular economy benefits is reducing the carbon footprint of construction materials. Using fly ash in the creation of geopolymer mortar or in replacing cement drastically reduces carbon emissions. Fly ash-based geopolymer concrete can cut CO_2 emissions by up to 80% compared to traditional Portland cement, supporting climate change mitigation efforts. In industries like ceramics, rubber, and glass, incorporating fly ash reduces the energy intensity and carbon emissions associated with processing virgin materials. The reuse of waste fly ash closes the loop in industrial cycles, reducing the need for energy-intensive raw material production (Kumar et al., 2022).

1.6.6 Innovative Material Recycling

The development of adsorbents from fly ash for water and air purification represents another example of upcycling waste materials into new, valuable products. These adsorbents can remove contaminants from industrial wastewater and reduce air pollution, contributing to environmental clean-up efforts and preventing additional waste. The use of fly ash as a filler in detergents not only reduces the need for synthetic chemical fillers but also provides a pathway for reusing industrial waste in everyday consumer products. This closes the loop by integrating waste materials into new supply chains (Al-Ghouti et al., 2021).

1.6.7 Fostering a Circular Industrial Ecosystem

The various uses of fly ash promote an industrial ecosystem where waste from one process (coal combustion) becomes a valuable input for other industries (construction, agriculture, manufacturing). This symbiotic relationship between different industries is a hallmark of the circular economy, where waste is minimized, and resource flows are optimized. Utilizing fly ash locally for construction, reclamation, and agriculture reduces transportation costs and energy use, contributing to localized circular economies. This also supports the development of industries that can process fly ash into useful products, creating new jobs and economic opportunities.

1.6.8 Extending Product Life Cycles

Many fly ash-based products, such as concrete, ceramics, and glass, have long life cycles, making them inherently more sustainable. Using fly ash in durable products extends the material's usefulness far beyond its original life as a combustion byproduct. This durability aligns with the circular economy goals of creating long-lasting materials that reduce the need for frequent replacement. The valorization of fly ash supports the circular economy by turning an abundant industrial byproduct into a resource that can be used across a wide range of industries, from construction to agriculture. By integrating fly ash into the production of roads, embankments, ceramics, soil conditioners, and more, we can reduce the demand for virgin materials, minimize waste, and decrease carbon emissions. This not only mitigates the environmental impact of fly ash disposal but also contributes to a more sustainable and resource-efficient industrial system. As industries continue to adopt circular economy practices, fly ash valorization will play an essential role in achieving zero-waste goals and transitioning toward more sustainable economic models (Tang et al., 2021).

1.7 ENVIRONMENTAL IMPACT AND SUSTAINABILITY

1.7.1 Environmental Concerns Associated with the Disposal of Coal Fly Ash

Fly ash disposal poses significant environmental challenges due to the large volumes generated by coal-fired power plants and the potential for harmful substances to leach into the environment. Coal fly ash contains various heavy metals and trace elements, such as arsenic, lead, mercury, and selenium, which can be toxic to both humans and wildlife. The improper disposal of fly ash can lead to contamination of soil and water resources, posing serious environmental and public health risks (Yao et al., 2015). One of the primary environmental concerns associated with fly ash disposal is the potential for leachate formation. When fly ash is disposed of in landfills or surface impoundments, rainwater or other moisture can percolate through the ash, leaching out heavy metals and other toxic substances. This leachate can then migrate into surrounding soil and groundwater, leading to contamination. Studies have shown that the leaching potential of fly ash is influenced by factors such as the chemical composition of the ash, the pH of the leachate, and the presence of other materials in the disposal site.

In addition to the risk of leachate formation, the disposal of fly ash in landfills or surface impoundments can also contribute to air pollution. Fine particles of fly ash can become airborne during handling, transportation, and disposal, leading to the release of PM into the atmosphere. These airborne particles can have serious health implications, particularly for individuals with respiratory conditions, as well as contribute to environmental issues such as haze and reduced visibility. Another significant environmental concern is the potential for the structural failure of fly ash disposal sites. In some cases, surface impoundments or ash ponds have failed, leading to catastrophic releases of fly ash into nearby waterways and ecosystems. One of the most notable incidents occurred in 2008, when the Tennessee Valley Authority's Kingston Fossil Plant in the United States experienced a massive ash spill, releasing over a billion gallons of coal ash slurry into the surrounding environment. This event highlighted the risks associated with fly ash disposal and the need for more stringent regulations and oversight (Li et al., 2021a, 2021b).

1.7.2 Strategies for Mitigating the Environmental Impact

Given the significant environmental risks associated with fly ash disposal, it is essential to implement strategies to mitigate these impacts and promote the safe and sustainable management of fly ash. Several approaches can be employed to reduce the environmental footprint of fly ash disposal and enhance its beneficial use.

1.7.2.1 Improved Landfill and Impoundment Design

One of the most effective ways to mitigate the environmental impact of fly ash disposal is to improve the design and operation of landfills and surface impoundments. This includes the use of liners and leachate collection systems to prevent the migration of contaminants into the soil and groundwater. Additionally, covering fly ash disposal sites with soil or synthetic materials can help reduce the risk of airborne PM and minimize the potential for erosion and surface runoff. Regular monitoring of fly ash disposal sites is also crucial for early detection of potential environmental issues. This includes monitoring groundwater quality, leachate composition, and air quality to ensure that any adverse effects are identified and addressed promptly. The use of advanced monitoring technologies, such as remote sensing and geophysical surveys, can enhance the ability to detect and respond to environmental risks associated with fly ash disposal (Chindaprasirt and Rattanasak, 2010).

1.7.2.2 Encapsulation and Solidification

Encapsulation and solidification are techniques used to stabilize fly ash and reduce its potential for leaching and airborne release. These methods involve mixing fly ash with binding agents, such as cement or lime, to form a solid, stable material that can be safely disposed of or used in construction applications. Encapsulation can significantly reduce the mobility of heavy metals and other contaminants in fly ash, thereby minimizing the risk of environmental contamination. In addition to reducing the environmental impact of fly ash disposal, encapsulation, and solidification can also enhance the potential for beneficial use of fly ash. For example, stabilized fly ash can be used as a construction material in road bases, embankments, and other infrastructure projects, providing a sustainable alternative to traditional materials.

1.7.2.3 Regulatory Frameworks and Guidelines

The implementation of robust regulatory frameworks and guidelines is essential for ensuring the safe management of fly ash and mitigating its environmental impact. Governments and environmental agencies have developed regulations governing the disposal, handling, and use of fly ash to protect public health and the environment. These regulations often include requirements for the design and operation of disposal sites, limits on permissible contaminant levels, and guidelines for the beneficial use of fly ash in various applications (Wang et al., 2020). In many countries, the classification of fly ash as either hazardous or non-hazardous waste determines the regulatory requirements for its disposal and use. For example, in the United States, the Environmental Protection Agency (EPA) has established regulations under the Resource Conservation and Recovery Act (RCRA) that classify coal combustion residuals (CCRs), including fly ash, as non-hazardous waste, subject to specific management standards. However, the classification of fly ash can vary depending on its chemical composition and the presence of hazardous substances.

1.7.2.4 Promoting Beneficial Use and Recycling

One of the most effective strategies for reducing the environmental impact of fly ash disposal is to promote its beneficial use and recycling in various industrial applications. By finding valuable uses for fly ash, such as in cement production, concrete manufacturing, and construction materials, the need for disposal is reduced, and the environmental footprint of coal-fired power plants is minimized (Wei et al., 2022). The beneficial use of fly ash not only reduces the volume of waste requiring disposal but also contributes to resource conservation and sustainability. For example, the use of fly ash in cement and concrete production reduces the demand for raw materials, such as limestone and clay, and lowers energy consumption and CO_2 emissions associated with the production of Portland cement. Additionally, the use of fly ash in construction applications can improve the durability and performance of infrastructure, extending its lifespan and reducing the need for maintenance and repairs.

1.8 SUSTAINABLE MANAGEMENT PRACTICES

1.8.1 Role of Fly Ash in Promoting Circular Economy Principles

The concept of a circular economy emphasizes the importance of resource efficiency, waste minimization, and the recycling and reuse of materials to create a closed-loop system that reduces environmental impact and promotes sustainability. Fly ash, as a byproduct of coal combustion, has the potential to play a significant role in advancing circular economy principles by serving as a valuable resource in various industrial processes.

1.8.1.1 Resource Efficiency and Waste Minimization

Fly ash contributes to resource efficiency by providing a secondary raw material that can replace virgin resources in industrial applications. For example, in the construction industry, fly ash can be used as a SCM in cement and concrete production, reducing the need for natural resources such as limestone, clay, and gypsum (Jin et al., 2023). The use of fly ash in this way not only conserves natural resources but also reduces the environmental impact of raw material extraction and processing. Waste minimization is another key aspect of the circular economy, and the beneficial use of fly ash is an effective way to achieve this goal. By finding valuable applications for fly ash, the need for disposal in landfills or surface impoundments is reduced, leading to a decrease in the environmental footprint of coal-fired power plants. This approach aligns with the principles of circular economy by turning a waste product into a valuable resource, thereby closing the loop and reducing the need for new raw materials.

1.8.1.2 Recycling and Reuse

Recycling and reuse are central components of the circular economy, and fly ash offers numerous opportunities for both. In addition to its use in cement and concrete production, fly ash can be recycled into a variety of other products, such as bricks, tiles, and lightweight aggregates. These recycled products can then be used in construction projects, reducing the demand for traditional building materials and contributing to a more sustainable built environment. The reuse of fly ash in environmental remediation is another important aspect of its role in the circular economy. Fly ash can be used to stabilize contaminated soils, neutralize acidic mine drainage, and encapsulate hazardous waste, providing a sustainable solution for managing environmental pollutants. By recycling fly ash in these applications, the need for new materials and resources is reduced, and the environmental impact of waste disposal is minimized.

1.8.2 Case Studies on Successful Fly Ash Utilization and Recycling

Several case studies demonstrate the successful utilization and recycling of fly ash, showcasing its potential to contribute to a circular economy and promote sustainability in various industries.

1.8.2.1 Cement and Concrete Production

The use of fly ash in cement and concrete production is one of the most well-established and successful examples of its beneficial use. In the United States, the production of blended cements containing fly ash has been widely adopted, particularly in large infrastructure projects such as highways, bridges, and dams. For example, the Hoover Dam, one of the most iconic infrastructure projects in the United States, incorporated large quantities of fly ash in its concrete mix, resulting in a more durable and resilient structure (Townsend et al., 2020). In India, the use of fly ash in cement production has gained momentum due to government regulations requiring the use of fly ash in certain construction projects. This has led to the widespread adoption of fly ash-based cements, known as Portland Pozzolana Cement (PPC), which offer improved durability and environmental performance compared to traditional Portland cement. The use of PPC in large-scale projects, such as the construction of highways and urban infrastructure, has demonstrated the benefits of fly ash utilization in reducing CO_2 emissions and promoting sustainable construction practices.

1.8.2.2 Geopolymer Technology

Geopolymers, which are synthesized from aluminosilicate materials such as fly ash, represent an innovative and sustainable alternative to traditional Portland cement. Several case studies highlight the successful use of fly ash-based geopolymers in construction and environmental remediation. For example, in Australia, fly ash-based geopolymers have been used to construct precast concrete elements for infrastructure projects, offering enhanced durability, chemical resistance, and reduced CO_2 emissions compared to conventional concrete. In addition to construction applications, fly ash-based geopolymers have been successfully used to encapsulate hazardous waste, providing a safe and stable solution for long-term waste management. The use of geopolymers in this context not only reduces the environmental impact of waste disposal but also promotes the circular economy by recycling fly ash into a valuable material for environmental protection (Naik et al., 2023).

1.8.2.3 Soil Stabilization and Land Reclamation

Fly ash has been successfully used in soil stabilization and land reclamation projects, providing a sustainable solution for improving soil properties and restoring degraded land. For example, in China, fly ash has been used to stabilize expansive soils in highway construction, reducing the risk of soil shrinkage and swelling, which can lead to structural damage. The use of fly ash in this application not only improves the durability and performance of the infrastructure but also promotes the reuse of industrial waste in environmentally beneficial projects. In the United Kingdom, fly ash has been used in land reclamation projects to restore former industrial sites and create new habitats for wildlife. By blending fly ash with soil and other materials, degraded land can be transformed into valuable green spaces, contributing to biodiversity conservation and sustainable land use. These case studies demonstrate the potential for fly ash to play a key role in sustainable land management and environmental restoration (Andavan and Pagadala, 2020).

Fly ash offers significant potential for promoting environmental sustainability and advancing circular economy principles through its beneficial use and recycling. By addressing the challenges of fly ash disposal and implementing strategies for its sustainable management, industries can reduce their environmental impact and contribute to a more sustainable future.

1.9 FUTURE PROSPECTS AND RESEARCH DIRECTIONS

1.9.1 Emerging Technologies and Research in the Utilization of Fly Ash

As the global focus shifts toward sustainability and resource efficiency, the utilization of fly ash is gaining renewed attention in various research and industrial applications. Emerging technologies are exploring innovative ways to harness the properties of fly ash, moving beyond traditional uses in cement and concrete production. These new approaches aim to enhance the value of fly ash and reduce its environmental footprint, contributing to more sustainable industrial practices.

1.9.1.1 Advanced Material Synthesis

One of the most promising areas of research is the use of fly ash in the synthesis of advanced materials, such as zeolites, ceramics, and nanomaterials. Zeolites, which are microporous aluminosilicate minerals, have a wide range of applications in catalysis, adsorption, and ion exchange. Researchers have developed methods to convert fly ash into zeolites through processes such as alkaline hydrothermal treatment. These fly ash-derived zeolites can be used in environmental applications, such as wastewater treatment and air purification, providing a valuable use for this industrial byproduct. Another innovative application is the use of fly ash in the production of ceramics and glass ceramics. Fly ash contains silica, alumina, and other oxides that can be sintered or melted to form ceramic materials. These ceramics can be used in construction, electronics, and other high-performance applications. The development of fly ash-based ceramics not only provides a sustainable alternative to traditional raw materials but also offers a way to recycle large volumes of fly ash, reducing the need for disposal. Nanotechnology is another emerging field where fly ash is being explored as a precursor material. Researchers are investigating the production of fly ash-based nanoparticles, such as silica nanoparticles, which have potential applications in coatings, adhesives, and composite materials. The development of nanomaterials from fly ash could open up new markets and applications, further enhancing the value of this byproduct (Li et al., 2021a, 2021b).

1.9.1.2 Carbon Capture and Storage

The role of fly ash in CCS technologies is another area of active research. Fly ash can be used as a sorbent material to capture carbon dioxide (CO_2) from flue gases in power plants and industrial processes. Researchers are developing methods to enhance the CO_2 capture capacity of fly ash through chemical modification and surface treatment. The captured CO_2 can then be stored in geological formations or used in various industrial applications, such as the production of carbonates or enhanced oil recovery (Provis and Bernal, 2014). The integration of fly ash in CCS technologies not only helps reduce greenhouse gas emissions but also provides a valuable use for fly ash, contributing to the circular economy. This approach aligns with global efforts to mitigate climate change and transition toward a low carbon economy.

1.9.1.3 Geopolymer Technology

As discussed earlier, geopolymer technology is an emerging area of research that leverages the pozzolanic properties of fly ash to produce environmentally friendly alternatives to Portland cement. Geopolymers are synthesized by activating fly ash with alkaline solutions, resulting in materials with excellent mechanical properties and chemical resistance. Researchers are exploring the use of fly ash-based geopolymers in various applications, including construction, waste encapsulation, and high-temperature materials (Naik et al., 2023). The development of geopolymer technology has the potential to revolutionize the construction industry by providing a sustainable alternative to traditional cement. Fly ash-based geopolymers offer significant environmental benefits, including lower CO_2 emissions, reduced energy consumption, and the recycling of industrial waste. As research in this area continues to advance, the adoption of geopolymers in large-scale construction projects is expected to increase.

1.9.2 Potential of Bio-Coal Fly Ash in Future Sustainable Practices

The growing interest in bio-coal as a renewable energy source has led to the generation of BFA, which presents unique opportunities for sustainable practices. BFA, derived from the combustion of biomass, has different chemical and physical properties compared to traditional coal fly ash, offering potential advantages in various applications.

1.9.2.1 Sustainable Construction Materials

BFA has the potential to be used as a sustainable alternative to traditional fly ash in the production of construction materials. Due to its lower carbon content and renewable origin, BFA can be used to produce greener concrete and cement with a reduced environmental footprint. Researchers are exploring the use of BFA in the synthesis of geopolymers, as well as its incorporation into blended cements and concrete mixes (Gil, 2022). The use of BFA in construction materials aligns with the principles of the circular economy by turning a renewable waste product into a valuable resource. As the demand for sustainable building materials continues to grow, BFA has the potential to play a key role in reducing the carbon footprint of the construction industry.

1.9.2.2 Agricultural Applications

Another promising area of research is the use of BFA in agriculture. BFA contains nutrients such as potassium, phosphorus, and calcium, which can be beneficial for soil health and crop growth. Researchers are investigating the use of BFA as a soil amendment or fertilizer, particularly in regions with nutrient-deficient soils. The application of BFA in agriculture could enhance soil fertility, reduce the need for chemical fertilizers, and promote sustainable farming practices. However, the use of BFA in agriculture requires careful consideration of its chemical composition, particularly the presence of heavy metals and other contaminants. Ongoing research is focused on assessing the environmental safety of BFA and developing guidelines for its use in agricultural applications.

1.9.2.3 Environmental Remediation

BFA also holds potential for use in environmental remediation, particularly in the stabilization and remediation of contaminated soils and water. The high surface area and reactive properties of BFA make it suitable for the adsorption and immobilization of heavy metals and other pollutants. Researchers are exploring the use of BFA in the treatment of industrial effluents, mine tailings, and contaminated land, offering a sustainable solution for managing environmental pollutants. The use of BFA in environmental remediation not only provides a valuable use for this byproduct but also contributes to the protection and restoration of ecosystems. As research in this area continues to evolve, BFA could become an important tool in sustainable environmental management.

1.10 POLICY AND REGULATORY FRAMEWORK

1.10.1 Overview of Existing Policies and Regulations Governing Fly Ash Management

The management of fly ash is subject to a complex regulatory framework that varies by region and is influenced by environmental, health, and safety considerations. Governments and environmental agencies have developed policies and regulations to ensure the safe disposal, handling, and beneficial use of fly ash, with the goal of minimizing its environmental impact and promoting sustainability.

1.10.1.1 United States

In the United States, the EPA plays a key role in regulating the management of CCRs, including fly ash. The EPA's regulations are governed by the RCRA, which classifies CCRs as non-hazardous waste, subject to specific management standards. These standards include requirements for the design and operation of disposal sites, groundwater monitoring, dust control, and the closure and post-closure care of disposal units (Kravchenko and Ruhl, 2021). In addition to federal regulations, states may implement their own rules governing fly ash management, which can be more stringent than federal requirements. The EPA also encourages the beneficial use of fly ash through initiatives such as the Coal Combustion Products Partnership (C2P2), which promotes the recycling of fly ash in construction and other applications.

1.10.1.2 European Union

The European Union (EU) has established a comprehensive regulatory framework for the management of fly ash under the Waste Framework Directive (2008/98/EC) and the Landfill Directive (1999/31/EC). The Waste Framework Directive classifies fly ash as waste and sets out requirements for its handling, treatment, and disposal, with an emphasis on promoting recycling and recovery. The Landfill Directive imposes strict limits on the disposal of fly ash in landfills, including requirements for landfill design, leachate management, and monitoring (Ummik et al., 2024). The EU also promotes the use of fly ash in construction through European standards (EN 450-1) (Adu-Amankwah et. al., 2016), which specifies the requirements for fly ash used as a component in concrete. This standard ensures that fly ash meets specific quality criteria, including limits on the content of harmful substances, to ensure its safe use in construction.

1.10.1.3 India

India has implemented a robust regulatory framework for fly ash management, driven by the country's rapid industrialization and the need to manage large volumes of fly ash generated by coal-fired power plants. The Ministry of Environment, Forest and Climate Change (MoEFCC) has issued several notifications under the environment Protection Act, 1986, mandating the use of fly ash in construction activities within a certain radius of coal-based power plants. These regulations aim to promote the utilization of fly ash in the production of cement, concrete, bricks, and other construction materials (Darmansyah et al., 2023). The MoEFCC's regulations also include provisions for the disposal of fly ash, requiring power plants to ensure the safe storage and disposal of fly ash in designated areas. The regulations encourage the reuse of fly ash in infrastructure projects, such as road construction and land reclamation, to minimize the need for disposal and promote resource efficiency.

1.10.2 Recommendations for Future Policy Directions

As the utilization of fly ash continues to evolve, there is a need for policies and regulations that can adapt to emerging technologies and practices while ensuring the protection of public health and the environment (Li et al., 2024). The following recommendations outline key areas for future policy development.

1.10.2.1 Harmonization of Standards

One of the challenges in fly ash management is the variability in standards and regulations across different regions. Harmonizing standards at the national and international levels would help ensure consistent quality and safety of fly ash used in various applications. This could involve the development of global guidelines for fly ash classification, quality control, and environmental testing, facilitating the trade and use of fly ash across borders.

1.10.2.2 Promotion of Research and Innovation

Governments and environmental agencies should promote research and innovation in the utilization of fly ash, particularly in emerging areas such as advanced materials, carbon capture, and environmental remediation. This could involve providing funding and incentives for research and development projects, as well as supporting pilot projects that demonstrate the feasibility and benefits of new technologies. By fostering innovation, policymakers can help unlock the full potential of fly ash in sustainable practices.

1.10.2.3 Encouragement of Circular Economy Practices

Policies should encourage the adoption of circular economy principles in fly ash management, emphasizing the recycling and reuse of fly ash in various industrial applications. This could include setting targets for the beneficial use of fly ash, providing incentives for companies that incorporate fly ash into their products, and establishing regulations that prioritize recycling over disposal. By promoting circular economy practices, policymakers can help reduce the environmental impact of fly ash and contribute to resource conservation.

1.10.2.4 Strengthening of Environmental Safeguards

As new uses for fly ash are developed, it is important to ensure that environmental safeguards keep pace with these innovations. This includes updating regulations to address the potential risks associated with new applications, such as the use of fly ash in agriculture or environmental remediation. Policymakers should also ensure that environmental impact assessments are conducted for large-scale fly ash utilization projects, to identify and mitigate any potential adverse effects on the environment and public health.

1.10.2.5 Global Collaboration and Knowledge Sharing

Finally, there is a need for greater global collaboration and knowledge sharing in fly ash management. This could involve the establishment of international forums or networks where policymakers, researchers, and industry stakeholders can exchange information and best practices. By working together, countries can learn from each other's experiences and develop more effective strategies for managing fly ash in a sustainable and environmentally responsible manner.

The future prospects for fly ash utilization are promising, with numerous opportunities for innovation and sustainability. However, realizing this potential will require ongoing research, the development of supportive policies, and a commitment to environmental stewardship. By adopting a forward-looking approach to fly ash management, industries and governments can contribute to a more sustainable future.

1.11 CONCLUSION TOP OF FORM

The role of fly ash, both from coal and bio-coal combustion, in industrial applications continues to evolve as a sustainable solution to waste management and resource efficiency. This chapter underscores the potential of fly ash, not just as a byproduct of industrial processes, but as a critical resource for innovations in construction, agriculture, and environmental remediation. The unique chemical properties of fly ash open doors for its use in a variety of industries, contributing to a circular economy by reducing waste and the consumption of virgin materials. In particular, BFA presents exciting possibilities due to its renewable nature, highlighting the importance of developing tailored strategies for its use. While the environmental benefits are evident, ongoing research and advancements in technology will further enhance the value of fly ash across industries. The transition from waste to resource aligns with global sustainability goals, and its integration into industrial systems holds significant promise for reducing environmental impacts while promoting industrial resilience. Through the continued exploration of its applications, fly ash is poised to play a pivotal role in advancing sustainable industrial practices.

REFERENCES

Adeoye, A. O., Lawal, O. S., Quadri, R. O., Malomo, D., Aliyu, M. T., Dang, G. E.,..., & Hikon, B. N. (2023). Sustainable energy via thermochemical and biochemical conversion of biomass wastes for biofuel production. In: Sharma, S.K., Upadhyay, R.K., Kumar, V., Valera, H. (eds) *Transportation Energy and Dynamics* (pp. 245–306). Singapore: Springer Nature Singapore.

Adu-Amankwah, S., Khatib, J. M., Searle, D. E., & Black, L. (2016). Effect of synthesis parameters on the performance of alkali-activated non-conformant EN 450 pulverised fuel ash. *Construction and Building Materials*, *121*, 453–459.

Al-Ghouti, M. A., Khan, M., Nasser, M. S., Al-Saad, K., & Heng, O. E. (2021). Recent advances and applications of municipal solid wastes bottom and fly ashes: Insights into sustainable management and conservation of resources. *Environmental Technology & Innovation*, *21*, 101267.

Alrefaei, Y., Wang, Y. S., & Dai, J. G. (2021). Effect of mixing method on the performance of alkali-activated fly ash/slag pastes along with polycarboxylate admixture. *Cement and Concrete Composites*, *117*, 103917.

Andavan, S., & Pagadala, V. K. (2020). A study on soil stabilization by addition of fly ash and lime. *Materials Today:* Proceedings, 22, 1125–1129.

Cao, H., Gao, Q., Zhang, X., & Guo, B. (2022). Research progress and development direction of filling cementing materials for filling mining in iron mines of China. *Gels*, *8*(3), 192.

Chindaprasirt, P., & Rattanasak, U. (2010). Utilization of blended fluidized bed combustion (FBC) ash and pulverized coal combustion (PCC) fly ash in geopolymer. *Waste Management*, *30*(4), 667–672.

da Costa, T. P., Quinteiro, P., Arroja, L., & Dias, A. C. (2022). Environmental performance of different end-of-life alternatives of wood fly ash by a consequential perspective. *Sustainable Materials and Technologies*, *32*, e00411.

Darmansyah, D., You, S. J., & Wang, Y. F. (2023). Advancements of coal fly ash and its prospective implications for sustainable materials in Southeast Asian countries: A review. *Renewable and Sustainable Energy Reviews*, *188*, 113895.

Fernández-Jiménez, A., & Palomo, A. (2003). Characterisation of fly ashes. Potential reactivity as alkaline cements. *Fuel, 82*(18), 2259–2265.

Gil, A. (2022). Challenges on waste-to-energy for the valorization of industrial wastes: Electricity, heat and cold, bioliquids and biofuels. *Environmental Nanotechnology, Monitoring & Management, 17,* 100615.

Hoe-Woon, T., Cheng-Yong, H., Yun-Ming, L., Qi-Hwa, N., Wei-Ken, P., Chin-Yii, Y.,..., & Yong-Jie, H. (2024). Assessing viability and leachability in fly ash geopolymers incorporated with rubber sludge. *Journal of Industrial and Engineering Chemistry,* 142, 499–511.

Hu, X., Huang, X., Zhao, H., Liu, F., Wang, L., Zhao, X., ... & Ji, P. (2021). Possibility of using modified fly ash and organic fertilizers for remediation of heavy-metal-contaminated soils. *Journal of Cleaner Production, 284,* 124713.

Huang, B., Gan, M., Ji, Z., Fan, X., Zhang, D., Chen, X.,..., & Fan, Y. (2022). Recent progress on the thermal treatment and resource utilization technologies of municipal waste incineration fly ash: A review. *Process Safety and Environmental Protection, 159,* 547–565.

Jin, L., Chen, M., Wang, Y., Peng, Y., Yao, Q., Ding, J.,..., & Lu, S. (2023). Utilization of mechanochemically pretreated municipal solid waste incineration fly ash for supplementary cementitious material. *Journal of Environmental Chemical Engineering, 11*(1), 109112.

Kravchenko, J., & Ruhl, L. S. (2021). Coal combustion residuals and health. In: Siegel, M., Selinus, O., Finkelman, R. (eds) *Practical Applications of Medical Geology*, 429–474.

Kumar, M., Prashant, S., & Kamath, M. V. (2022). Enhancing the sustainability of high strength concrete in terms of embodied energy and carbon emission by incorporating sewage sludge and fly ash. *Innovative Infrastructure Solutions*, *7*(4), 240.

Li, G., Hu, R., Hao, Y., Yang, T., Li, L., Luo, Z.,..., & Shen, G. (2023). CO_2 and air pollutant emissions from bio-coal briquettes. *Environmental Technology & Innovation, 29,* 102975.

Li, W., Gu, K., Yu, Q., Sun, Y., Wang, Y., Xin, M.,..., & Zhang, D. (2021a). Leaching behavior and environmental risk assessment of toxic metals in municipal solid waste incineration fly ash exposed to mature landfill leachate environment. *Waste Management, 120,* 68–75.

Li, X., Sun, Y., Li, W., Nie, Y., Wang, F., Bian, R.,..., & Lu, C. (2024). Solidification/stabilization pre-treatment coupled with landfill disposal of heavy metals in MSWI fly ash in China: A systematic review. *Journal of Hazardous Materials*, 478, 135479.

Li, Z., Fei, M. E., Huyan, C., & Shi, X. (2021b). Nano-engineered, fly ash-based geopolymer composites: an overview. *Resources, Conservation and Recycling, 168,* 105334.

Naik, S. S., Pandey, S., Pawar, S. N., Shinde, B. H., & Prakash, C. (2023). Innovative and interactive methodology for development of geopolymer mortar using fly ash of agricultural waste briquettes. *International Journal on Interactive Design and Manufacturing (IJIDeM),* 18(8), 1–9.

Provis, J. L., & Bernal, S. A. (2015). Milestones in the analysis of alkali-activated binders. *Journal of Sustainable Cement-Based Materials, 4*(2), 74–84.

Rawat, S., & Kumar, S. (2022). Critical review on processing technologies and economic aspect of bio-coal briquette production. *Preparative Biochemistry & Biotechnology, 52*(8), 855–871.

Sahay, D. K., & Bansal, S. (2022). Use of fly ash—A resourceful byproduct in road embankment: A review. *Advances in Construction Materials and Sustainable Environment: Select Proceedings of ICCME 2020*, (LNCE,volume 196) 539–550.

Saikia, B. K., Hower, J. C., Islam, N., Sharma, A., & Das, P. (2021). Geochemistry and petrology of coal and coal fly ash from a thermal power plant in India. *Fuel, 291,* 120122.

Tang, W., Pignatta, G., & Sepasgozar, S. M. (2021). Life-cycle assessment of fly ash and cenosphere-based geopolymer material. *Sustainability, 13*(20), 11167.

Townsend, S. W., Spreadbury, C. J., Laux, S. J., Ferraro, C. C., Kari, R., & Townsend, T. G. (2020). Blending as a strategy for reusing municipal solid waste incinerator ash in road-base construction. *Journal of Environmental Engineering, 146*(9), 04020106.

Ummik, M. L., Järvik, O., Reinik, J., & Konist, A. (2024). Ecotoxicity assessment of ashes from calcium-rich fuel combustion: contrasting results and regulatory implications. *Environmental Science and Pollution Research, 31*(35), 48523–48533.

Varshney, A., Dahiya, P., Sharma, A., Pandey, R., & Mohan, S. (2022). Fly ash application in soil for sustainable agriculture: An Indian overview. *Energy, Ecology and* Environment, 7(4), 340–357.

Wang, N., Sun, X., Zhao, Q., Yang, Y., & Wang, P. (2020). Leachability and adverse effects of coal fly ash: A review. *Journal of Hazardous Materials, 396*, 122725.

Wei, Y., Liu, S., Yao, R., Chen, S., Gao, J., & Shimaoka, T. (2022). Removal of harmful components from MSWI fly ash as a pretreatment approach to enhance waste recycling. *Waste Management, 150*, 110–121.

Yao, X., Mao, J., Li, L., Sun, L., Xu, K., Ma, X.,..., & Xu, K. (2021). Characterization comparison of bottom ash and fly ash during gasification of agricultural residues at an industrial-scale gasification plant–Experiments and analysis. *Fuel, 285*, 119122.

Yao, Z. T., Ji, X. S., Sarker, P. K., Tang, J. H., Ge, L. Q., Xia, M. S., & Xi, Y. Q. (2015). A comprehensive review on the applications of coal fly ash. *Earth-Science Reviews, 141,* 105–121.

Zhang, Z., Li, Z., Yang, Y., Shen, B., Ma, J., & Liu, L. (2022). Preparation and characterization of fully waste-based glass-ceramics from incineration fly ash, waste glass and coal fly ash. *Ceramics International, 48*(15), 21638–21647.

2 Analysis of Bio-Coal and Coal-Derived Fly Ash

Suranjana V. Mayani, Shivam M. Mod, Pooja M. Pancholi, Rucha N. Thaker and Rajkumar Patel

2.1 INTRODUCTION

Fly ash (FA), a hazardous byproduct, is produced when pulverized coal is burned in a coal power station to generate electricity. Particularly, FA did not offer any benefits; in former times, it was discharged directly into the atmosphere. Due to this, a negative effect was created in the environmental condition. FA is the complex, heterogeneous residue left over from burning coal. Approximately 2,300 years before, FA was utilized in the development of the very initial hydraulic cement. The first ever product was produced in Pozzuoli, a small Italian town, ultimately the word pozzolan arose from it. Globally, coal ash production is currently projected at nearly 600 million tons. The amount of coal waste being discharged by the industry and thermal power plants has been rising globally. India's emissions rose to 226.13 million tons ($22{,}613 \times 10^7$ kg) of FA in the period of 2019–2020 and by 2025, production is projected to be in the range of 300–400 million tons (Alterary *et al.*, 2021). Coal-based electrical power generation offers energy for everyday CONSUMPTION and also for industry, additionally inducing the degradation of atmosphere and water. The producers of thermal electricity are therefore searching for ways to utilize coal-based FA, which is seen as a waste material from the standpoint of power generation but as a resource that hasn't been fully utilized from the standpoint of coal utilization. The cement industry might use it as an initial component for manufacturing concrete. In addition to being used as a byproduct, coal fly ash released by power stations is also being researched for use in recycling materials for engineering and agriculture. Among the numerous byproducts of coal-fired power plants include coal ash, various heavy metals, gypsum, CO_2, NO_2, SO_2, and volatile organic compounds like benzene, toluene, aldehydes, methylene chloride, etc. For instance, coal ash was regarded as hazardous trash until the ministry deemed it to be non-hazardous waste in 2000 after realizing its value in numerous economic areas. Some of these byproducts like gypsum and coal ash can be sold and used for commercial purposes. Coal ash is produced when fly ash and bottom ash are combined. The coal ash that rises with the exhaust gases is known as fly ash.

 DOI: 10.1201/9781003498124-2

Demand toward sources of green energy has increased in recent years due to challenges of both energy and the environment security on a worldwide basis. Among these resources, biomass has the potential to play a significant role because it is regarded as a sustainable and CO_2 neutral source of energy that can eventually supply energy for heat, electricity, and transport once utilization rates are significantly lower than the rate of growth. The forms of solid fuel that can easily be fired which consist of organic matter from woodlands, farms, or the food industry can be referred to as biomass. The fly ash generated when any type of biomass undergoes combustion must be known as "bio-ash." Different kinds of biomass used in a biomass power plant, such as wood, bark, or agricultural wastes, have a significant impact on the amount and quality of ashes that are produced. Bark, agricultural wastes, and plant-based biomass all have larger amounts of ash than wood, so burning wood produces fewer ashes which have to be maintained. Biomass has begun to take up the place of coal in power resources. As a result, new fly ash types have developed, such as coal bio-ash (CBA), which has been produced from low-alkali coal fly ash (CFA), that gets introduced into the boiler furnace during the combustion of wood pellets to prevent corrosion and sustain the fuel's essential thermal efficiency. In order to make optimal use of this resource, an evaluation of the characteristics of a new type of ash is necessary (Nina *et al.*, 2019). The filler behavior and the pozzolanic activity of CBA when utilized as a partial substitute for cement are examined in this chapter. When CBA was compared to traditional CFA, it turned out to be a more beneficial alternative. This chapter has demonstrated that wood FA might be shifted from a waste byproduct to a useful asset for the manufacturing of concrete via the addition of CFA to the boiler furnace. Particular differences between biomass and coal ashes relate to their chemistry and mineralogy.

The features of the ashes produced by burning biomass can differ greatly and are determined by the following factors: (i) the characteristics of the biomass (such as wood, bark, or herbaceous material); (ii) the type of combustion technology used (such as fixed bed or fluidized bed); and (iii) the location of the ashes collection (such as bottom ashes or fly ashes). Industrial byproducts and other waste materials that can be used as binder substitutes in cement compositions have been the subject of extensive research. Using these materials has several advantages, such as lower costs and less of an adverse effect on the environment from cement manufacture and waste disposal. One major contributor to the atmospheric release of carbon dioxide (CO_2) is the cement industry (Rejini *et al.*, 2009). The utilization of agricultural waste ashes, such as rice husk ash, sawdust ash, bagasse ash, and palm oil fuel ash, as ingredients in concrete has been investigated by numerous researchers. These agricultural waste ashes have the potential to replace cement in pozzolanic materials because they contain a significant amount of silica in an amorphous state. The morphological and elemental investigations found that the electrical and aerodynamic characteristics of bio-ash and CFA generated in electrostatic precipitators differ at each successive stage of electrostatic precipitation (ESP) and that the participation of various constituents differs based on the FA source. The primary constituents of FA from coal-fired boilers are silicon dioxide, Al_2O_3 (1/4), and Fe_2O_3 (1/8), as well as oxides or chlorides of Ca, Na, Mg, K, Ti, and S, which typically make up higher than 1 mg/g each. These findings have been the subject of numerous investigations. Depending on the coal, the proportion of these chemicals may change. The majority of the elements found in particles smaller than 1 mm are heavy metals in the form of oxides, as well as minor amounts of components containing K, S, Mn, and Cl. Because SiO_2, Al_2O_3, and Fe_2O_3 are not present in large amounts in biomass, their percentage in FA from cofired biomass boilers is substantially less. However, the ratio of P_2O_5 and alkaline compounds like CaO, MgO, Na_2O, and K_2O that are produced as a result of burning biomass is increased (Arrigoni *et al.*, 2020). Generally speaking, bio-ash has a higher alkali and alkaline earth metal composition than CFA. These metals are primarily present as oxides or salts. This could lead to lower collection efficiency and increased resistivity of bio-ash due to low water content; however, biomass-derived flue gas typically has a comparatively elevated moisture content, which causes salts to dissociate and lower bio-ash resistivity throughout all ESP phases. Even while the amount of sulfur (S) in bio-ash is magnitudes more than that in CBA, the proportion of sulfur, which also influences for both varieties of FA, the ability to resist becomes greater with each ESP stage.

2.2 COMPOSITION AND PROPERTIES OF FA

The mineralogical composition of FA contains partially unburned black carbon and the crystalline and glass mineral elements. FA is divided into primary, secondary, and tertiary phase based on the formation time. There is no change in primary phases. Oxides and silicates are among the secondary stage that are created when coal burns. Furthermore, the transportation of FA containing gypsum and portlandite creates the tertiary phases (Alterary et al., 2021). The majority elements of FA are oxides of Ca, Fe, and Si. In smaller amounts, there is also S, Ti, K, Na, and Mg. When used as a mineral additive in concrete, FA can be separated into two types of uses: class C or class F ash, depending on its chemical makeup (Bhatt et al., 2019). Burning lignite or sub-bituminous coal provides class C ashes, which are cementitious and pozzolanic. The American Society for Testing and Materials (ASTM) states that the overall concentration of silicon dioxide, iron oxide, aluminum dioxide must be more than 50% (Bentz et al., 2011). Class F ash is described as having more than 70% weight percentage of silicon dioxide, aluminum oxide, and iron oxide.

2.2.1 Crystalline Phase

Its bulk is composed of 5%–50% FA crystalline regions. There are ten phases in the crystalline phases. Class C FA includes anhydrite ($CuSO_4$), merwinite, periclase, and C3A except from multiple lime, hematite (Fe_2O_3), magnetite (Fe_3O_4), quartz, lite, and mellite. Anhydrite is produced in the furnace by the presence as O_2, SO_2, and CaO, and it grows as SO_3 levels rise. Furthermore, the primary factor in the creation of ettringite is anhydrite. Mullite is more prevalent in class F FA and is not reactive. But every kind of FA involves quartz. Since lime is available in all FA types, it fulfills a critical function in being involved in hydration reactions. By dissolving, lime also raises the pH by more than 12. The expansion of sulfate is caused by C3A. X-ray diffraction (XRD) is used to identify the crystalline phases in FA and distinguish between the reactive and stable phases.

2.2.2 Amorphous Phase

The particles in the amorphous phase are rounded, ranging in diameter between 1 and 5 μm, and are created when particles are quenched, which leads to their disorganization. This process prevents the formation of well-defined crystal structures, making them hard to characterize because they lack long-range atomic order and distinct diffraction patterns. Aluminosilicate glass is created when parent coal has a low Ca (calcium) percentage, whereas calcium aluminosilicate glass is created when the calcium content is high. In FA, the latter is recognized as a recognized reactive phase. Its greater sensitivity in comparison to class F is due to class C. Total solid is subtracted from the total FA mass to determine the total amorphous phases (Alterary *et al.*, 2021).

2.3 Few Characterization Techniques for Coal-Based FA

FA is a material that has been synthetically produced by volcanic eruptions. "Silicate/Aluminosilicate material, which is not liable to self-cementing or has inadequate durability" is the definition of volcanic FA. It reacts with calcium hydroxide [Ca $(OH)_2$] at room temperature and in a suitably split state to produce a gel product. XRD and X-ray fluorescence (XRF) technique has been employed in mineralogical analysis via scanning electron microscopy (SEM) technique, structure of the surface of the CFA has been analyzed, and the chemical composition has been determined through EDS. For identification of function groups, Fourier-transform infrared (FTIR) technique has been employed. Particle size distribution (PSD) has been used in analysis for the study of particles' size distribution (Apua et al., 2018).

2.3.1 X-Ray Diffraction and Energy Dispersive X-Ray Spectroscopy

XRD analysis was implemented for assessing the total content of element and chemical constituents. Oxide composition of FA can be determined by the XRF technique. The principal elements observed in coal and FA with different concentrations were typically in oxide form. Phase determined by XRD analysis and components obtained correspond well, in particular, with the majority of silicon and Al composition (quartz and mullite); additionally, XRF results revealed that the sample's low calcium oxide (CaO) levels include it in class F CFA. The ideal results from the XRF and XRD analysis suggest that the primary element was Al, Fe, Si, Ca, Ti, etc. Figure 2.1 depicts the XRD of coal-based FA.

This technique relies on the interaction between the source of X-ray emission and FA. An energy dispersive spectrometer is employed to measure the intensity and frequency of X-ray emitted from fly ash with an objective of extracting

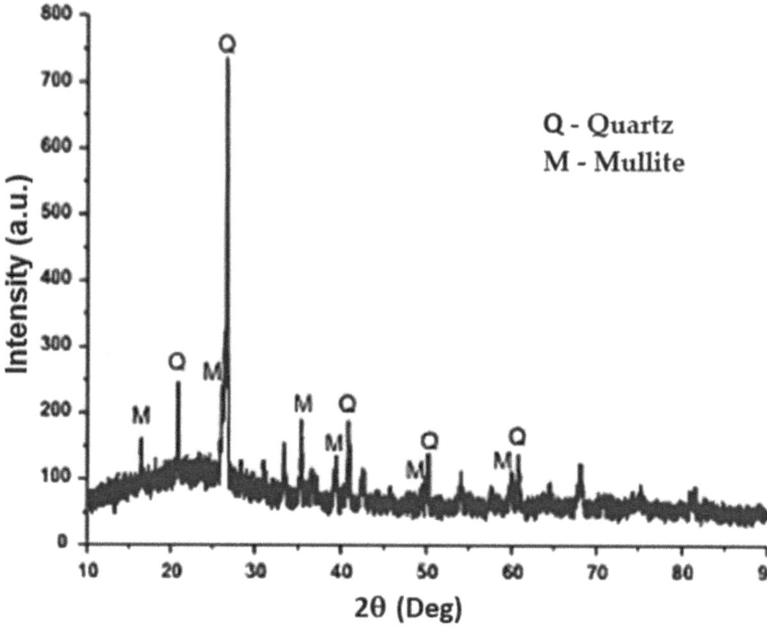

FIGURE 2.1 XRD pattern of coal-based fly Ash.

characteristic X-ray from FA. Energy dispersive X-ray (EDX) spectroscopy contributes in elemental analysis. Despite this, when an element's proportion is smaller than one percentage (1%), EDX lacks the ability to determine the elements of FA (Panda *et al.*, 2020).

2.3.2 FTIR (Fourier-Transform Infrared) Spectroscopy

This technique consists of using a CFA sample and applying it to infrared (IR) radiation. The spectrum of FTIR can be obtained between the range of 4,000 $cm-^{1}$ and 500 $cm-^{1}$. Wave number vertical axis derived from the spectrometer represents the degree of transmitted light, whereas the horizontal plane shows absorbance of infrared spectrum which indicates the absorption and transmittance of the molecule, functioning as a molecular fingerprint of the sample (Apua *et al.*, 2018).

2.3.3 Particle Size Distribution

The analysis of CFA indicates that the sample has P80 fragments and have a diameter of 49 μm. Particles smaller than 49 μm are suitable for extracting Fe and Al in CFA, because the smaller dimension of the particle improves recovery efficiency (Apua *et al.*, 2018). The most common method for measuring FA particle size is laser diffraction. The analyzer examines the interaction between light and material particles according to the diffraction pattern, matching the examined pattern to a model that is computed theoretically. A Fourier-transform lens is used to focus the diffracted light beam in order to extract electric signals from the incident energy. The angle between the incident and diffracted light is proportional to particle size. Isopropyl alcohol is the preferred dispersive agent because of its high viscosity and improved consistency. A wider range of particle sizes distribution signifies a higher packing density of the blend, which in turn leads to increased workability.

2.3.4 NMR Spectroscopy

Analysis using only XRD technique sometimes fails to accurately identify the structure of amorphous particles in FA. Enhanced knowledge of the amorphous structure of FA can be given by NMR spectroscopy. The embraces and shift of silicon-29 (^{29}Si) and Aluminum-27 atom might be identified through the NMR spectroscopy.

2.3.4.1 Silicon-29 NMR Spectra of FA

From the spectrum, the amorphous or partially crystalline compound exhibits a chemical shift between –50 and –150 ppm, which can be attributed to fluctuation in the surrounding environment of nominally equivalent $Q_n(_mAl)$ – type silicon (Figure 2.2). In FA, crystalline phase mullite is obligated for a peak at the chemical shift which appears around -90 ppm in spectra. The crystalline phase in FA is responsible for chemical shift near –110 ppm; in addition, the fly ash contains less concentrated monomer components which might be the cause of chemical shift at around –80 ppm. The intense alternate bands observed between 0 and –50 ppm and –150 and –200 ppm is achieved by paramagnetic elements like Fe_3O_4 (iron (II, III) oxide) (Alehyen *et al.*, 2017).

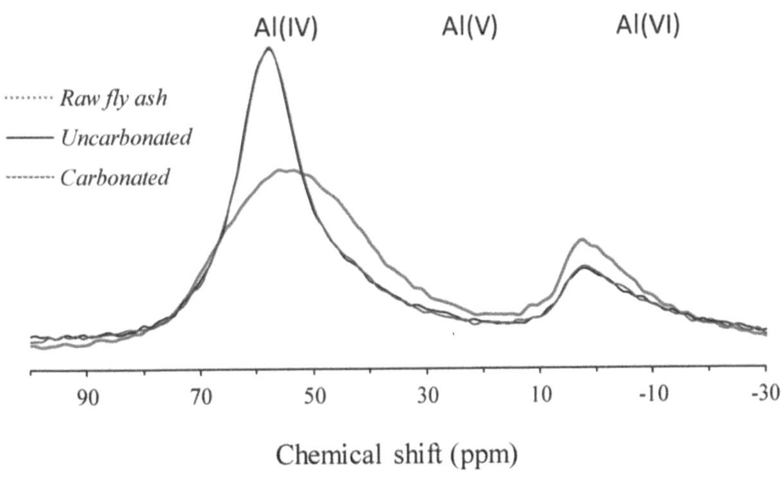

FIGURE 2.2 ^{29}Si MAS NMR spectra of (a) all; (b) raw fly ash; (c) uncarbonated; and (d) carbonated alkali-activated fly ash.

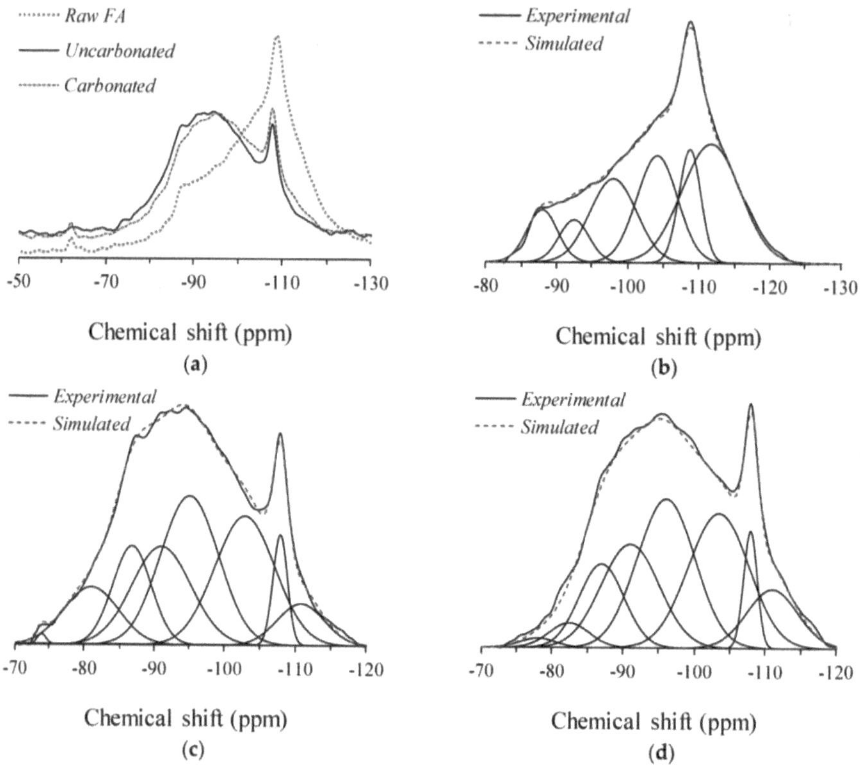

FIGURE 2.3 ^{27}Al MAS NMR spectra of raw fly ash and uncarbonated and carbonated alkali-activated fly ash.

2.3.4.2 Aluminum-^{27} NMR spectra of FA

The spectra from ^{27}Al-NMR of FA. NMR spectra of FA indicate significant peaks from –200 to 200 ppm. Figure 2.3 indicates two peaks at –3.4 and 43.7 ppm because of octahedrally coordinated Al (Vl) and terahedrally coordinated Al (IV).

2.3.5 Thermogravimetric Analysis

Currently, CFA carbon content has been determined by the thermogravimetric analysis (TGA) method to determine both quantity as well as quality. TGA is often used to investigate any physical or chemical processes which result in a variation in weight through the adsorption of gases, such as oxygen (O_2). In terms of science, the carbon content in FA acts as an indication for the combustion mechanism efficiency. The higher the percentage of carbon in the coal, greater heat is released during combustion, making the coal sample less efficient (Panda *et al.*, 2020).

2.4 UTILIZATION OF FA

FA is still disposed of in large quantities in ash dams and landfills, even if some of it is being used productively. FA may leak harmful metals into ponds or landfills because it contains trace levels of these elements, including selenium, chromium, boron, and in certain situations, mercury and barium. Because toxic materials can contaminate groundwater, they have the potential to have a huge impact on both terrestrial and aquatic ecosystems. With every passing day, disposing of waste is getting harder and more expensive because of tighter environmental rules. In light of this, it will benefit the society if such waste materials are transformed into beneficial goods. Some of the beneficial effects of the utilization of FA are covered here.

2.4.1 Soil Properties Improvement

Many studies have been conducted on the application of FA as a fertilizer, insecticide, plant nutrient, soil amendment, stabilizing erosion-prone soils, normalizing acidic soil, and other applications. FA has the power to influence the soil's biological characteristics, pH, bulk density, texture, and capacity to hold water. It is suggested that FA should be used as a potential material for landfills rehabilitation and soil amendment. Comprehensive research indicates that the incorporation of FA to soil enhances its texture, ability to retain water, and fertility. Its pH is the primary determinant of whether a given type of soil is suitable or not for a specified species. For the best growth conditions for crops, a pH range of 6.5–7.0 is often preferred. According to the necessary quantity of calcium, CFA can have a basic or acidic pH. Consequently, when a certain amount of FA is put to soil, when applied carefully and regularly examined, FA can significantly improve the nutritional content of soil. The use of mineral lime in agriculture contributes to warming the earth by releasing carbon dioxide into the atmosphere – as claimed by the Intergovernmental Panel on Global Climate Change. As a soil improver, substituting FA for lime lowers overall carbon dioxide emissions and contributes to a reduction in climate change (Panda *et al.*, 2020).

2.4.2 As Pesticides

In order to combat pests, the majority of farmers currently employ hazardous chemical pesticides. Nonetheless, it is discovered that most of the time, pesticides have a negative impact on the surrounding area in addition to their intended target. They enter through food chains into the human body. Many beneficial pests are also eliminated by harmful chemical insecticides. In such cases, natural source-based biopesticides that control pests without harming the ecosystem might be taken into evaluation. Fly-ash-based pesticide formulations are widely used because they are inexpensive and readily available. As a result, farmers can simply access these biopesticides at the ground level. Consequently, using FA as a biopesticide is a safe and effective technique to get rid of FA (Panda *et al.*, 2020).

2.4.3 As Catalyst

Coal fly ash's potential as a catalyst is currently the topic of extensive research. CFA can be used as a heterogeneous catalyst due to its massive surface area, heterogeneity, superior heating stability, and ease of availability at a low cost. Because they are easier to recover from the reaction process, heterogeneous catalysts are preferred to homogeneous catalysts. Nickel (Ni) is supported by FA during the carbon dioxide's methane reforming process. The catalytic reduction of nitric oxide by ammonia (NH_3) and catalytic oxidation of volatile organic molecules are two further instances of FA catalysis (Panda *et al.*, 2020).

2.4.4 As Ceramic and Glasses

Oxides such as silicon dioxide (SiO_2), iron (III) oxide, calcium oxide (CaO), and aluminum oxide make up CFA in different amounts. FA is a good beginning material for the production of ceramics, glass ceramics, and glass materials because of its composition. Because of its chemical and mineralogical makeup, CBA is also used in the manufacturing of structural glass ceramics. Highly durable structural glass ceramics that resist acids can be made with class C CFA, which contains iron (III) oxide, silicon dioxide, and aluminum oxide in excess of 50%.

2.5 MICROSTRUCTURE AND MINERALOGICAL PROPERTIES OF BIOMASS FA

A range of techniques for analysis have been utilized to examine the microstructure and mineralogy of biomass ash that comes from various sources. These methods are thought to be helpful in learning about the morphology, crystallographic arrangement, and content of biomass ash particles. The biomass ash particles from various feedstocks exhibit notable morphological differences, as shown by the SEM. The presence of phytoliths and amorphous silica in remaining ash from agriculture, including millet and rice husk ash (RHA) and wheat straw ash, causes the development of particles with coarse textures and irregular forms (Mayumi *et al.*, 2003). As the wood has a variety of cellular structures and burns under different conditions, the ash from wood biomass has a heterogeneous look that is caused by irregularly shaped and sized particles. The mineral phases found in biomass ash were identified by XRD technique (Milan *et al.*, 2014) which showed that crystalline K_2SiO_3 and $CaSiO_3$, as well as amorphous SiO_2, make up the residual ash produced by agricultural activities. It is shown that crystalline calcium species, such as CaO and $CaCO_3$, and potassium make up wood biomass ash (Alehyen *et al.*, 2017). The pattern of different plant-based ash samples by the XRD technique can be seen in Figure 2.4.

2.6 UTILIZATION AND LIMITATIONS OF BIO-CFA

Studies reveal that using biomass ash to make concrete gives a useful purpose to a waste product, making the process of producing biomass energy more circular. Adding biomass ash to other SCMs, for instance, was found to enhance concrete properties and lower the risk of an alkali-silica reaction (ASR). The fresh and hardened properties of biomass-ash-containing concrete were assessed and compared to concrete made with cement or FA by Nagaraju *et al.* (2023) conducted an investigation into the impact of substituting silica fume and ground granulated blast furnace slag (GGBS) for RHA in order to increase the compressive strength of the geopolymer. To optimize the utilization of biomass ash, a number of aspects need

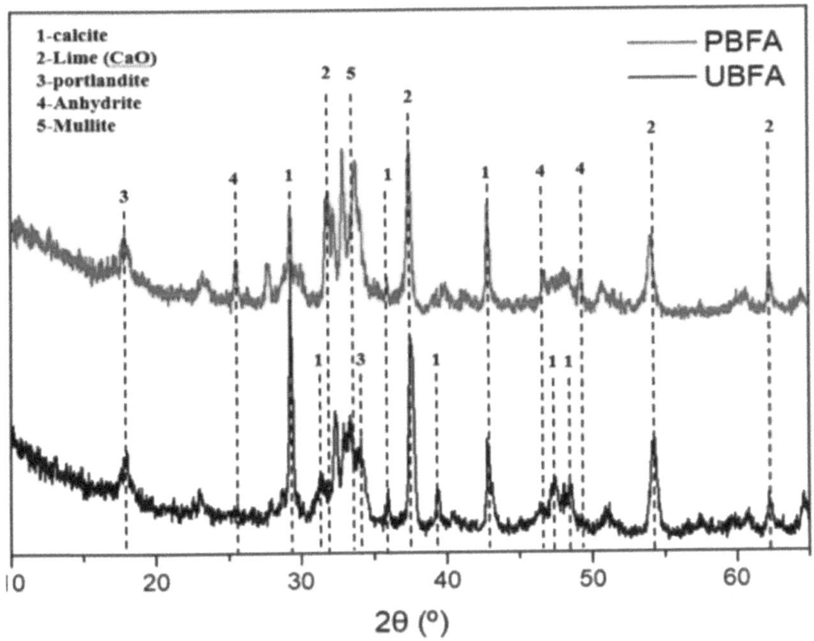

FIGURE 2.4 The pattern of different plant-based ash samples by XRD technique.

to be carefully taken into account. Ash produced by burning plant-based materials can have a wide variety of physical and chemical properties, depending on the source. Applying wood ash and other agricultural wastes can result in a variety of behaviors and qualities since the temperature at which they burn affects a number of these properties. Its pozzolanic and cementitious properties are thereby enhanced by a higher percentage of amorphous silica and calcium oxide, which also has a higher reactivity and density. According to a study, concrete's workability is enhanced by larger particle sizes, while its strength is developed by smaller ones because of their higher reactivity. Because of the proportioning of cement to biomass in concrete, a thorough evaluation is necessary to preserve the material's strength and effectiveness. Reducing the adverse effects of ASR requires the use of pozzolanic materials and alkali-resistant aggregates. The possible ecological issue associated with using biomass waste to make concrete has drawn attention. The characteristics of biomass ash supply and the combustion process affect how the ecosystem reacts to the introduction of bio-ash. The environmental benefits from FA are illustrated in Figure 2.5.

Further study is required in a number of areas in order to get over the drawbacks of using biomass ash in concrete. There are several ways to reduce the alkalinity of biomass ash. The pace at which FA is replaced with biomass ash can be tuned for a variety of applications, and the durability of biomass-ash concrete over time can be evaluated. To provide uniform guidelines for their use in the manufacturing of concrete, it is advised to look at the effects of various types of bio-ash and burning methods on the characteristics of concrete. There aren't many long-term stability and

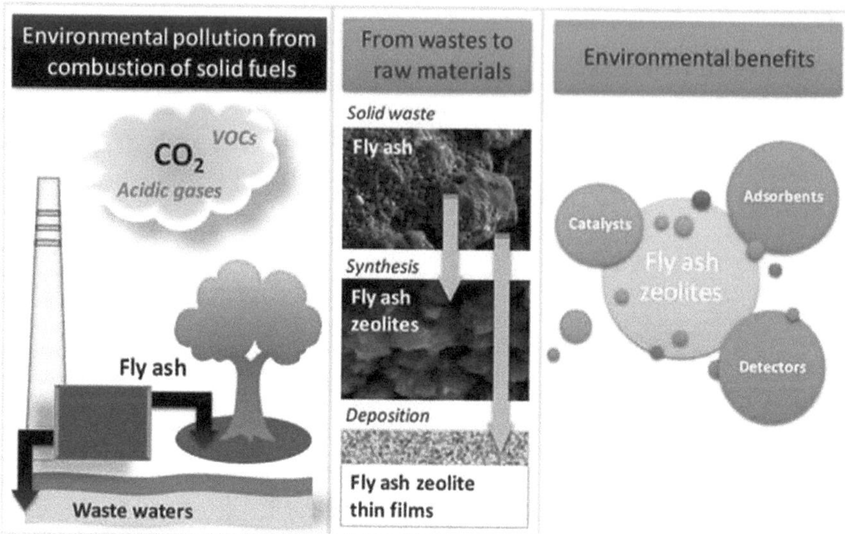

FIGURE 2.5 The process of utilizing coal fly ash.

sustainability studies on concrete made with biomass ash. Concrete that contains biomass ash should be tested over an extended period of time for chemical assault, wetting-drying cycles, and freeze-thaw cycles. When it comes to managing, storing, and processing biomass ash during the concrete-making process, there are no set criteria. This hinders the replacement of bio-ash and FA, making it challenging to compare study findings. Future research should standardize the use of biomass ash in the production of concrete in order to boost its use in construction (Oladele *et al.*, 2023).

2.7 CONCLUSION

The biomass ash has different qualities and can be included with other cementitious materials to create amalgamated biomass ashes and improve the consistency and quality of concrete; more research is needed to maximize the effectiveness of biomass ash in concrete. Examining the viability of adding biomass ash to concrete as an additive offers a viable line of inquiry that can help develop sustainable construction practices and reduce the environmental effect of concrete. Based on filler behavior and pozzolanic activity, CBA is a superior cement alternative than CFA. As a result, the technique of combining wood FA with CFA in the stream has allowed wood FA, a byproduct that was previously useless, to become a valuable resource for making concrete. FA offers high porosity, extensive surface area, suitable pore size, percentage of calcium oxide, and surface function groups that make it suitable as the basis material for various adsorbents that remediate air and water pollutants. It can be produced into fiber membranes and zeolite, among other composite material. Nowadays, a variety of engineering applications consume more than 20 million metric tons of FA annually. In conclusion, both bio-coal and coal-derived FA play key roles in the ongoing transition to more sustainable energy and material practices. The challenge lies in balancing the environmental benefits of bio-coal with the practical utility of FA, ensuring that both materials contribute positively to future energy and environmental goals.

REFERENCES

Alehyen S, Achouri MEL, Taibi M, Characterization, microstructure and properties of fly ash-based geopolymer. *Journal of Materials and Environmental* Science Volume 8, Issue 5, 2017, 1783–1796.

Alterary SS, Marei NH, Fly ash properties, characterization, and applications: A review. *Journal of King Saud University – Science* Volume 33, 2021, 101536.

Apua C, Simate G, Characterization of coal fly ash for the production of coagulant for usage in wastewater treatment, 2018, https://doi.org/10.7449/2018/MST_2018_1512_1521.

Arrigoni A, Panesar DK, Duhamel M, Opher T, Saxe S, Posen ID, MacLean HL, Life cycle greenhouse gas emissions of concrete containing supplementary cementitious materials: cut-off vs. substitution. *Journal of Cleaner Production*, Volume 263, 2020, 121465, ISSN 0959-6526, https://doi.org/10.1016/j.jclepro.2020.121465.

Bentz, D., Durán-Herrera, A., & Galvez-Moreno, D. (2011). Comparison of ASTM C311 strength activity index testing versus testing based on constant volumetric proportions. *Journal of ASTM International*, 9(1), JAI104138.

Bhatt A, Priyadarshini S, Mohanakrishnan AA, Abri A, Sattler M, Techapaphawit S, Physical, chemical, and geotechnical properties of coal fly ash: A global review 2019. *Case Studies in Construction Materials* Volume 11, 2019, e00263.

Geršl M, Mareček J, Matysek D, Tomaš Vitěz, Pavol Findura quantitative determination of mineral phases by the x-ray powder diffraction method in organomineral substrates (digestate, compost and ash from biomass) acta universitatis agriculturae et silviculturae mendelianae brunensis, Volume 62, Issue 5, 2014, https://doi.org/10.11118/actaun201462050905.

Nagaraju VT, Bahrami A, Azab M, Naskar S, Geopolymer concrete and mortar using agricultural biomass—A strength performance and sustainability analysis. *Frontiers in Materials* Volume 10, 2023, Article 1128095, https://doi.org/10.3389/fmats.2023.1128095.

Olatoyan OJ, Kareem MA, Adebanjo AU, Olawale SOA, Alao KT, Potential use of biomass ash as a sustainable alternative for fly ash in concrete production: A review. *Hybrid Advances* Volume 4, 2023, 100076, ISSN 2773-207X, https://doi.org/10.1016/j.hybadv.2023.100076.

Panda L, Dash S, Characterization and utilization of coal fly ash: A review. Volume 9, Issue 3, 921–934 (Issue publication date: September 2020), https://doi.org/10.1680/jemmr.18.00097.

Rajamma R, Ball RJ, Tarelho LAC, Allen GC, Labrincha JA, Ferreira VM, Characterisation and use of biomass fly ash in cement-based materials, *Journal of Hazardous Materials* Volume 172, 2009, 1049–1060, ISSN 0304-3894, https://doi.org/10.1016/j.jhazmat.2009.07.109.

Sigvardsen NM, Ottosen LM, Characterization of coal bio ash from wood pellets and low-alkali coal fly ash and use as partial cement replacement in mortar. *Cement and Concrete Composites* Volume 95, 2019, 25–32, ISSN 0958-9465, https://doi.org/10.1016/j.cemconcomp.2018.10.005.

Tsukada M, Yamada H, Kamiya H, Analysis of biomass combustion ash behavior at elevated temperatures, *Advanced Powder Technology* Volume 14, Issue 62003, 2003, 707–717, ISSN 0921-8831, https://doi.org/10.1163/15685520360731990.

3 Enhancing Mechanical and Chemical Properties of Coal and Bio-Coal Fly Ash

Seetharam Shreemathi Giri, Natrajan Giri, Giri Pranavanathan and Arti Thanki

3.1 INTRODUCTION

Coal is the primary global source of electricity, contributing over 36% of the world's power supply. In the United States alone, coal-fired plants generate nearly a quarter of all electricity consumed. Despite its significant role, coal's share in electricity generation has diminished due to the rise of zero-emission technologies like nuclear and renewables, which combat climate change. However, coal remains crucial in addressing global energy poverty, offering reliable and affordable energy that meets demands consistently. Beyond electricity, coal supports diverse industrial applications. It's essential in steel and cement production and finds use in creating activated carbon for purification systems, carbon fiber for aerospace and automotive sectors, and silicon metal for lubricants and chemical processes. Moving forward, the coal industry's challenge lies in advancing technologies that minimize emissions, particularly carbon dioxide, to mitigate its environmental impact while sustaining its vital industrial contributions. Fly ash, a byproduct of coal combustion in thermal power plants, is an alumino-silicate material composed primarily of silicon dioxide and aluminum oxide (Panda and Dash, 2020). Fly ash poses environmental challenges due to its potential toxicity and the large land areas required for disposal. Fly ash is a fine, powdery byproduct generated from the combustion of coal in thermal power plants. Composed primarily of silica, alumina, and oxides of iron and calcium, fly ash is often collected from the flue gases using electrostatic precipitators or filter bags. While it presents environmental challenges due to its potential toxicity and the vast land areas required for its disposal, fly ash also offers significant opportunities for recycling and reuse. In construction, it is widely used as a supplementary cementitious material, enhancing the strength and durability of concrete while reducing the demand for Portland cement. Additionally, fly ash is utilized in the production of bricks, road embankments, and other construction materials, contributing to sustainable building practices. Despite its benefits, the

 DOI: 10.1201/9781003498124-3

management of fly ash remains a critical environmental concern, requiring careful handling and innovative solutions to mitigate its impact on ecosystems and human health.

3.2 IMPORTANCE OF BIO-COAL

High efficiency, low emissions (HELE) coal technologies, together with carbon capture and sequestration or storage (CCS), are critical to meeting energy needs and climate goals. According to World Coal Association (WCA), "Moving the current average global efficiency rate of coal-fired power plants from 37.5% to 47.5% by deploying more advanced technology could cut 2 gigatons of CO_2 emissions while allowing affordable energy for economic development and poverty reduction" (Figueroa et al., 2008). Deploying HELE coal-fired power plants is also a key first step along a pathway to near-zero emissions from coal with CCS. The challenge lies in making these cutting-edge technologies commercially viable.

Bio-coal is the only solution for all the environmental issues to replace the fossil coals in the industry, and specifically as an energy source. Bio-coal is a carbon-neutral fuel that can replace fossil coal in industry specifically as an energy source. It is produced within the process of biogreen pyrolysis and carbonization of raw biomass performed within controlled temperature and residence time conditions. Thermal conversion of biomass, which is done under the oxygen-free conditions process, allows the removal of volatile organic compounds and cellulose components from the feedstock and creates a uniform, solid biofuel with characteristics similar to the ones in fossil coal. Such fuels are gaining increasing attention in the carbon-intensive industries and the launch of new bio green production facilities dedicated to bio-coal production is heralding a new approach toward the use of fossil-free solid fuels.

3.3 OBJECTIVES AND SCOPE OF THE CHAPTER

The objectives and scope of this chapter are to provide a comprehensive analysis of coal and bio-coal fly ash, focusing on their characterization, mechanical, and chemical properties, as well as enhancement techniques. This chapter aims to explore the physical, chemical, and structural attributes of fly ash, assess the impact of particle morphology on mechanical properties, and examine various surface modification and functionalization approaches to improve chemical properties. Additionally, it compares the characteristics of bio-coal fly ash (BFA) with conventional coal fly ash (CCFA), considering environmental implications and market applications. This chapter also discusses experimental results of enhancement techniques, highlights regulatory challenges, and identifies emerging trends and research gaps, offering recommendations for industry and future research.

3.4 NATURE AND COMPOSITION OF FLY ASH

It is primarily composed of harmful heavy metals such as arsenic, mercury, and lead. These micron-sized earth elements consist primarily of silica, alumina, and iron. When mixed with lime, fly ash forms a cementitious compound with properties

TABLE 3.1
Typical chemistry of coal fly ash (in wt. %)

	Class F*	Class F*	Class C*	Class C*
	Low-Fe	High-Fe	High-Ca	Low-Ca
SiO_2	46–57	42-54	25–42	46–59
Al_2O_3	18–29	16.5–24	15–21	14–22
Fe_2O_3	6–16	16–24	5–10	5–13
CaO	1.8–5.5	1.3–3.8	17–32	8–16
MgO	0.7–2.1	0.3–1.2	4–12.5	3.2–4.9
K_2O	1.9–2.8	2.1–2.7	0.3–1.6	0.6–1.1
Na_2O	0.2–1.1	0.2–0.9	0.8–6.0	1.3–4.2
SO_3	0.4–2.9	0.5–1.8	0.4–5.0	0.4–2.5
LOI	0.6–4.8	1.2–5.0	0.1–1.0	0.1–2.3
TiO_2	1–2	1–1.5	<1	<1

very similar to that of Portland cement. It is generally categorized into two: Class F and Class C based on chemical composition, which is influenced by the type of coal burned during its production (Vassilev and Vassileva, 2007). Class F fly ash is produced from burning bituminous coal and is pozzolanic, requiring a cementitious material to react. Class C fly ash comes from sub-bituminous or lignite coal and is both pozzolanic and cementitious, meaning it can set and harden on its own. A standard specification for coal fly ash is ASTM International (American Society for Testing and Materials). It outlines the chemical and physical properties of fly ash and natural pozzolans, along with guidelines for their use in concrete mixtures (Table 3.1).

3.5 CHARACTERIZATION OF FLY ASH

3.5.1 Physical Characteristics

The color of fly ash varies from tan to gray or black, primarily depending on its carbon content: Darker shades indicate higher carbon levels. Consequently, fly ash derived from lower-grade coals (like lignite and sub-bituminous) tends to be lighter in color, ranging from tan to buff, while fly ash from higher-grade coals (such as anthracite and bituminous) appears darker, from gray to black. Additionally, the calcium oxide content influences color, with lower-grade coals containing more calcium, which imparts a white shade to the fly ash. The specific surface area and specific gravity of fly ash typically range from 2,000 to 6,800 cm^2/g and 2.1 to 3.0 g/cm^3, respectively (Pandian et al., 1998). Particle sizes of fly ash vary geographically (Goodarzi, 2006). U.S. fly ash has sandy particles ranging from 2 to 0.5 mm and silt particles from 0.05 to 0.002 mm, while Indian fly ash has sandy particles from 4.75 to 0.075 mm and silt particles from 0.075 to 0.002 mm. Clay particles are less than 0.002 mm in both U.S. and Indian fly ashes. U.S. fly ash contains about 32.4% sandy particles, while Indian fly ash has 35.69%, showing a 2%–4% higher sandy content. Silty particles exceed 60% in both, but U.S. fly ash has slightly more at 63.2%, compared to 62.39% in Indian fly ash. Clay particles constitute 4.3% of U.S. fly ash versus 1.91% of Indian fly ash, indicating a 2%–3% higher clay content in the U.S. fly ash. Additionally, fly ash variation is influenced by the structural properties of particles such as cenospheres, plerospheres, ferrospheres, and angular-shaped carbon particles. Cenospheres, with a bulk density of 0.4–0.6 ton/m^3, make up to 5% of the total fly ash weight (Moreno et al., 2005).

3.5.2 Morphology of Fly Ash

The scanning electron microscope (SEM) images the sample surface by scanning it with a high-energy beam of electrons in a faster scan pattern. A wide range of magnification ranging from 10 to 500,000 times is carried out, about 250 times magnification limit of the best light microscopes. Back-scattered electrons (BSE) are often used in analytical SEM along with the spectra made from the characteristic X-rays. Because the intensity of the BSE signal is strongly related to the atomic number (Z) of the specimen, BSE images can provide information about the distribution of different elements in the sample. The characteristic X-rays are used to identify the composition and measure the abundance of elements in the sample. Chemical analysis in the SEM is performed by measuring the energy or wavelength and intensity distribution of X-ray signals generated by a focused electron beam on the specimen (Sun et al., 2020).

With the attachment of the energy dispersive spectrometer (EDS) or wavelength dispersive spectrometer (WDS), the precise elemental composition of the material can be obtained with high spatial resolution. From bulk specimens in the SEM, very precise accurate chemical analysis (relative error – 1%–2%) can be obtained from larger areas of the solid (0.5–3 μm dia) using an EDS or WDS. Using the above techniques, researchers were able to study the morphology of fly ash particles. It was found that fly ash is a sphere-shaped, micron-sized (0.01–100 μ) heterogeneous material, having depositions of mainly aluminum, silicon, iron, and carbon in variable compositions on its surface, and closely resembles the volcanic ashes. The fly ash particles may be either rough or smooth surfaced based on the type of depositions on their surface. Morphologically, fly ash particles may have differently shaped particles, which also vary in their elemental composition viz. ferrospheres (ferrous-rich spherical particles), cenospheres or aluminosilicate spheres (Al- and Si-rich particles), plerospheres (larger spherical particles encapsulate smaller particles), and carbon nanomaterial—i.e., soots, buck balls, fullerenes, and unburned carbon, including both organic and inorganic materials. Cenospheres are more dominant structures in the fly ash, followed by the ferrospheres, which are spherical-shaped ferrous-rich particles, whose sizes fall in the micron range. The ferrospheres have high depositions of ferrous or Fe, which could be either rough, smooth, elliptical, or molten drop-shaped. Besides cenospheres and ferrospheres, there is the third type of micron-sized spherical-shaped particles, called plerospheres, which are less frequent in fly ash in comparison to the other two forms. These plerospheres encapsulate several small fly ash particles, minerals, and gases inside them during the formation from the molten slag at high temperatures in the furnace. Due to the combustion of organic and inorganic carbon minerals present in the coal, there are a large number of carbonaceous nanomaterials, such as fullerenes, graphene, soots, and unburned irregular-shaped carbon particles are also present in fly ash (Inkson, 2016).

3.5.3 Surface Area Analysis

The specific surface area is defined as the ratio of the absolute surface area (A) of a solid to its mass (m), with the unit of measurement being m^2/g. The surface area includes all parts mainly pore wall inner surfaces (mainly pore wall surfaces) (Dollimore et al., 1976). The specific surface area of fly ash is estimated from the amount of nitrogen adsorbed in relationship with its pressure, at the boiling temperature of liquid nitrogen under normal atmospheric pressure.
The Brunauer-Emmett-Teller (BET) method is widely used for calculating surface areas of porous materials through gas adsorption analysis (Dollimore et al., 1976). It has been applied to study fly ash, a coal combustion byproduct, revealing specific surface areas ranging from 2 to 40 m^2/g. BET analysis of water adsorption on fly ash from various sources demonstrated that water uptake occurs primarily on the particle surface, with adsorption behavior influenced by the source region and power plant efficiency. The method's applicability extends to metal-organic frameworks (MOFs), providing useful surface area estimations for most MOFs, although some exceptions exist. BET analysis has shown that fly ash can adsorb more than a monolayer of water under normal atmospheric conditions, suggesting its potential environmental impact (Agarwal et al., 2016). These studies highlight the BET method's importance in characterizing porous materials and understanding their adsorption properties.

3.6 CHEMICAL COMPOSITIONAL ANALYSIS

3.6.1 Fourier Transform Infrared Analysis

Fourier Transform Infrared (FTIR) analysis is crucial for characterizing fly ash by identifying its functional groups, assessing amorphous and crystalline phases, and evaluating surface chemistry. It detects trace elements and organic compounds, aiding in environmental assessments. Additionally, FTIR supports quality control, helps understand reactivity, and ensures safe utilization of fly ash in various applications.

The FTIR spectrum of fly ash from Malaysian power plants has been investigated to assess its suitability for geopolymer synthesis and radioactive waste immobilization. Studies have shown that Malaysian fly ash contains essential components like silica, alumina, and calcium oxide, with particle sizes predominantly in the 1–15 μm range (Siyal et al., 2016). The chemical composition analysis showed that class F fly ash has more Si and Al content compared to Class C. There are six band ranges in the FTIR analysis. The band area detail of 4,000–3,500 cm^{-1} indicates the loss of $Ca(OH)_2$, 3,500–1,600 cm^{-1} indicates the stretching of the (–OH) bond and bending of the (H-O-H) vibrations, 1,600–1,000 cm^{-1} indicates the gains of the Si-O-Si bond typical of quartz, 1,000–800 cm^{-1} indicates the loss of the $CaCO_3$ gel, 800–500 cm^{-1} indicates the symmetric stretching of the Si-O-Si and Al-O-Si bonds, and the last band is less than 500 cm^{-1} indicates the bending of the Si-O-Si and O-Si-O bonds studied the FTIR spectrum of four coal-based fly ashes. The results reveal that $SiO_2$is the most common mineral in all samples. The asymmetric stretching vibration of bridging oxygen (Si-O-Si) appears between 1,086 and 1,102 cm^{-1}, with an additional peak at 1,058 cm^{-1} observed in the FTIR spectra of fly ash. This vibration suggests the presence of intrinsic impurities of other cations or the multiphase presence of silica in the tetrahedral arrangement of SiO_2. The symmetric stretching vibration of Si-O-Si (quartz) is identified as a double peak band within the range of 776–778 cm^{-1} and 794–798 cm^{-1} in all four coal-based fly ashes. Peaks at wavenumbers between 692 and 694 cm^{-1} represent the symmetric vibrations of Si-O bonds, while peaks between 456 and 462 cm^{-1} indicate asymmetric vibrations of Si-O bonds in fly ashes. A high-intensity peak at 434 cm^{-1} is observed along with other shoulder peaks at 450-470 cm^{-1}. These may indicate either a shift in the Si-O bond peak or the presence of another metal oxide bond in the amorphous phase. The presence of quartz can be identified from the peaks at 456–462 cm^{-1}, 692–694 cm^{-1}, 776–778 cm^{-1}, 1,086–1,102 cm^{-1}, and 1,058 cm^{-1}. Anhydrite is indicated by peaks at 594, 678, and 1,098 cm^{-1}. A broad peak between 800 and 1,200 cm^{-1} in the FTIR spectrum of coal fly ash suggests peak degeneration due to overlapping.

Mullite is present in coal fly ash with peaks at 596, 736, 826, and 886 cm^{-1} (Gupta et al., 2020) Additional peaks in the FTIR spectra, such as 564 cm^{-1} in fly ash, 564 and 612 cm^{-1} and 414, 486, 506, 524, 562, and 584 cm^{-1} indicate the presence of organic or inorganic bonds, which are crucial for the reactivity of fly ash and the presence of organic or inorganic crystalline bonds (Zhang et al., 2012). FTIR spectroscopy in combination with particle size analysis is used to predict the reactivity of fly ash. The above bands provide the fingerprint of the composition of different types of fly ash (Figures 3.1 and 3.2).

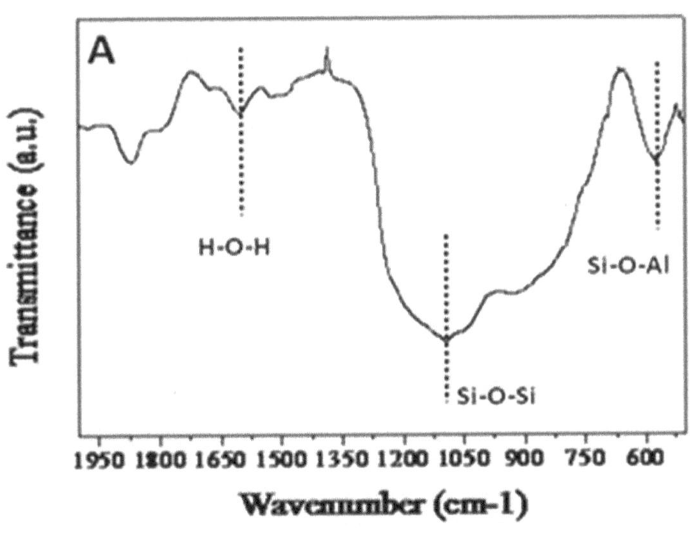

FIGURE 3.1 FTIR spectrum of fly ash from India.

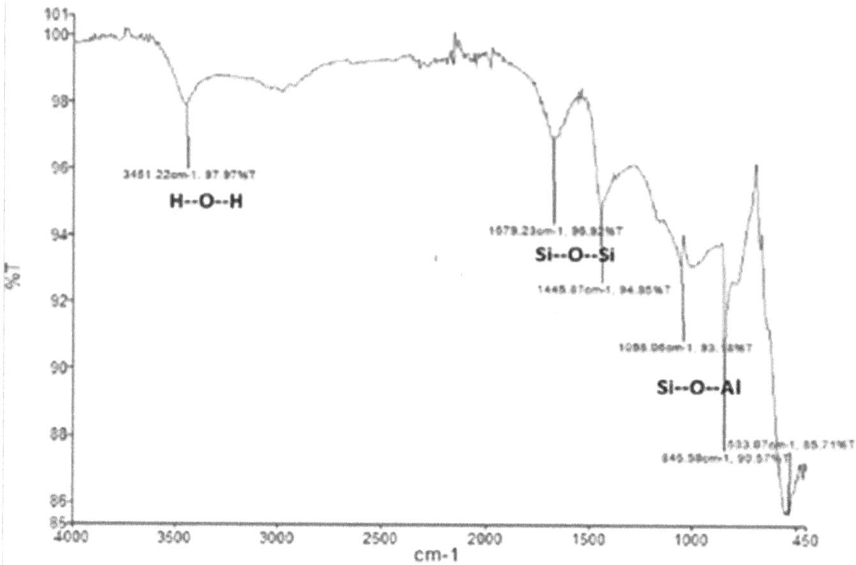

FIGURE 3.2 FTIR spectrum of fly ash from Pakistan.

3.6.2 Raman Spectroscopy

Raman spectroscopy has emerged as a valuable tool for characterizing fly ash and cement-based materials. It can identify mineral compositions in raw materials and hydration products in fly ash paste (Garg and Wang, 2012). The technique has been used to analyze unburned carbon in fly ash, revealing graphitization and crystallization processes. Tai et al. (2010) observed bands at 1,350 cm^{-1} (D-band) and 1,580 cm^{-1} (G-band) for carbonaceous material such as amorphous carbon. In waste-to-energy fly ashes, photobleaching for two minutes significantly reduces fluorescence, enabling the detection of calcite, calcium sulfate, zincite, and carbon. Raman spectroscopy can also assess the pozzolanic reactivity of fly ashes by studying the evolution of sulfates and hydroxides in cement-fly ash pastes over time (Garg, 2012). The method has shown good agreement with traditional strength activity index tests (Garg and Wang, 2012). These studies demonstrate the potential of Raman spectroscopy for rapid monitoring of fly ash mineralogy and reactivity in various applications. The "Raman bands" refer to the specific wavenumber ranges of 950–1,050 cm^{-1} for studying the evolution of sulfates, and 3,600–3,700 cm^{-1} for studying the evolution of hydroxides, as used in the Raman spectroscopy analysis of the cement and fly ash mixtures. Common Raman bands in fly ash include a strong peak at 464 cm^{-1} which is characteristic of quartz, peaks at 225 cm^{-1} and 500–700 cm^{-1} for hematite, and bands between 400–500 cm^{-1} and 800–1,100 cm^{-1} for aluminosilicates like mullite. Additionally, calcite exhibits a prominent peak at 1,086 cm^{-1}, while magnetite is identified by a characteristic peak at 670 cm^{-1}. Proper sample preparation, involving finely grinding and uniformly spreading the fly ash, is crucial for obtaining clear Raman spectra. The choice of laser wavelengths, such as 532 nm or 785 nm, can affect the Raman signal and fluorescence background. Data interpretation requires comparison with reference spectra of known minerals and compounds.

3.6.3 Nuclear Magnetic Resonance Spectroscopy

Nuclear Magnetic Resonance (NMR) spectroscopy has been extensively used to analyze fly ash and its activation products. Studies have employed various NMR techniques, including ^{1}H, ^{13}C, ^{29}Si, ^{27}Al, and ^{23}Na MAS NMR, to investigate the structural properties and hydration kinetics of fly ash (Palomo et al., 2004). NMR spectroscopy has proven to be a valuable tool for analyzing fly ash and its activation products. Studies have shown that the alkaline activation of fly ash leads to the formation of amorphous tectosilicates, with the Si/Al ratio increasing over time. NMR analysis reveals that activation changes the fly ash structure from high to low polymerization, producing silicon-oxide bodies with end groups (Wen-xun, 2004). Multinuclear NMR techniques have been used to study hydration kinetics, structural properties, and adsorption characteristics of coal fly ash, demonstrating differences in hydration behavior compared to Portland cement. Advanced geo polymerization techniques, such as mechanochemical co-grinding, have been explored using multinuclear MAS NMR, revealing changes in Si/Al/Na coordination environments and framework connectivity. These findings contribute to a deeper understanding of fly ash activation processes and the potential development of improved cementitious materials.

3.6.4 X-Ray Diffraction Studies

X-ray Diffraction (XRD) is one of the primary methods used to characterize the structural properties of coal fly ash. This technique allows the identification of the crystalline phases present in the fly ash, which can provide information about the mineralogical composition and the degree of amorphization. XRD is a valuable technique for studying the mineralogy and composition of fly ash. Research has shown that fly ash typically contains a high glass content, with diffuse scattering maxima observed in X-ray diffractograms (McCarthy and Johansen, 1988). The crystalline phase assemblage varies depending on the source coal and calcium content, with low-calcium (Class F) fly ashes generally containing quartz, mullite, hematite, and magnetite, while high-calcium (Class C) fly ashes may have a more complex composition including anhydrite, tricalcium aluminate, and lime (McCarthy, 1987; McCarthy and Johansen, 1988). Quantitative XRD analysis using the Rietveld method has been employed to determine the glass content and crystalline phase composition of fly ash samples, with glass content ranging from 65% to 78% by weight. The presence of aluminous cement in some fly ashes can explain their high strength development after hydration. Particle size is an important parameter in processing reinforcement particles utilized as particulate metal matrix composites (MMCs). These particulate reinforcements provide advantages such as high stiffness, high strength, wear resistance properties, corrosion, and thermal resistance to MMCs. The study of the effects of ball milling on fly ash's density, microstructure and morphology was conducted. The milling times were varied as 0, 15, 30, 45, 60 and 180 minutes. The density of the fly ash samples was measured as 2.44 g/cm^3 using Le Chatlier's principle. The microstructural analysis showed a change in morphology with increasing milling times. The SEM micrographs showed an agglomeration of the fly ash particles due to the dry milling conditions. The EDS analysis showed the presence of oxygen, aluminum, silicon, and iron as the major elements common to all the fly ash samples at different milling times. The XRD analysis identified mullite ($3Al_2O_3SiO_2$), silica (SiO_2), alumina (Al_2O_3), wustite (FeO), and hematite (Fe_2O_3) as the major phases.

3.7 MECHANICAL PROPERTIES ENHANCEMENT

3.7.1 Influence of Particle Morphology and Size Distribution

Fly ash, a byproduct of coal combustion, has diverse applications, particularly in MMCs and concrete production. The particle size and morphology of fly ash are crucial parameters influencing its effectiveness and the properties of the final products in the following applications.

3.7.1.1 Concrete Production

In concrete mixtures, finer fly ash particles enhance workability and compressive strength, as reported by researchers. The increased surface area provided by finer particles contributes to better packing density and pozzolanic reactions, leading to improved strength and durability of the concrete. Ball milling of fly ash can significantly alter its physical and chemical properties, enhancing its reactivity and potential applications in concrete mixtures. Also milling reduces particle size, increases surface area, and modifies morphology, leading to improved pozzolanic activity and was used to study the effect of milling times on the microstructural morphology and density of fly ash at different milling times. The analysis revealed an acceptable value for density which makes fly ash an ideal candidate for producing lightweight materials. The SEM analysis revealed the presence of agglomeration due to cold welding at different milling times. It also revealed a slight change in the shapes of the fly ash particles. The efficiency of milling depends on the equipment type and duration, with attrition mills showing better results than vibration mills in terms of particle size reduction and reactivity (Chu et al., 2019). The mechanical activation of fly ash can increase its industrial usage beyond the typical 25 wt% limit by enhancing its reactivity and early strength development. These modifications in fly ash properties lead to improved compressive strength and overall performance in concrete applications.

3.7.1.1 Unburned Carbon Content

Fly ash often contains unburned carbon, which is typically found in the larger particles. This unburned carbon content has a significant impact on the fly ash's surface area and sorption capacity, making it suitable for contaminant removal applications (Zhu et al., 2013). The sorption capacity is crucial for environmental applications where fly ash is used to remove heavy metals and other pollutants from wastewater.

3.7.1.3 Metal Matrix Composites

In MMCs, the particle size and distribution of fly ash influence the mechanical properties, tribological behavior, and thermodynamic characteristics of the composites. The incorporation of fly ash and basalt ash as reinforcements in aluminum MMCs significantly influences their mechanical properties. Studies on Al6061 and A356 MMCs reinforced with varying weight fractions of fly ash and basalt ash show that tensile strength and impact strength increase with higher fly ash content, while hardness improves with increased basalt ash (Kumaravel and Mohanraj, 2017). By varying the weight fraction of the reinforcements (2.5% basalt ash + 7.5% fly ash, 5% basalt ash + 5% fly ash, 7.5% basalt ash + 2.5% fly ash), three sets of hybrid MMC were prepared. The result reveals that impact strength and tensile strength increased with increasing weight fraction of fly ash, whereas hardness increased with increasing weight fraction of the basalt ash. Smaller, uniformly distributed fly ash particles can enhance the reinforcement effect by improving the load transfer between the matrix and the particles. The rheological behavior of the matrix melts, methods for incorporating the particles, and the interactions between the particles and the matrix are critical factors that determine the effectiveness of the reinforcement (Dey and Pandey, 2016). Particle size of fly ash also affects mechanical properties, with smaller particles enhancing tensile strength, compressive strength, and hardness. Stir casting is identified as a cost-effective manufacturing process for producing these composites. Microstructural analysis reveals uniform dispersion of reinforcements in the matrix, thereby contributing to the enhanced properties of the composites (Kumaravel and Mohanraj, 2017).

3.7.2 Strength and Durability Assessments

Fly ash, a byproduct of coal combustion, is increasingly recognized for its potential as a partial cement replacement in concrete, offering significant improvements in durability, particularly in sulfate and alkali resistance, as well as reducing chloride ion penetration and water absorption (Nokken et al., 2014). While higher replacement levels of fly ash generally lead to reduced compressive strength, optimal strength is typically observed at 5%–15% replacement, with hydration time further enhancing strength (Reddy and Reddy, 2015). Studies suggest that 40% fly ash replacement provides the best overall durability, contributing to both enhanced performance and environmental sustainability by reducing CO_2 emissions and utilizing industrial waste. Additionally, fly ash has shown promise in the production of geopolymer bricks with strong acid resistance. The collective findings underscore fly ash as a viable and sustainable cement alternative in concrete applications.

3.7.3 Testing Methodologies for Mechanical Properties of Fly Ash Composites

Research has shown that incorporating fly ash as a reinforcement in materials such as epoxy resin, High-Density Polyethylene (HDPE), natural rubber, and concrete generally enhances mechanical properties, including tensile and flexural strength, impact resistance, hardness, and compressive strength (Ahmad and Mahanwar, 2010). Smaller fly ash particles are particularly effective in improving strength and elongation in HDPE composites. Thermal analysis techniques like TG and differential scanning calorimetry (DSC) indicate improved thermal stability. Additionally, fly ash acts as an efficient flame retardant, enhancing fire resistance in epoxy composites. Various manufacturing techniques, including stir casting and hand lay-up, are used to produce fly ash composites (Manimaran et al., 2018).

3.8 CHEMICAL MODIFICATION TECHNIQUES

3.8.1 Surface Properties Modification

Surface modification of fly ash to enhance its suitability for the plastic industry involves various techniques that aim to improve its whiteness, frictional properties, and specific surface area. Common methods include chemical treatments, such as the use of alkyl silane coupling agents, which enhance adhesion between fly ash and the polymer matrix, thereby improving mechanical properties and compatibility. Titanate and zirconate coupling agents are also used to enhance the dispersibility of fly ash within the polymer matrix. Additionally, acid and alkali treatments can clean the fly ash surface, remove impurities, and increase surface activity, further enhancing its performance in plastic composites.

3.8.2 Chemical Treatment Methods

Chemical treatment methods for fly ash have been extensively studied to enhance its properties and applications. Acid treatment with 1M HCl has been shown to increase the specific surface area of fly ash, improving its adsorption capabilities for organic compounds in wastewater treatment. XRD pattern of fly ash. (a) As-received fly ash; (b) heat-treated fly ash; (c) acid-treated fly ash are shown in the figure (Bada and Potgieter-Vermaak, 2008). Various characterization techniques, including microscopy, spectroscopy, and chemical analyses, are employed to assess fly ash composition and properties. Chemical fixation using ferrous sulfate solutions has demonstrated effectiveness in reducing the mobility of trace elements like As, B, Cr, Mo, Ni, Se, Sr, and V in fly ash, potentially improving its environmental safety and utilization potential. Additionally, fly ash containing 10%–12% CaO has been successfully used to neutralize acid mine water, offering a cost-effective treatment method that addresses both fly ash disposal and acid mine drainage issues (Pradhan and Deshmukh, 2008) (Figure 3.3).

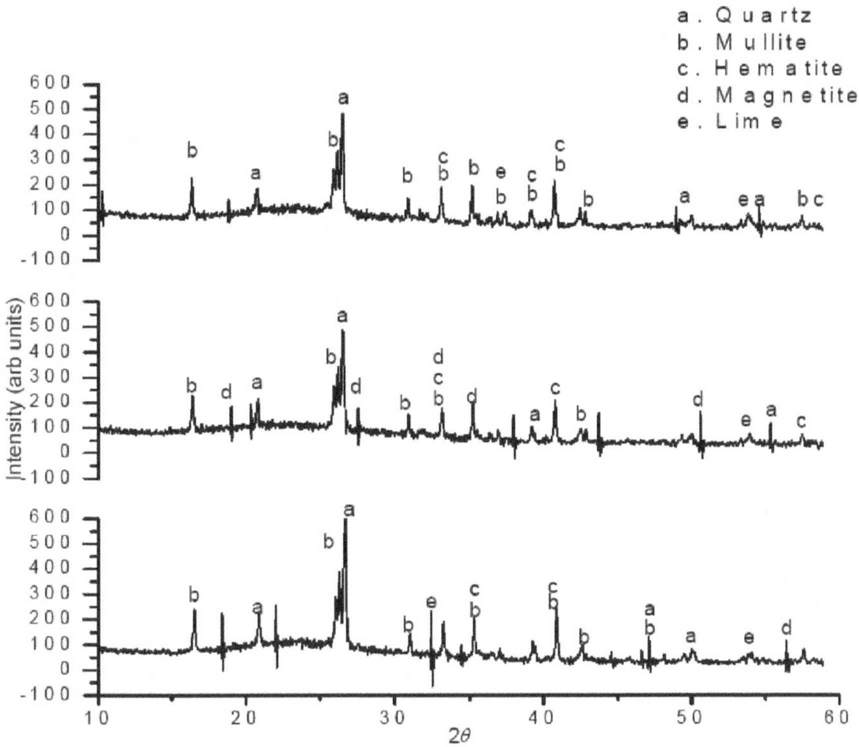

FIGURE 3.3 X-ray diffraction pattern of fly ash. (a) As-received fly ash; (b) heat-treated fly ash; (c) acid-treated fly ash.

3.8.3 Combined Methods

Combining chemical and physical methods can significantly enhance fly ash properties, making it a viable filler in the plastic industry. Mechanical milling followed by chemical treatment improves surface area, whiteness, and frictional properties, leading to better performance and cost-effectiveness in polymer composites. Various techniques for surface modification, such as mechano-chemical (MC) treatment, isothermal heating, planetary ball milling, and chemical treatments, have been explored to optimize fly ash characteristics (Kato et al., 2019). MC treatment, for example, involves analyzing changes in size, surface morphology, and crystal structure of fly ash, leading to insights into "grinding" and "activation" effects. While acid and base treatments have shown limited effectiveness in improving reinforcement in polyisoprene compounds, chemical modifications with silane coupling agents have proven effective in enhancing dispersion and mechanical properties of fly ash in high-temperature polyether ketone composites. These surface modification methods enhance fly ash's performance in diverse applications, including geopolymers, composites, and wastewater treatment adsorbents.

3.9 BIO-COAL FLY ASH – PROPERTIES AND COMPARATIVE ANALYSIS

BFA differs significantly from CCFA in terms of composition and properties. Kjersgaard et al. (2007) describe a project called Bio Crete that aims to facilitate and increase the reuse of bio ash (sludge ash) for the production of concrete, and the key findings include that bio ash can be used to partially substitute fly ash in concrete up to 50%, but bio ash has less pozzolanic effect than fly ash, and aluminum-based bio ash is better for concrete production than iron-based bio ash. BFA generally has lower pozzolanic activity compared to CCFA, limiting its potential as a cement replacement. Cement was replaced with 20%, 40%, and 60% fly ash, and biomass fly ash replaced 0%, 50%, and 100% of coal fly ash. The compressive strength of concrete made with biomass fly ash was similar to that of concrete made with coal fly ash. Concrete made with biomass fly ash showed similar or better abrasion resistance compared to concrete made with coal fly ash. BFA typically contains lower concentrations of heavy metals compared to literature values for CFA, but still exceeds regulatory limits for use as a soil conditioner. Despite these challenges, BFA has potential for metal recovery and can be used as a partial replacement for CCFA in concrete mixtures, with a 50% substitution rate being a common maximum (Kjersgaard et al., 2007).

3.9.1 Environmental Impact

When it comes to environmental impact, traditional coal combustion releases large quantities of greenhouse gases, sulfur dioxide, nitrogen oxides, and particulate matter into the atmosphere, contributing to acid rain, air pollution, and climate change. In contrast, bio-coal combustion produces fewer emissions of sulfur and nitrogen compounds. Reduced levels of greenhouse gases (carbon dioxide) are produced. Moreover, bio-coal offers the potential for carbon sequestration when produced from sustainably managed biomass sources. By locking carbon into a stable solid form, bio-coal can help balance carbon emissions from other sources and contribute to efforts to lessen climate change.

Both bio-coal and fly ash are important in the context of sustainability and waste management. Bio-coal represents a way to utilize biomass residues and reduce greenhouse gas emissions, while fly ash can be recycled into useful products, reducing the environmental impact of coal combustion. BFA, a byproduct of biomass pyrolysis, has a distinctive elemental composition that reflects its origins and processing methods. Typically, BFA contains a high carbon content, ranging from 50% to 90% by weight, which results from the carbonization of biomass during pyrolysis. The hydrogen content is relatively low, generally between 2% and 6%, as much of it is lost in the process. Oxygen levels vary widely, from 5% to 40%, and influence the chemical reactivity of the bio-coal. Nitrogen is present in small amounts, usually 0.5%–2%, which impacts its role as a soil amendment. The mineral content, or ash, in BFA ranges from 5% to 20% and includes key nutrients such as calcium (1%–10%), potassium (1%–5%), and magnesium (1%–3%), all of which contribute to soil fertility and structure. Phosphorus and sodium are also present but in smaller amounts. Sulfur content is typically less than 1%, and trace elements like iron, manganese, zinc, and copper are found in minimal concentrations. The composition can vary based on the biomass feedstock and pyrolysis conditions but overall, BFA is known for its lower levels of heavy metals compared to coal fly ash, making it a more environmentally friendly option for applications such as soil enhancement and carbon sequestration.

Under Europe's REACH (Registration, Evaluation, Authorization, and Restriction of Chemicals) regulations, biomass ash has been classified as a UVCB substance (Unknown or Variable Composition, Complex Reaction Product, or Biological Origin), containing components such as oxides of calcium, sodium, potassium, silicon, iron, manganese, magnesium, aluminum, phosphorus, titanium, and sulfur. The associated chemical safety assessment determined that biomass ash does not require hazardous classification under REACH. Extensive research has focused on characterizing "clean" biomass ash, particularly from wood-fired boilers in Scandinavia, where it is permitted for use as a forestry fertilizer. Various studies and databases have compiled data on the macro and trace element composition of biomass ash, with comparisons made to coal ash and soil. International Energy Agency collected data from 560 ash samples from biomass facilities across six countries. In Europe, additional data have been compiled by the (Swedish University of Agricultural Sciences) and The Energy Centre of the Netherlands, although much of this information overlaps. The data reflect biomass boilers of varying types, sizes, and fuels, leading to a wide range of trace element values. It's important to note that metal concentrations can vary between different ash streams, such as bottom ash and fly ash, with metals often being more concentrated in fly ash. Generally, ash from burning straw, cereal, and grasses contains lower metal levels compared to ash from woody material and bark. Trace metal concentrations in ash from waste wood, commonly used for electricity generation in Europe, are higher than in clean wood ash. Compared to coal ash, clean wood ash has lower levels of arsenic, chromium, and nickel but higher levels of manganese, cobalt, cadmium, and zinc. Despite these differences, these metals make up less than 2% of the total ash composition, and, as long as occupational exposure limits for dust are met, exposure to these metals is unlikely to exceed limit values. Jokiniemi et al. (2008) studied rats exposed to fine fly ash from coal, biomass, and coal/biomass co-firing and found no significant impact on lung inflammation from biomass-derived fly ash, unlike coal and coal/biomass ash, which did elicit effects. The study suggests that the higher soluble salt content in biomass ash may contribute to its reduced impact, as it can dissolve and be eliminated from the lungs more readily than the less soluble coal ash.

3.9.2 Radioactivity

Concerns about the potential radioactivity of biomass ash arise from the possibility that natural or artificial radioactivity present in plant material may become concentrated in the ash after combustion. While natural radiation is generally considered negligible, there is greater concern about anthropogenic radionuclides, which may be present in higher concentrations in plants and soils in regions affected by nuclear fallout (Pitman, 2006). The primary radionuclides of concern are cesium-137, with a half-life of 30.17 years, and strontium-90, with a half-life of 28.8 years. The long half-lives of these isotopes mean that contamination can persist for many decades following a nuclear event, such as the Chernobyl or Fukushima disasters, where significant amounts of both isotopes were released.

A limited number of studies have investigated the potential occupational radiation risks associated with biomass fuel. For instance, a study at a peat-fired power station in Ireland assessed exposure through on-site monitoring of airborne dust, radon levels, and ambient gamma dose rates. The findings indicated that workers involved in various plant operations did not receive a radiation dose above the concern level set by the Irish government, with a calculated dose of 0.3 mSv per year compared to the action level of 1 mSv per year (Cocchi, 2011). Additionally, radiation exposure risks to workers have been explored in areas where the biomass fuel stock may be contaminated with radioactivity.

3.10 MARKET APPLICATIONS AND TRENDS

Bio ash, a byproduct of the increasing use of renewable resources in energy production, presents ecological and economic challenges for disposal. However, due to its favorable chemical and physical properties, bio ash has potential applications in the construction industry, particularly as raw materials. Research has highlighted the potential of BFA, which is chemically similar to Class C fly ash, in road construction and as a supplementary cementitious material (Rajamma et al., 2009). Though BFA can be used in lower dosages (<20%) due to concerns about workability and chemical content, studies have shown that a blend of 20% BFA and 10% metakaolin effectively inhibits alkali-silica reaction (ASR) in cement mortars and concrete. Furthermore, research on biomass bottom ash (BBA) has explored its use in various construction applications, including the manufacture of roller-compacted concrete (RCC), paver-compacted concrete (PCC), and conventional concrete. BBA has been found to increase porosity and decrease mechanical strength due to its high absorption and rough, porous nature. However, despite these drawbacks, BBA has demonstrated stabilizing properties, particularly in expansive soils, and meets the minimum requirements for lightweight structural concrete. Market trends show a growing focus on sustainability, with green policies and corporate responsibility driving interest in bio-coal and BFA as cleaner alternatives to traditional materials. Technological advancements in pyrolysis and bio-coal production are improving efficiency and product quality, leading to increased awareness and adoption in sectors such as agriculture and energy. The market for BFA is expanding globally, with diverse applications emerging, including specialty products and construction materials. Economic factors, such as the cost-effectiveness of BFA production and government incentives for renewable energy, play a significant role in market growth (Table 3.2).

TABLE 3.2
Enhancement of techniques and outcomes

Enhancement technique	Specific outcomes	Associated costs	Cost range/ per ton	Applicability
Alkali treatment (Antoni et al., 2012)	**Increased reactivity**: Activates silica and alumina, enhancing pozzolanic activity.	**Chemicals**: Moderate to high	\$100–\$2,000	Suitable for cement and concrete products.
	- **Improved mechanical strength**: Boosts compressive and tensile strength in cementitious materials	**Process**: Moderate	\$500–\$1,500	Effective in high-performance concrete.
	Enhanced workability: Reduces water demand, making the mix easier to handle	**Overall**: Moderate to high	\$500–\$2,000	Often used in concrete with strict performance criteria.
Acid treatment (Sharma et al., 2015)	**Surface purification**: Removes impurities, such as carbonates, improving reactivity	**Chemicals**: High	\$500–\$5,000	Ideal for applications requiring high durability.
	Increased durability: Enhances resistance to chemical attacks and ASR.	**Environmental Management**: High	\$1,000–\$5,000	Best for environments with exposure to aggressive chemicals.
	Potential environmental impact: Risks associated with acid handling and waste management	**Overall**: High	\$1,500–\$5,000	Use is limited due to environmental concerns.
Silane coupling agents (Alkadasi et al., 2004)	**Improved polymer compatibility**: Enhances adhesion between fly ash and polymers.	**Chemicals**: High	\$1,000–\$5,000	Specialized use in polymer composites.
	Hydrophobic surface: Imparts water-resistant properties, reducing moisture-related issues	**Application process**: High	\$2,000–\$5,000	Used in moisture-sensitive applications.
	Enhanced mechanical properties: Improves tensile and flexural strength in composites	**Overall**: High	\$1,500–\$5,000	Applied in advanced composite materials.

(Continued)

TABLE 3.2
(Continued)

Enhancement technique	Specific outcomes	Associated costs	Cost range/ per ton	Applicability
Titanate and zirconate agents (Kulkarni and Mahanwar, 2016)	**Superior dispersion**: Ensures even distribution of fly ash in polymer matrices	**Chemicals**: Very high	\$2,000–\$10,000	High-end applications like specialty composites.
	Improved thermal stability: Enhances the material's performance under high temperatures	**Application process**: Very high	\$4,000–\$10,000	Suited for products requiring high thermal resistance.
	Enhanced mechanical strength: Provides superior mechanical properties compared to other treatments.	**Overall**: Very high	\$5,000–\$10,000	Limited to advanced engineering applications.
Carbonation (Yuan et al., 2021)	**Carbon sequestration**: Captures CO_2, contributing to environmental sustainability	**Process**: Low	\$50–\$500	Broad applicability in green construction materials.
	Improved durability: Reduces permeability, enhancing concrete longevity	**Integration**: Low to moderate	\$100–\$300	Common in cement plants aiming for sustainable practices.
	Increased resistance to chemical attacks: Effective in environments with acidic exposure.	**Overall**: Low to moderate	\$100–\$500	Widely applicable in eco-friendly projects.

3.11 RESEARCH GAPS

The review of biomass fly ash underscores its potential for diverse industrial applications, particularly in construction, soil stabilization, and sustainable material development. However, to fully harness this potential, several specific research gaps need to be addressed. Advanced characterization techniques, such as XRD and SEM, are essential to gain a deeper understanding of the microstructure and nanostructure, which directly influence the material's properties.

Standardized chemical treatment protocols are critical for achieving consistent and reproducible results, thereby improving the reliability of research findings and their applicability in industry. Additionally, long-term performance data, particularly concerning durability under various environmental conditions, is necessary to assess the material's real-world lifespan and suitability for different applications. Environmental and health impact studies, with a focus on heavy metal leaching and long-term exposure risks, are crucial for ensuring that the use of biomass fly ash is both safe and sustainable (Gollakota et al., 2019). Economic feasibility studies must also be expanded to include comprehensive cost-benefit analyses that account for indirect costs, potential savings, and the benefits of waste reduction.

In the realm of material optimization, research should focus on tailoring the properties of biomass fly ash for specific applications, balancing factors such as reactivity, workability, and durability. Geo polymerization techniques, in particular, present a promising avenue for converting biomass fly ash into high-performance construction materials. These techniques involve activating the fly ash with alkaline solutions to form geopolymer binders, which can result in materials with superior strength, durability, and resistance to chemical attacks. Innovation in chemical treatment technologies, including bio-based treatments, nanotechnology, and advancements in geo polymerization processes, could significantly enhance the efficiency, cost-effectiveness, and performance of biomass fly ash. Addressing these research gaps through targeted studies will not only improve the understanding and application of biomass fly ash but also support the development of more sustainable industrial practices, contributing to a circular economy and reducing environmental impact.

3.12 SUMMARY AND CONCLUSION

The analysis coal fly ash and BFA underscores its significant potential for use in various industries, particularly in construction, soil stabilization, and the development of sustainable materials. However, to fully capitalize on this potential, several research gaps need to be addressed. Advanced characterization techniques, standardized processing protocols, and long-term performance data are essential to better understand and utilize the unique properties of biomass fly ash. Additionally, comprehensive studies on environmental and health impacts, especially regarding heavy metal leaching, are crucial for ensuring safe and sustainable use. Economic feasibility remains a key consideration, highlighting the need for thorough cost-benefit analyses that include both direct and indirect costs, as well as potential savings from waste reduction. Optimizing the properties of biomass fly ash for specific applications—balancing reactivity, workability, and durability—is vital to expanding its industrial use. Furthermore, innovation in chemical treatment technologies, focusing on more efficient and cost-effective methods, could significantly enhance the material's performance and broaden its applicability. By addressing these research gaps, future studies can enhance the understanding and application of bio-coal fly ash, supporting the development of more sustainable industrial practices, reducing waste, minimizing environmental impact, and advancing the principles of the circular economy.

REFERENCES

Agarwal, A.K., Kadu, M.S., Pandhurnekar, C.P., & Muthreja, I.L. (2016). Brunauer-Emmett-Teller (B.E.T.), Langmuir and freundlich isotherm studies for the adsorption of nickel ions onto coal fly ash. *Asian Journal of Water, Environment and Pollution*, 13, 49–53.

Ahmad, I., & Mahanwar, P.A. (2010). Mechanical properties of fly ash filled high density polyethylene. *Journal of Minerals and Materials Characterization and Engineering*, 09, 183–198.

Alkadasi, N.A., Hundiwale, D., & Kapadi, U.R. (2004). Studies on the effect of silane coupling agent (2.0 per cent) on the mechanical properties of flyash filled polybutadiene rubber. *Journal of Scientific & Industrial Research*, 63, 603–609.Antoni, A., Wattimena, O.K., & Hardjito, D. (2012). Improving surface durability of high volume fly ash concrete with application of alkali solution. *Advanced Materials Research*, 626, 636–640.

Bada, S.O., & Potgieter-Vermaak, S. (2008). Evaluation and treatment of coal fly ash for adsorption: Application. *Leonardo Electronic Journal of Practices and Technologies*, 7(12), 37–48.

Chu, Y., Davaabal, B., Kim, D.S., Seo, S.K., Kim, Y.L., Ruescher, C.H., & Temuujin, J. (2019). Reactivity of fly ashes milled in different milling devices. *Reviews on Advanced Materials Science*, 58, 179–188.

Cocchi, M. (2011). Global wood pellet industry market and trade study, IEA bioenergy task 40 sustainable international bioenergy trade. Available online: https://www.bioenergytrade.org/downloads/t40-global-wood-pellet-market-study_final.pdf (accessed on 13 May 2011).

Dey, A., & Pandey, K. M. (2016). Characterization of fly ash and its reinforcement effect on metal matrix composites: A review. *Reviews on Advanced Materials Science*, 44(2), 168–181.

Dollimore, D., Spooner, P., & Turner, A. (1976). The bet method of analysis of gas adsorption data and its relevance to the calculation of surface areas. *Surface Technology*, 4, 121–160.

Figueroa, J., Fout, T., Plasynski, S., Mcilvried, H., & Srivastava, R. (2008). Advances in CO2 capture technology—The U.S. Department of Energy's carbon sequestration program ☆. *International Journal of Greenhouse Gas Control*, 2, 9–20. https://doi.org/10.1016/S1750-5836(07)00094-1.

Garg, N. Raman spectroscopy for characterizing and determining the pozzolanic reactivity of fly ashes. Available online: http://lib.dr.iastate.edu/ cgi/viewcontent.cgi?article=3589&context=etd

Garg, N., & Wang, K. (2012). Applying Raman spectroscopy for studying cement-based materials. In Ye, G., Van Breugel, K., Sun, W., Miao, C. (Eds.), *Microstructural-Related Durability of Cementitious Composites*, 275–282.

Gollakota, A.R., Volli, V., & Shu, C. (2019). Progressive utilisation prospects of coal fly ash: A review. *The Science of the Total Environment*, 672, 951–989.

Goodarzi, F. (2006). Characteristics and composition of fly ash from Canadian coal-fired power plants. *Fuel*, 85, 1418–1427. https://doi.org/10.1016/J.FUEL.2005.11.022.

Inkson, B. (2016). Scanning electron microscopy (SEM) and transmission electron microscopy (TEM) for materials characterization. 17–43. https://doi.org/10.1016/B978-0-08-100040-3.00002-X.

Jokiniemi, J., Hytonen, K., Tissari, J., Obernberger, I, Brunner, T, Barnthaler, G., et al. (August 2008). Biomass combustion in residential heating: Particulate measurements, sampling, and physicochemical and toxicological characterization. Final report, University of Kuopio, Fine Particle Aerosol Technology Laboratory.

Kato, K., Xin, Y., Hitomi, T., & Shirai, T. (2019). Surface modification of fly ash by mechanochemical treatment. *Ceramics International*, 45(1), 849–853.

Kjersgaard, D., Jacobsen, B. N., Rindel, K., Andreasen, L., Larsen, F., Nyegaard, P., ... & Bodker, J. (2007). The reuse of bio ash for the production of concrete: a Danish case study [Held jointly with the 4. Canadian organic residuals and biosolids management conference].

Kulkarni, M.B., & Mahanwar, P.A. (2016). Studies on effect of titanate-coupling agent (0.5, 1.5, and 2.5%) on the mechanical, thermal, and morphological properties of fly ash–filled polypropylene composites. *Journal of Thermoplastic Composite Materials*, 29, 344–365.

Kumaravel, S., & Mohanraj, D. (2019). Channankaiah. Production and Mechanical Properties of Fly ash and Basalt ash reinforced Al 6061 composites. *Indian Journal of Engineering*, 16, 276–283.

Manimaran, R., Jayakumar, I., Mohammad Giyahudeen, R., & Narayanan, L.A. (2018). Mechanical properties of fly ash composites—A review. *Energy Sources, Part A: Recovery, Utilization, and Environmental Effects*, 40, 887–893.

McCarthy, G.I. (1987). X-Ray powder diffraction for studying the mineralogy of fly ash. *MRS Proceedings*, 113, 75.

McCarthy, G., & Johansen, D.M. (1988). X-Ray powder diffraction study of NBS fly ash standard reference materials. *Powder Diffraction*, 3, 156–161.

Moreno, N., Querol, X., Andrés, J., Stanton, K., Towler, M., Nugteren, H., Janssen-Jurkovicová, M., & Jones, R. (2005). Physico-chemical characteristics of European pulverized coal combustion fly ashes. *Fuel*, 84, 1351–1363. https://doi.org/10.1016/J.FUEL.2004.06.038.

Nokken, M. R., Salloum, T., Idle, A., & Martinez Ramos, L. A. (2014). Standardized testing for determining the durability of high-volume fly ash mixtures. *Journal of Materials in Civil Engineering*, 26(1), 206–210.

Palomo, A., Alonso, S., Fernández-Jiménez, A., Sobrados, I., & Sanz, J. (2004). Alkaline activation of fly ashes: NMR study of the reaction products. *Journal of the American Ceramic Society*, 87, 1141–1145.

Panda, L., & Dash, S. (2020). Characterization and utilization of coal fly ash: A review. *Emerging Materials Research*, 9, 921–934.

Pandian, N.S. et al. (1998). Studies on the specific gravity of some Indian coal ashes. *Journal of Testing and Evaluation, ASTM*, 26, 177–186.

Pitman, R.M. (2006). Wood ash use in forestry—A review of the environmental impacts. *Forestry*, 79, 563–588.

Pradhan, A., & Deshmukh, J. P. (2008). Utilization of fly ash for treatment of acid mine water. *Journal of Environmental Research and Development*, 3(1), 137–142.

Rajamma, R., Ball, R., Tarelho, L., Allen, G., Labrincha, J., & Ferreira, V. (2009). Characterisation and use of biomass fly ash in cement-based materials. *Journal of Hazardous Materials*, 172(2–3), 1049–1060. https://doi.org/10.1016/j.jhazmat.2009.07.109.

Reddy, A.K., & Reddy, D.K. (2015). Effect of fly ash on strength & durability parameters of concrete of high-volume fly ash mixtures. *Journal of Materials in Civil Engineering*, 26, 206–210.

Sharma, A., Kabra, S.P., Katara, S., & Rani, A. (2015). Variation of surface morphology and physico-chemical properties of the fly ash through mechanical and thermal activations. *Journal of Advanced Chemical Sciences*, 1 (2), 70–74.

Siyal, A.A., Azizli, K.A., Ismail, L.B., Man, Z., & Khan, M. (2016). Suitability of Malaysian fly ash for geopolymer synthesis. *Advanced Materials Research*, 1133, 201–205.

SLU. Swedish University of Agricultural Sciences Wood Ash Database, 2014. Available online: https://woodash.slu.se/eng/ (accessed on 13 May 2011).

Sun, C., Lux, S., Müller, E., Meffert, M., & Gerthsen, D. (2020). Versatile application of a modern scanning electron microscope for materials characterization. *Journal of Materials Science*, 55, 13824–13835. https://doi.org/10.1007/s10853-020-04970-3.

Tai, F.C., Wei, C., Chang, S.H., & Chen, W.S. (2010). Raman and X-ray diffraction analysis on unburned carbon powder refined from fly ash. *Journal of Raman Spectroscopy*, 41, 933–937.

Vassilev, S.V., & Vassileva, C.G. (2007). A new approach for the classification of coal fly ashes based on their origin, composition, properties, and behavior. *Fuel*, 86, 14901512. https://doi.org/10.1016/j.fuel.2006.11.020.

Wen-xun, Q. (2004). NMR research on silicon-oxide polyhedron structure in activated fly ash. *Journal of Materials Science & Engineering*. 22(4), 561–563.

Yuan, Q., Yang, G., Zhang, Y., Wang, T., Wang, J., & Romero, C. E. (2022). Supercritical CO2 coupled with mechanical force to enhance carbonation of fly ash and heavy metal solidification. *Fuel*, 315, 123154.

Zhang, Z., Wang, H., & Provis, J.L. (2012). Quantitative study of the reactivity of fly ash in geopolymerization by FTIR. *Journal of Sustainable Cement-Based Materials*, 1, 154–166.

Zhu, Z., Wang, X., Dai, S., Huang, B., & He, Q. (2013). Fractional characteristics of coal fly ash for beneficial use. *Journal of Materials in Civil Engineering*, 25, 63–69.

4 Utilizing Fly Ash for Environmental Remediation

A Sustainable Alternative

Shalini Shashni, Rahul Gautam and Simran Kaur Dhillon

4.1 INTRODUCTION

The significant reliance on fossil fuels to fulfil the increasing energy demand has resulted in a concerning rise in environmental pollution. Increased energy consumption is required due to the growing population, with electricity significantly meeting this demand worldwide (Umar et al., 2021). Most electricity is produced by thermal power plants, which burn solid municipal trash, biomass, and coal. Consequently, the world produces and releases over 1 billion tonnes of fly ash (FA) (Dindi et al., 2019). When coal is burned, the lucrative byproduct substance known as fly ash (FA) is produced in power plants. The spherical glassy particles that make up this fine grey powder contain pozzolanic elements, which react with lime to form cementitious compounds. Fly ash (FA) is used in concrete, mines, landfills, and dams. It is an essential component of the construction sector. Its distinctive qualities and heterogeneous nature account for its usefulness. FA is a valuable byproduct of burning coal in power plants, and its use depends on its production procedures. To produce FA for different purposes, these processes entail multiple phases, such as separation, grinding, granulation, roasting, and drying. As the building industry moves towards more environmentally friendly materials, FA is a viable replacement for Portland cement, which has a significant carbon footprint. Nonetheless, there are still issues, most notably the variation in FA composition and its environmental effects. Notwithstanding these obstacles, the FA sector offers bright prospects because of the growing need for housing, the construction of infrastructure, and government programmes that support the production and use of FA.

However, some obstacles need to be overcome, like the switch to renewable energy sources, uneven fly ash quality that affects building projects, and environmental worries about contaminants and hazardous materials in fly ash. Putting innovative processing methods and safety procedures into place are essential steps in reducing

DOI: 10.1201/9781003498124-4

FIGURE 4.1 Scope of fly ash utilization in remediation technologies.

environmental effects and guaranteeing FA's safe and effective use in various applications (Figure 4.1). It is impossible to overlook FA's potentially dangerous environmental and public health effects. The two main issues are the leaching of toxicants into soil and water and the entrainment of FA particles in the air (Dindi et al., 2019). Consequently, FA is the most popular way to overcome its detrimental environmental effects. Over time, FA has been used increasingly in the concrete industry for making bricks, building roads, and other projects. However, only 30% of the world's production is utilized globally (Dindi et al., 2019). Due to its hazardous heavy metals and soluble components, another portion is frequently stored in ponds or landfills, which raises worries about water and soil pollution (Sharma et al., 2020). Direct disposal or storage of FA is, therefore, not permitted.

On the other hand, there is growing interest in converting FA into high-value goods that may benefit air, water, and soil quality. FA has the potential to be a low-cost precursor material made of metal oxides and silicates that can be used to create adsorbents, catalysts, zeolites, and membranes that are required for environmental remediation. The recycling of FA for the benefit of the environment, the building industry, the agricultural sector, the ceramic industry, and the recovery of valuable components has been the subject of numerous reviews in the literature (Blissett and Rowson, 2012).

FA is abundantly produced due to coal combustion, making it readily available for remediation projects. Its widespread availability ensures that it can be easily sourced for environmental cleanup efforts, thus facilitating timely interventions in contaminated sites. Utilizing FA for remediation offers a cost-effective solution compared to traditional methods. It repurposes FA for remediation instead of disposal as waste, saves on many costs, and ensures optimal use of resources such as financial resources allocated to environmental cleanup. Because FA has natural adsorption qualities, it can effectively remove and arrest environmental pollutants. This ability to immobilize pollutants makes it an effective tool in reducing the mobility and exposure of economic exploitation. As a result, remediation activities in polluted areas or sites can be further validated. The chemical treatment of FA can increase its surface area, which significantly improves the capacity to capture environmental pollutants, such as soil, water, and air. FA, therefore, creates more surface area for pollutants to occupy, increasing the process's efficiency.

FA is incorporated into the solid-state fermentation and sustainable life cycle concept and promotes the application of biosurfactants for the remediation of bio-polluted areas. Biosurfactants can be produced and used to disperse petroleum hydrocarbon contaminants and mobilize heavy metals while cleaning contaminated soil and groundwater. The abundance of FA and its availability makes it an ideal resource for disposing of a bio-polluted zone, especially when the area has been significantly contaminated. Additionally, FA may also help in bioremediation and soil washing over contaminated land with Petroleum Hydrocarbons (PHC) and electroplating in a cosponsored strategy due to the high phytoremediation potential. The effectiveness of FA incorporation in biosurfactant production is confirmed as necessary, as biosurfactant-based fungal efficiency will include the utilization of FA in considerably bio-polluted spots. The methods previously discussed, in which FA is used, provide a source of subsidy. Inventive methods can be used to extract substances from FA, supporting an economy's principles and lessening the environmental consequences of waste management. By retrieving resources, like metals from FA, efforts to clean up the environment can be more sustainable economically and environmentally. This chapter highlights the possibility of incorporating waste materials such as FA into cleanup solutions, providing financial and ecological benefits (Table 4.1).

4.2 APPLICATIONS OF FA IN ENVIRONMENTAL REMEDIATION

4.2.1 Soil Remediation Techniques

The toxic trash present in the soil in the form of pollutants is always a significant concern for public health and environment (Derakhshan et al., 2018; Varol et al., 2019; Alengebawy et al., 2021). The long-established soil treatment techniques were quite expensive, as were resource intake processes, which led to resourceful and sustainable approaches (Hoang et al., 2021; Lee et al., 2020, 2021). FA readily reacts with water and lime, thereby improving the texture and soil-bearing capacity necessary for application in soil remediation techniques (Renjith et al., 2021).

Studies showed that adding FA to contaminated soil improves soil stability, solidification, and other remediation methods. Further, FA can improve the bearing capacity and lower the plasticity of soft, fine-grained soils used for road construction (Edil et al., 2006). Ahmaruzzaman (2010) emphasized the environmental and cost benefits of using FA for soil stabilization, aiding waste management and reducing greenhouse gas emissions. Al-Rkaby et al. (2022) assessed using geopolymer derived from high-calcium class C FA as an alternative to the traditional soil stabilization methods

TABLE 4.1
Application of fly ash-based materials for environmental applications

Sl.no	Adsorbent	Adsorbate	Removal	Comment	Reference
1	Fly ash-based geopolymer	Cu	152 mg/g	Wastewater treatment	Al-Harahsheh et al. (2015)
2	Fly ash-based zeolite	Zn	53.5 mg/g	Wastewater treatment	Buema, Trifas and Harja (2021)
3	NaOH-modified fly ash	Cd	56.31 mg/g	Wastewater treatment for heavy metals removal	Huang et al. (2020)
4	NaOH-modified fly ash	Pb	126.55 mg/g	Wastewater treatment for heavy metals removal	Huang et al. (2020)
5	Coal fly ash	Ni	47 mg/g	Wastewater treatment for heavy metals removal	He et al. (2020)
6	Biomass fly ash	Methylene blue	20.5 mg/g	Dye removal	Novais et al. (2018)
7	Fly ash-derived nano-zeolite	Methylene blue	0.164 g/100 ml	Dye removal	Sivalingam et al. (2019)
8	Zinc oxide nanorods loaded on activated carbon	Crystal violet	113.64 mg/g	Dye removal	Dil et al. (2016)
9	magnetized biocarbon	Fuchsin dye	99.35%	Dye removal	Renita et al. (2021)
10	Edible fungi residue	Reactive black 5	172.43 mg/g	Dye removal	Xiao et al. (2012)
11	Fly ash mixed with $CaCO_3$	H_2S gas	99.1%	Gas removal	Asaoka et al. (2021)
12	Fly ash mixed with CaO	SO_2 gas	95%	Gas removal	Li et al. (2007)
13	Zeolite-carbon composites from high-carbon fly ash	Engine oil, used engine oil and kerosene	1.14–1.39 g/g	Wastewater treatment for oil contamination	Bandura et al. (2021)
14	Fly ash (FA100%)	Increase metal concentration in soil for rice crop	–	Soil reclamation	Nayak et al. (2015)

(Continued)

TABLE 4.1
(Continued)

Sl.no	Adsorbent	Adsorbate	Removal	Comment	Reference
15	Fly ash-derived zeolite	Cd removal from leachate	2.2mg/kg	Leachate treatment	Xu et al. (2021)
16	Fly ash-derived zeolite	Pb removal from leachate	176.21 mg/kg	Leachate treatment	Xu et al. (2021)
17	NaOH-modified fly ash	Cd	9.18–48.5 mg/g	Wastewater treatment for heavy metals removal	Buema, Lupu, et al. (2021)
18	Fly ash-derived zeolite	Pb^{2+}	25.5 mg/g	Wastewater treatment for heavy metals removal	Kobayashi et al. (2020)
19	Fly ash-derived zeolite	Hg^{2+}	7.5 mg/g	Wastewater treatment for heavy metals removal	Kobayashi et al. (2020)
20	TiO_2 ultrafiltration membrane from flash	Dye removal	–	Wastewater treatment and dye removal	Bouazizi et al. (2017)
21	Fly ash-derived zeolite	Traces of strontium and caesium cations	–	Wastewater treatment for heavy metals removal	Feng et al. (2018)
22	TiO_2 membrane with fly ash ceramic support	Cu(II),	9.56 mg/g	Wastewater treatment for heavy metals removal	Zhang et al. (2021)
23	TiO_2 membrane with fly ash ceramic support	Cr(VI)	97.09%	Wastewater treatment for heavy metals removal	Zhang et al. (2021)
24	Fly ash-derived zeolite	Cd(II)	130.68 mg/g	Wastewater treatment for heavy metals removal	Yao et al. 2024
25	Fly Ash-Based Geopolymer	Building materials	–	Sustainable construction	Eze et al. (2020)
26	Coal fly ash	Cs^{+}	2.12 mmol/g	Wastewater treatment for heavy metals removal	Tian et al. (2019)
27	Coal fly ash	Sr^{2+}	1.93 mmol/g	Wastewater treatment for heavy metals removal	Tian et al. (2019)

such as ordinary Portland cement. Kodaz (2022) investigated the positive impact of FA on highly plastic expansive clays, reporting significant improvements in strength properties with the addition of FA, particularly a 66% increase in unconfined compressive strength (UCS) and California bearing ratio (CBR) values with 20% FA content.

Soil stabilization by FA can be broadly categorized into physical and chemical techniques. FA alters the soil stabilization property without any chemical interaction. Some approaches in this direction include mechanical mixing, spreading and compacting FA, layering to facilitate uniform distribution, use of fly ash-embedded geosynthetic materials, and FA injection into the soil. A study by Wen et al. (2020) examined the impact of FA addition in soil slurry on its wettability. Mechanical pressure and electro-osmosis were employed to lower the moisture content in the soil slurry. Incorporating 5% FA increased the amount of water drained to 350 mL. Further, the water levels in the treated slurry fell to 20.5% and 19.1% for the anode and cathode sections, respectively. Following this, the compressibility coefficient and shear strength were 0.170 MPa^{-1}, 0.152 MPa^{-1}, 81.2 kPa, and 85.1 kPa, respectively.

Yoobanpot et al. (2020) explored a composite of cement and FA gravel as an alternative to deep cement mixing columns. This offered more efficient supporting embankments for soft clays, offering high strength and permeability. Turan et al. (2022) compared class C and F FA for stabilizing clay soil, evaluating their effects on mechanical behaviour and microstructure. Mypati et al. (2022) used low-calcium FA to stabilize expansive soil with the in situ deep soil mixing technique, achieving improved strength and controlling swell-shrink behaviour. Lu et al. (2022) tested cement-soil composites with various binders, finding that adding FA and sisal fibre improved strength and reduced permeability.

Stabilization with FA mitigates leaching risk, preventing groundwater contamination. Guan et al. (2023) optimized foamed concrete backfilling in urban areas, adjusting the water-solid ratio and soil content for enhanced stability. Increased foam, soil, and FA bolstered ductility and soil quality, with optimal FA amounts refining pore size and roundness. Mohanty et al. (2021) worked on stabilizing dispersive soil erosion susceptibility triggered by higher sodium content. Soil stabilization occurs when soil is modified by adding soil stabilization agents, commonly known as FA, ground granulated blast furnace slag (GGBS), or cement clinker, to improve its strength and durability.

FA can also be an auxiliary agent to improve soil properties. When added, it can function as a thermal insulator, even increase soil-resistance frost, or support soil conservation in custodial or highly salty environments. In Li et al.'s (2020) *study, soil modification with clay, waste fly ash and oil shale ash increases* the silt clay thermal insulation capacity. Its purpose is to assess the microstructure, and Clay Linear Thermal Conductivity (CLTC) changes on the clay after Freeze Thaw Cycles (FTS). Furthermore, Cheng et al. (2021) investigated the salinized soil of the Suihua to Daqing expressway, China. This study aimed to identify the consequences caused by freeze-thaw cyclic and FA content on the strength properties of the soil. They found that applying freeze-thaw cycles weakened the saline soil while adding FA strengthened cohesion. The optimum amount of FA, 15%, exhibited the maximum cohesion and internal friction angle. Cui et al. (2024) mixed FA and cement into the soil beneath roads to improve its resistance to freezing and thawing. They used physical methods to combine these materials. The study investigated how these additives affected the soil's mechanical properties and behaviour during freezing and thawing cycles. The results showed that the FA and cement changed the soil's physical structure.

Another technique involves adding different proportions of chemicals to the soil mixture, including FA. This changes the soil's composition and properties by forming cement-like compounds. This technique improves the soil's strength and durability. Commonly used chemicals include lime, cement, and specialized additives. Mateos et al. (1962) proved this in their study by adding granular soil with 3.6% lime and 10.25% FA, but for clayey soils, they added 5.9% lime with 10.25% FA. Calcitic hydrated lime performed better than monohydrate lime at lower lime levels, while the opposite was true at higher concentrations. Barman et al. (2022) provided a comprehensive review on soil stabilization using chemical additives, recommending lime and cement combination as the most effective stabilizer, supplemented by FA or slags for soils lacking pozzolans. In a report by Silvani et al. (2022), through experimental shear tests on sand FA lime soil samples, they observed how the orientation of stress and deviatoric linear stress path influenced the mechanical response. The research revealed that changes in the stress ratio significantly impacted stress-strain behaviour volume properties and strength characteristics, showing that strength decreased as this ratio increased.

Different stress paths indicated a more isotropic response, suggesting that cementation may mitigate material anisotropy. Mohammadi et al. (2023) investigated using low-concentration fly ash-based alkali cement in recycled alkaline solutions for soil stabilization. Their research showed increased soil stiffness and strength, with high-calcium FA performing better in an environment with less alkalinity. Bazarbekova et al. (2024) addressed the need for environmentally sustainable soil stabilization methods in the construction industry by proposing a one-part alkali-activation approach using dry powdered alkaline activators. Their study evaluated the effectiveness of silty soil in conjunction with class C and F FA and GGBS, assessing engineering properties through UCS testing. They found that dry SM shows promise for creating alkali-activated mixtures, offering the potential for soil stabilization in transportation infrastructure systems.

Geopolymer could be a greener alternative to cement or lime, according to Maheepala et al. (2023) offering cost-effective, eco-friendly, and long-lasting benefits. Class F-grade FA is extensively used in geopolymer technology for soil stabilization.

A study by Rađenović et al. (2024) examined the pros of FA, hydrated lime, and lime/FA mixtures on the geotechnical characteristics of soft soils in Germany Hallecity region. This reduced the plasticity of soil and increased the moisture content. Among the additives, lime significantly improved the UCS. While lime/FA mixture transformed the nature of soil from ductile to brittle.

Saride et al. (2024) investigated if columns made of FA geopolymer without cement could effectively control the swelling and shrinkage of expansive soil in a confined setting. They developed the "area improvement ratio," which considers the ratio of column area to untreated soil area within confined boundaries. They installed the columns in triangular and square patterns. Results showed that cement-free Deep soil mixing columns effectively mitigated swelling and shrinkage. The square pattern with an area improvement ratio of 0.16 reduced these behaviours by approximately 68%, while the triangular pattern had a reduction of 72%. The binder ratio Alkali activator/Fly ash of 1.5 performed best in improving UCS. Hamed et al. (2024) studied how geopolymer made from FA and GGBS affected the strength of very soft clay. They ran tests using different amounts of FA, slag, and an alkaline solution. After the mixture hardened for seven and 28 days, they tested its strength by squeezing it (UCS), its ability to support weight (California bearing ratio, CBR), and how well water flowed through it (permeability). They found the best mixture was 38% geopolymer (24% FA, 14% slag) with a 25.75% alkaline activator that was 12 M. The study showed that geopolymer with FA and slag is a green way to make soil stronger than cement.

As soil solidifies, FA, acting as a bonding agent, traps pollutants, reducing their movement. A study by Gondek et al. (2023) tested the effectiveness of willow biochar and fly ash-based zeolite in immobilizing heavy metals and boosting enzyme activity in contaminated sandy soil. They used biochar and zeolite alone and together, which led to decreased cadmium, lead, and zinc mobility in the soil. The analysis also revealed that these materials boost soil enzyme activity by altering the acidity (pH) and electric conductivity (EC). By that, the content of metal in the soil changes overtime. Kumar et al.'s (2022a, 2022b) main concern is the pollution caused by industries that directly affect the soil. They inspected that incorporating black cotton soil with an adequate amount of this industrial waste (FA and silica fume) had better efficiency than the same. It maximized its dry density and CBR% value by subsiding black soil's free swelling index.

Cui et al. (2023) use the same concept of modifying the soil with FA, resulting in degraded heavy metal content and increased pH levels. Apparently, it is harder for plants to adsorb heavy metals, which improves soil quality for agriculture, making it more fertile and resistant to erosion. Researchers (Munda et al., 2016; Dash & Sahoo, 2017; Ijaz et al., 2022; Khalid et al., 2023) observed that combining modified fly ash-based soil with other materials adds nutrients and increases water retention. In the long run, it is essential to regularly track soil health, chemical contamination levels, and environmental effects for fly ash-based remediation's long-term success and sustainability.

4.2.2 Water Treatment Processes

The physicochemical properties of FA include a porous structure with a large surface area and metal oxide components. Generally, the impurities in water bodies are present as metal or organic matter components. On treating contaminated water with fly ash-based materials, they show a phenomenon where they get adsorbed on the surface of FA or form coagulants with the metal oxides present in FA. These properties make it an efficient water purifier.

4.2.3 Adsorption

Despite being an environmental pollutant, FA is extensively used as an adsorbent for water treatment. The adsorbent property is due to the porous structure with a large surface area in FA. This makes an effective adsorbent for removing contaminants like dyes, organic compounds, and heavy metals from water. These adsorbents form bonds (via hydrogen bonding, π-π interactions, and van der Waals forces) with the suspended or dissolved contaminants in water and are drawn out. Singh et al. (2022) recommended it as a versatile material for water remediation. The efficiency of these FA adsorbents can be strengthened by altering their surface characteristics or activating them with chemical or physical processes. They were incorporated with nanoporous materials like zeolites and mesoporous silica. Furthermore, they can be converted into geopolymers by treating them with alkali activators. They serve as adsorbents or photocatalysts, showcasing their potential in addressing water quality issues and waste management challenges. Miricioiu et al. (2020) point out the high silica content of FA sampled from power plants using different combustion methods for synthesizing valuable materials like zeolites or mesoporous silica. They were used for CO_2 capture from emissions and wastewater treatment, promoting reprocessing and sustainability with minimal environmental impact. Various contaminants such as heavy metals, organic pollutants, and dyes can be effectively adsorbed onto FA particles, improving water quality:

4.2.3.1 Removal of Organic Pollutants as Contaminants from Water

Contaminants in water, such as volatile organic substances (VOCs), compounds found in smoke and tar, and pesticides, jeopardize our well-being. In 2022, Sadiya Mushtaq removed organic pollutants from wastewater through adsorption technology. To change its surface, they created a surfactant-coated coal fly ash (CFA) using sodium dodecyl sulphate (SDS). The study showed that the fly ash-based adsorbent was able to adsorb chlorobenzene (CB) and nitrobenzene (NB) from water and showed better surface features and thermal stability for adsorption. Teng et al. (2021) explored the effectiveness of fly ash-red mud particle electrodes (FRPEs) in a 3D electrochemical system for treating wastewater. They targeted the breakdown of Atrazine (ATZ) and identified hydroxyl radicals (OH) as the primary active agent in ATZ breakdown. The reaction was primarily occurring on FRPE surfaces. The optimized conditions showed a maximum degradation of 90.1%. Additionally, FRPEs in 3D electrochemical systems exhibited stability and recyclability, providing a fresh approach to wastewater treatment and demonstrating potential for environmental cleanup.

Other water purification techniques include FA to create a Fenton-like catalyst and employing it to treat organic wastewater. This approach incorporates adsorptions and photocatalysis where pollutants are removed by binding to surfaces by advanced oxidation processes (AOPs) by breaking down pollutants using highly reactive molecules.

Microwave pre-enhancement significantly boosted radical OH generation, influenced by FA leaching behaviour. Wang et al. (2022) developed a reusable CFA catalyst for a Fenton-like process under microwave (MW) assistance to treat organic wastewater. They prepared the catalyst (acidic FA (AFA)) by impregnating CFA with H_2SO_4 and optimized the preparation conditions. Characterization techniques were used to analyse AFA's properties. The study optimized dynamic treatment conditions and demonstrated AFA's effectiveness in removing p-nitrophenol (PNP) from wastewater, achieving over 90% removal efficiency. The AFA showed potential for reuse and applicability in treating natural coking wastewater. Wang et al. (2021) examined the performance of modified CFA based catalyst for treating wastewater from polymer flooding operations. The raw CFA was treated with sulphuric acid (ACFA), which increased the specific surface area and active site density, thereby rendering it beneficial for the catalytic potential of the catalyst.

4.2.3.2 Removal of Heavy Metals as Contaminants from Water

Heavy metals, such as copper, nickel, lead, and zinc, are common pollutants in wastewater from industrial activities such as mining, metal processing, and urban runoff. Medykowska et al. (2023) investigate the adsorption capabilities of hydrothermal reaction products of fly ash (HC FA), including zeolites (Na-X, NA-P1) and their carbon composites (Na-X(C), Na-P1(C)), for removing heavy metals (Pb(II), Zn(II)) and organic compounds (diclofenac, poly(acrylic acid)) from mixed solutions. Na-X zeolite adsorbed 322.1 mg/g of Pb(II), 332.5 mg/g of Zn(II), and 6.68 mg/g of diclofenac, while Na-P1 adsorbed 332.9 mg/g of Pb(II), 103.6 mg/g of Zn(II), and 21.19 mg/g of diclofenac. The representative adsorption values showed the potential of these materials for environmental remediation.

CFA particle size significantly impacted the decontamination potential in water samples. Police et al. (2020) investigated five different size fractions of FA to examine the adsorption capacity of each size fraction. Size fraction <20 μm demonstrated the highest uranium adsorption capacity, where the adsorbent dose and initial concentration of uranium positively affected both the adsorption capacity and removal efficiency. Freundlich and Dubinin–Radushkevich (D-R) isotherm models were established to analyse. The isotherms suggested multilayer adsorption. Further, thermodynamic studies revealed the adsorption of FA as an endothermic and spontaneous process. Recently, Tan et al. (2020) concluded that geopolymer could be a potential green and effective alternative to clean up heavy metals in water. Zhang et al. (2021) used a TiO_2 nanofiber membrane supported by a ceramic FA substrate for wastewater treatment. The membrane removed up to 90.15% of copper (II) and 97.09% of chromium (VI) from wastewater. Lekgoba et al. (2021) examined CFA to remove copper and nickel from contaminated water in batch and fixed bed columns. Operational parameters, including acidity (pH), FA content, reaction time, temperature, column height, and flow rate, were examined to evaluate the application of FA as an adsorbent. FA effectively removes both copper and nickel. The removal of copper could be attributed to its small size and high electronegativity. Batch experiments followed the pseudo-second-order kinetic model and the modified Langmuir isotherm. In contrast, Thomas's model fitted the fixed bed column experiments iron oxide nanoparticles (IONPs) from CFA and exhibited the ability to remove hazardous lead and chromium ions from a medium with 20% CFA. Yadav et al. (2023) offered an affordable and environmentally friendly way to remove heavy metals from wastewater.

4.2.3.3 Removal of Dyes as Contaminants

Dyes are widely employed in textiles, paper production, and dye manufacturing industries. These dyes are problematic because they readily dissolve in water and are challenging to break down, resulting in water pollution. To some extent, FA can effectively remove these dyes as a contaminant. Rodwihok et al. (2023) prepared a low-cost photocatalytic adsorbent from FA incorporating metal oxides like zinc oxide (ZnO) and reduced graphene oxide (rGO). The pyrolysis-hydrothermal techniques were used to prepare these FA-based ternary ZnO/FA-zeolite/rGO nanocomposites. These nanocomposites exhibited negative surface charges with large surface areas. They increased the efficiency by breaking down and eliminating the organic dye (methylene blue) from wastewater through light-activated processes by adsorbing it on its surface. They attain this by generating reactive oxygen species that attack the dye, preventing the charges from recombining and increasing the material's reusability. Rozek et al. (2020), to eliminate methylene blue, created composite materials combining geopolymer and expanded glass to increase the adsorbing property of the materials. Eteba et al. (2022) used chemically altered CFA (TCFA) to remove direct blue 78 dye from water. TCFA was used in a pilot facility to treat textile wastewater, batch tests were conducted to optimize dye removal variables, and removal efficiency up to 99.7% was recorded.

4.2.3.4 Coagulation

Calcium oxide and silicon oxide can react with water and coagulate. Its high content in FA and the apparent behaviour of oxides make it an invaluable resource for water treatment. Maila et al. (2020) performed a jar test to show the competence of fly ash-prepared coagulants in wastewater treatment. Initially, FA and sodium hydroxide-FA were calcined at 800°C and rinsed with hydrochloric acid before being used as coagulants. The coagulants were tested across different turbidity ranges of low (19 NTU), medium (49 NTU), and high (80 NTU) with varying doses. The result showed that an increase in coagulant dose from 0.05 to 0.25g/l decreased residual turbidity. Calcined FA was more effective at low and medium turbidity at a dosage of 0.20 g/l, while complex coagulant displayed a 99.8% reduction at higher turbidity for the same dose. Kumar et al. (2022a, 2022b) developed a liquid coagulant using fly ash (FAC) through acid-alkali leaching and polymerization techniques. FAC effectively removed trihalomethanes (THMs) causing substances from water by neutralizing their charge and capturing them. Additionally, FAC reduced the likelihood of natural organic matter in water to produce THMs, indicated by a high ratio of biodegradable material to total organic carbon (72.86%).

Incorporating FA in water treatment processes contributes to addressing environmental concerns. It effectively removes suspended solids from industrial wastewater and other sources of polluted runoff, reducing their detrimental impact on the ecosystem and overall environmental health. Wang et al. (2020) performed a lab experiment where they combined FA with de-alkalized red mud in treating simulated wastewater with diatomite under murky conditions. The multi-flocculant prepared under optimal conditions was efficient enough to reduce turbidity from 683.6 NTU to 0.67 NTU with a removal rate of 98.76% at a 90 μL/L dosage. The optimal synthetic condition of the coagulant prepared from waste FA was initially treated in a ratio of 1:1 with sodium carbonate, and this modified FA was further reacted with dealkalized red mud in a 1:3 ratio.

Wang et al. (2022) synthesized magnetic coagulants (Fe-AFA, Fe-BFA) from acid-modified fly ash (AFA) and base-modified fly ash (BFA) for treating desulphurization wastewater. Fe-AFA exhibited superior Chemical Oxygen Demand (COD) adsorption performance (5.69 mg/g) compared to raw FA, with superparamagnetic behaviour. The COD removal process followed quasi-second-order kinetics and Langmuir adsorption isotherm. Thermodynamic tests showed COD removal as spontaneous, endothermic, and irreversible. Fe-AFA maintained COD removal efficiency after five cycles, suggesting the potential for FA as a low-cost adsorbent in wastewater treatment. A study by Gao et al. (2023) developed an eco-friendly coagulant using waste FA to treat chemical pollutants in coking wastewater. They extracted aluminium (Al) and iron (Fe) from FA using acids, achieving high leaching efficiencies (up to 96.75% for Al and 97.64% for Fe). The optimal molar ratios of A13+/SO42− and A13+/Fe3+ were determined to make the polymeric aluminium ferric chloride sulphate (PAFCS) coagulant. This coagulant formed a complex 3D network structure that effectively removed turbidity and reduced organic pollutants in the wastewater. Nguyen et al. (2023) developed a coagulant from CFA using sulphuric acid. The flash ash waste was collected from the Mong Duong power plant situated in Vietnam. In laboratory tests, the 5% solution coagulant effectively removed suspended solids from wastewater. Nearly 83.6% of solids from wastewater containing 60 g of solids per litre were removed.

4.2.3.5 Filtration

Scientists have proven FA's effectiveness in creating a membrane filter for water purification. The reason is FA's porous nature provides a large surface area. Therefore, the contaminants of water bind on its surface and work as a water purifier. In terms of energy savings and cost cutting, industrial waste as a water purifier is an environmental assent.

A recent assessment was based on the potential use of fly ash-based ceramic membrane filters for water purification by Goswami et al. (2022). The filters were optimized based on parameters such as particle size, membrane thickness, and water flow rate. They ensured the ability to remove impurities from water to provide clean drinking water or for industrial applications.

Sun et al. (2022) provide a general strategy for designing and utilizing FA porous materials for filtration and adsorption applications. Through the foaming process, they utilized FA to create porous materials with high interconnected porosity. These materials were boosted with layered double hydroxide (LDH) that exhibited a significant water permeation flux and suitable compressive strength. An excellent adsorption efficiency for Congo red from wastewater was observed. This showed a potential for industrial-scale wastewater treatment.

Diana et al. (2022) prepared a ceramic membrane filter from FA and clay for river water treatment. The membrane, sintered at low temperatures, exhibited effective filtration properties. Chemical characterization using XRD and EDX confirmed the composition of FA and clay. Filtration experiments, employing both dead-end and crossflow systems at varying pressures, demonstrated significant removal efficiencies for contaminants such as Pb, Fe, and *E. coli* bacteria, reaching up to 96.59%,

95.55%, and 99.29%, respectively, with a crossflow system at 0.75 bar. SEM analysis revealed pore size and membrane porosity changes before and after filtration, with the highest lead intensity observed in the membrane used for crossflow filtration at 0.75 bar pressure.

Another use of FA as a ceramic membrane was by Chathurappan et al. (2022) who prepared the membrane through the uniaxial compaction method. The membrane exhibited a minimum sintering temperature of 800°C. The porosity was 39.8%, with an average hydraulic pore radius of 41 nm. The pure water flux reached 11.31 m^3/m^2 sec with an applied pressure of 483 kPa. The separation efficiency of fly ash-based ceramic membranes was examined with rhodamine B, bovine serum albumin, and vanillin. At the pressure of 207 kPa, its efficiency for the participants was found to be 95.2%, 92.8%, and 39.4%, respectively. This study recommends using these membranes in industrial separation and purification applications. Rao et al. (2023) prepared ceramic membranes for microbial fuel cells (MFCs) by mixing them with potter's clay. The optimum results were shown by the membrane containing 10% FA. The physical characterization of the FA base membrane showed a large number of evenly distributed pores, which helped protons to pass through easily and retain water, where X-ray analysis showed the quartz nature of the fly ash-based membrane. The MFCs of this membrane worked well in electrochemical tests and efficiently removed organic components. The testing states that the MFCs incorporated with fly ash-based membranes will be more affordable and widespread.

Oni et al. (2023) suggested using FA as a filtrating agent in drilling fluids by varying its concentration. Tests were conducted on ilmenite-derived densified water-based mud samples at reservoir conditions, and results were compared with API (American Petroleum Institute) standards. Results showed improved filtration properties of mud with FA particles sealing filter cake openings. At the same time, the filtrate volume was reduced by 75%–79%.

4.2.4 Air Pollution Control Technologies

FA, a byproduct of coal combustion, offers valuable applications in air pollution control technologies, particularly in developing fly ash-based sorbents and flue gas treatment processes. Nguyen et al. (2023) review the waste-to-wealth applications of FA for environmental remediation, highlighting its potential as an adsorbent for flue gas and polluted water treatment, catalyst support for organic pollutant degradation and wastewater reforming. They emphasize the superior heavy metal ion removal of fly ash-derived materials compared to conventional adsorbents and enhanced Ni catalyst activity for biogas reforming. Here's an overview of how FA is utilized in these methods.

4.2.5 Fly Ash-Based Sorbents

FA is a sorbent for pollutants like SO_2, NO_x, Hg, and volatile organic compounds (VOCs) from industrial emissions, mitigating environmental and health risks. Miricioiu et al. (2021) developed polysulfone-based mixed matrix membranes

(MMMs) using Mobil Composition of Matter (MCM-41-type) silica from CFA as a filler for natural gas purification, exhibiting high resistance to CO_2 plasticization. Liu et al. (2022) tested fly ash/hydrated lime sorbents for CO_2 capture under conditions similar to flue gas desulfurization (FGD) systems, finding higher capture with <50% FA content and emphasizing the importance of low SO_2 concentration for effective capture.

Xiao et al. (2022) review the diverse applications of FA, including its use as a catalyst and adsorbent for environmental remediation. They discuss preparation methods for fly ash-based materials, photocatalytic degradation of pollutants, characterization techniques, and the sorption of pollutants from flue gases. The review also outlines future research directions for developing low-cost materials for environmental remediation using FA. Fomenko et al. (2023) converted dispersed FA microspheres into high-strength glass ceramics with magnetic properties, offering a solution for solid waste pollution reduction. Kunecki et al. (2023) utilized fly ash-derived zeolites X and A modified with silver to capture elemental mercury from gas streams, showcasing improved adsorption compared to untreated zeolites. Their novel hybrid synthesis method offers the potential for efficient mercury removal, aligning with the principles of circular economy. Tan et al. (2023) synthesized solid amine sorbents from fly ash-derived mesoporous silica, impregnated with tetraethylenepentamine (TEPA), exhibiting high CO_2 adsorption capacity (5.17 mmol/g at 75°C) and rapid kinetics (5.08 mmol/g in 10 minutes), making it promising for carbon capture technologies.

4.2.6 Flue Gas Treatment

FA aids in flue gas treatment by capturing harmful gases like SO_2 in processes such as FGD while also trapping particulate matter in electrostatic precipitators (ESPs) and baghouse filters, enhancing air quality. Xin et al. (2022) modified fly ash-derived magnetospheres with H_2S, creating efficient sorbents (S-MS) for the removal of elemental mercury (HgO). Sulphidation at 150°C resulted in over 80% HgO adsorption efficiency, with low leaching and superior cost-effectiveness compared to activated carbon, showing promise for coal combustion flue gas treatment. Gadore et al. (2024) review FA's utilization as a catalyst and adsorbent for environmental remediation, detailing preparation methods and applications in photocatalytic degradation and flue gas treatment while outlining future research directions. Xu et al. (2023) investigate SO_3 removal from flue gas using modified FA adsorbents. They find increased adsorption efficiency post-modification, with the optimal performance achieved using a 1.5 mol/L NaOH solution. Temperature optimization is observed at 350°C, while SO_2 presence competes with SO_3 adsorption. Wang et al. (2023) investigated Pb ion removal from fly ash (FA) washing solution using simulated exhaust gases, achieving high efficiency under specific conditions. Thermodynamic analysis suggests a spontaneous, exothermic process, meeting regulatory thresholds for residual pollutants and pH. The study offers guidance for pollutant remediation in FA washing solutions and carbon emissions sequestration.

Kalisz et al. (2023) investigated dry sorbent injection (DSI) desulfurization using sodium bicarbonate to reduce SO_2 emissions from a grate boiler combustion chamber. Unique nozzle design allowed injection into high-temperature zones (~700°C), achieving an 81% reduction in SO_2 levels. Chemical analysis of resulting FA revealed altered composition, notably increased concentrations of S, Na, Zn, Cu, As, Mo, Pb, loss of ignition (LOI), and total organic carbon (TOC), affecting leachate quality with elevated Total Dissolved Solids (TDS) and concentrations of sulphates, chlorides, fluorides, As, Pb, Cu, and Zn. The study addresses safe storage and circular economy principles for fly ash utilization. Sheng et al. (2023) propose a novel catalytic adsorption approach using CFA and carbide slag (CS) as adsorbents and Na_2O/Na_2O_2 as a catalyst for efficient and low-cost De-SO_2 and De-NOx technology. The highest SO_2 and NOx adsorption capabilities were observed at 0.55 and 0.06 mg/g under 2% Na_2O/FA-CS adsorbent at 150°C. The synergistic adsorption process facilitated by Na_2O_2 loading enhanced catalytic activation and adsorption performance, offering implications for the high-efficiency desulfurization and denitrification of CFA and CS.

4.3 FUTURE DIRECTIONS AND RESEARCH NEEDS

The use of fly ash-derived materials, despite their exceptional physicochemical characteristics and efficient use for various applications, such as heterogeneous catalytic reactions and impurity adsorption, still needs to be solved. First, materials made from FA have several undesirable and complex components that may need to be thoroughly purified before being used as adsorbents or catalytic supports, which could be expensive. Remarkably, sulphur-containing components are recognized to be one of the primary causes of catalyst deactivation in thermochemical reactions and may cause significant equipment corrosion. Therefore, removing these undesirable chemicals to satisfy process safety and catalyst synthesis requirements is a difficult and expensive procedure. The second environmental problem preventing the industrial use of products obtained from FA is the possibility of harmful metals, including Pb, Cr, As, Hg, and V, leaking into land and groundwater. In addition to the previously mentioned factors, the high cost of FA transportation and the market competition from existing industrial products, such as popular adsorbents and catalysts, may cause the industrial application of waste FA materials to be delayed. There haven't been any thorough investigations into the intrinsic mechanism and related kinetics for fly ash-based materials, in contrast to conventional adsorbents in the adsorption process and catalysts in photodegradation and methane dry reforming reactions. These critical studies must be carried out to facilitate the design and scaling-up of adsorption columns and catalytic reactors for large-scale production. Furthermore, to increase the effectiveness and lower the cost of FA utilization, further research should be done on sophisticated synthesis techniques for transforming waste FA into valuable materials and recovery techniques for spent fly ash-derived adsorbents and catalysts.

4.4 CONCLUSION

The disposal of FA, one of the primary hazardous solid wastes that thermal power plants release in excess to meet the demand for energy in many nations, has received a lot of attention as an environmental concern. Converting waste FA into valuable materials could be a better option than storing it in landfills, which unnecessarily take up a lot of space. In particular, FA has been successfully changed for creating zeolite and geopolymer materials by various techniques, including multi-step modification, sonication assistance, and improved microwave modification. These compounds, generated from FA, are acknowledged as potential adsorbents for treating organic pollutants, such as industrial colours in aqueous solutions, heavy metal ions, and hazardous gases. Fly ash-based adsorbent for flue gas, heavy metals and dyes has undergone extensive testing of the adsorption circumstances, including pH value, contact time, initial adsorbate concentration, and adsorbent loading, as well as adsorption modes, such as batch and continuous column. Notably, geopolymer and zeolite adsorbents made from FA demonstrated a higher adsorption capacity for the removal of heavy metal ions when compared to other traditional carbon-based materials. Furthermore, FA works well as a dispersant for active metals in other heterogeneous catalytic processes, such as the reforming reaction for the formation of H_2 and the photodegradation of wastewater. FA also produces batteries, building materials, and soil amendments since it includes SiO_2, Al_2O_3, and other valuable rare metals like zirconium and vanadium.

REFERENCES

Ahmaruzzaman, M. (2010). A review on the utilization of fly ash. *Progress in Energy and Combustion Science*, 36(3), 327–363.

Alengebawy, A., Abdelkhalek, S. T., Qureshi, S. R., & Wang, M. Q. (2021). Heavy metals and pesticides toxicity in agricultural soil and plants: Ecological risks and human health implications. *Toxics*, *9*(3), 42.

Al-Harahsheh, M. S., Al Zboon, K., Al-Makhadmeh, L., Hararah, M., & Mahasneh, M. (2015). Fly ash based geopolymer for heavy metal removal: A case study on copper removal. *Journal of Environmental Chemical Engineering*, *3*(3), 1669–1677.

Asaoka, S., Saito, H., Ichinari, T., Hayakawa, S., & Oikawa, T. (2021). Removal of hydrogen sulfide gas using coal fly ash–blast furnace cement composite. *Journal of Water, Sanitation and Hygiene for Development*, *11*(5), 824–830.

ASTM, N. (2011). *C1157/C1157M, Standard Performance Specification for Hydraulic Cement, Pennsylvania: West Conshohocken, Pa. EE. UU.* American Society for Testing and Materials.

Bandura, L., Panek, R., Madej, J., & Franus, W. (2021). Synthesis of zeolite-carbon composites using high-carbon fly ash and their adsorption abilities towards petroleum substances. *Fuel*, *283*, 119173.

Barman, D., & Dash, S. K. (2022). Stabilization of expansive soils using chemical additives: A review. *Journal of Rock Mechanics and Geotechnical Engineering*, 14(4), 1319–1342.

Bazarbekova, A., Naik, S. R., Kim, Y. R., Little, D., Jung, J. S., & Park, Y. B. (2024). One-part alkali-activated soil stabilization with sodium metasilicate: Mechanical-geochemical-mineralogical characterization. *Transportation Geotechnics*, 44, 101163.

Blissett, R. S., & Rowson, N. A. (2012). A review of the multi-component utilisation of coal fly ash. *Fuel*, *97*, 1–23.

Bouazizi, A., Breida, M., Karim, A., Achiou, B., Ouammou, M., Calvo, J. I.,..., & Younssi, S. A. (2017). Development of a new TiO2 ultrafiltration membrane on flat ceramic support made from natural bentonite and micronized phosphate and applied for dye removal. *Ceramics International*, *43*(1), 1479–1487.

Buema, G., Lupu, N., Chiriac, H., Ciobanu, G., Bucur, R. D., Bucur, D.,..., & Harja, M. (2021). Performance assessment of five adsorbents based on fly ash for removal of cadmium ions. *Journal of Molecular Liquids*, *333*, 115932.

Buema, G., Trifas, L. M., & Harja, M. (2021). Removal of toxic copper ion from aqueous media by adsorption on fly ash-derived zeolites: Kinetic and equilibrium studies. Polymers, 13(20), 3468.

Chathurappan, R., & Jayapal, A. (2022). Synthesis and characterization of coal fly ash based multipurpose ultrafiltration membrane. *Materialia*, *25*, 101548.

Cheng, Z., Cui, G., Yang, Z., Gang, H., Gao, Z., Zhang, D., & Xi, C. (2021). Improvement of the salinized soil properties of fly ash by freeze-thaw cycles: An impact test study. *Sustainability*, *13*(5), 2908.

Cui, G., Cheng, Z., Sun, F., Ma, S., & Liu, Z. (2024). Mechanical properties and neural network prediction of cement fly ash-enhanced roadbed soil in seasonal frozen zones under short-term curing. *KSCE Journal of Civil Engineering*, *28*(1), 106–123.

Cui, H., Yin, L., Huang, X., Yu, Z., Zhang, Z., & Dai, Z. (2022). Zeolite fly ash-enhanced coagulation treatment of oil recovery wastewater from polymer flooding. *Environmental Science and Pollution Research*, *29*(60), 90318–90327.

Dash, S., & Sahoo, S. (2017). Effect of fly ash amendment on growth of mustard. *International Journal of Applied Environmental Sciences*, *12*(9), 1617–1629.

Derakhshan Nejad, Z., Jung, M. C., & Kim, K. H. (2018). Remediation of soils contaminated with heavy metals with an emphasis on immobilization technology. *Environmental Geochemistry and Health*, *40*, 927–953.

Diana, S., Munawar, M., Saifuddin, S., Nahar, N., Silmina, S., Razi, F., & Arahman, N. (2023). Ceramic membrane-based on fly ash-clay for river water treatment. *Biointerface Research in Applied Chemistry*, *13*(5), 430.

Dil, E. A., Ghaedi, M., Ghaedi, A., Asfaram, A., Jamshidi, M., & Purkait, M. K. (2016). Application of artificial neural network and response surface methodology for the removal of crystal violet by zinc oxide nanorods loaded on activate carbon: kinetics and equilibrium study. *Journal of the Taiwan Institute of Chemical Engineers*, *59*, 210–220.

Dindi, A., Quang, D. V., Vega, L. F., Nashef, E., & Abu-Zahra, M. R. (2019). Applications of fly ash for CO2 capture, utilization, and storage. *Journal of CO2 Utilization*, *29*, 82–102.

Edil, T. B., Acosta, H. A., & Benson, C. H. (2006). Stabilizing Soft Fine-Grained Soils with Fly Ash. *Journal of Materials in Civil Engineering*, *18*(2), 283–294.

Eteba, A., Bassyouni, M., & Saleh, M. (2022). Modified coal fly ash for textile dye removal from industrial wastewater. *Energy & Environment*, 0958305X221130536.

Eze, C. P. et al. (2020). Fly ash-based geopolymer building materials for green and sustainable development. *Materials*, *13*(24),5699, 1–17.

Feng, W., Wan, Z., Daniels, J., Li, Z., Xiao, G., Yu, J.,..., & Li, G. K. (2018). Synthesis of high quality zeolites from coal fly ash: Mobility of hazardous elements and environmental applications. *Journal of Cleaner Production*, *202*, 390–400.

Fomenko, E. V., Akimochkina, G. V., Knyazev, Y. V., Semenov, S. V., Yumashev, V. V., Solovyov, L. A., & Anshits, A. G. (2023). Characterization and magnetic properties of sintered glass-ceramics from dispersed fly ash microspheres. *Magnetochemistry*, *9*(7), 177.

Gadore, V., Mishra, S. R., Sillanpää, M., & Ahmaruzzaman, M. (2024). Emerging applications of waste fly ash for remediation of environmental contaminants: a high-value and sustainable approach towards utilization of waste materials. *Nanotechnology for Environmental Engineering*, *9*(2), 207–237.

Gao, Y., Sui, H., Yu, Z., Wu, J., Chen, W., & Liu, Y. (2023). Cemented waste rock backfill enhancement via fly ash-graphene oxide hybrid under different particle size distribution. *Construction and Building Materials*, *394*, 132162.

Gondek, K., Mierzwa-Hersztek, M., & Jarosz, R. (2023). Effect of willow biochar and fly ash-derived zeolite in immobilizing heavy metals and promoting enzymatic activity in a contaminated sandy soil. *Catena, 232,* 107429.

Goswami, K. P., Pakshirajan, K., & Pugazhenthi, G. (2022). Process intensification through waste fly ash conversion and application as ceramic membranes: A review. *Science of the Total Environment, 808,* 151968.

Guan, L. L., Chen, Y. G., Ye, W. M., Wu, D. B., & Deng, Y. F. (2023). Foamed concrete utilizing excavated soil and fly ash for urban underground space backfilling: Physical properties, mechanical properties, and microstructure. *Tunnelling and Underground Space Technology, 134,* 104995.

Hamed, E. A., & Demirmoöz, A. (2024). Optimization of geotechnical characteristics of clayey soils using fly ash and granulated blast furnace slag-based geopolymer. Available at SSRN 4755845.

He, X., Yao, B., Xia, Y., Huang, H., Gan, Y., & Zhang, W. (2020). Coal fly ash derived zeolite for highly efficient removal of Ni2+ inwaste water. *Powder technology, 367, 40*-46.

Hoang, S. A., Lamb, D., Seshadri, B., Sarkar, B., Choppala, G., Kirkham, M. B., & Bolan, N. S. (2021). Rhizoremediation as a green technology for the remediation of petroleum hydrocarbon-contaminated soils. *Journal of Hazardous Materials*, 401, 123282.

Huang, X., Zhao, H., Zhang, G., Li, J., Yang, Y., & Ji, P. (2020). Potential of removing Cd (II) and Pb (II) from contaminated water using a newly modified fly ash. *Chemosphere, 242*, 125148.

Ijaz, N., Rehman, Z., Ijaz, Z. (2022). Principles and Prospects of Using Lignosulphonate as a Sustainable Expansive Soil Ameliorator: From Basics to Innovations. In: Reddy, K.R., Pancharathi, R.K., Reddy, N.G., Arukala, S.R. (eds) *Advances in Sustainable Materials and Resilient Infrastructure* (pp. 103–116). Springer Transactions in Civil and Environmental Engineering. Springer, Singapore.

Kalisz, S., Wejkowski, R., Maj, I., & Garbacz, P. (2023). A novel approach to the dry desulfurization process by means of sodium bicarbonate: A full-scale study on SO2 emission and geochemistry of fly ash. *Energy, 279,* 128046.

Khalid, U., Rehman, Z. U., Ijaz, N., Khan, I., & Junaid, M. F. (2023). Integrating wheat straw and silica fume as a balanced mechanical ameliorator for expansive soil: A novel agri-industrial waste solution. *Environmental Science and Pollution Research, 30*(29), 73570–73589.

Kobayashi, Y., Ogata, F., Nakamura, T., & Kawasaki, N. (2020). Synthesis of novel zeolites produced from fly ash by hydrothermal treatment in alkaline solution and its evaluation as an adsorbent for heavy metal removal. *Journal of Environmental Chemical Engineering, 8*(2), 103687.

Kodaz, G., Demirci, H. E., & Pulat, H. F. (2022). Improvement of strength characteristics of a highly plastic expansive soil with fly ash. *Sakarya University Journal of Science, 26*(3), 448–458.

Kumar, A., Kumari, M., & Gupta, S. K. (2022a). Performance study of fly-ash-derived coagulant in removing natural organic matter from drinking water: Synthesis, characterization, and modelling. *Environmental Monitoring and Assessment, 194*(11), 821.

Kumar, B., Gupta, A., & Kasaudhan, A. (2022b). CBR characteristics and swelling behaviour of black cotton soil mixing with fly ash and silica fume. *SAMRIDDHI: A Journal of Physical Sciences, Engineering and Technology, 14*(04), 49–54.

Kunecki, P., Wdowin, M., & Hanc, E. (2023). Fly ash-derived zeolites and their sorption abilities in relation to elemental mercury in a simulated gas stream. *Journal of Cleaner Production, 391,* 136181.

Lee, S. H., Kim, S. O., Lee, S. W., Kim, M. S., & Park, H. (2021). Application of soil washing and thermal desorption for sustainable remediation and reuse of remediated soil. *Sustainability, 13*(22), 12523.

Lee, S. H., Lee, J. H., Jung, W. C., Park, M., Kim, M. S., Lee, S. J., & Park, H. (2020). Changes in soil health with remediation of petroleum hydrocarbon contaminated soils using two different remediation technologies. *Sustainability, 12*(23), 10078.

Lekgoba, T., Ntuli, F., & Falayi, T. (2021). Application of coal fly ash for treatment of wastewater containing a binary mixture of copper and nickel. *Journal of Water Process Engineering, 40*, 101822.

Li, Q., Wei, H., Zhang, Y., Han, L., Han, S., & Ding, N. (2020). The variations on thermal conductivity and structures of silty clay modified by waste fly ash and oil shale ash after freeze–thaw cycles. *Construction and Building Materials, 260,* 119954.

Li, Y., Qi, H., You, C., & Xu, X. (2007). Kinetic model of CaO/fly ash sorbent for flue gas desulphurization at moderate temperatures. *Fuel, 86*(5–6), 785–792.

Liu, C. F., Yang, C. H., & Shih, S. M. (2022). CO_2 capture by fly ash/hydrated lime sorbents at low temperatures. *Industrial & Engineering Chemistry Research, 61*(14), 4774–4783.

Lu, L., Liu, C., Qu, S., & Zhang, M. (2022). Experimental study on the mechanical and hydraulic behaviour of Fibre-Reinforced cemented soil with fly ash. *Construction and Building Materials, 321,* 126374.

Maheepala, M. M. A. L. N., Nasvi, M. C. M., Robert, D. J., Gunasekara, C., & Kurukulasuriya, L. C. (2023). Mix design development for geopolymer treated expansive subgrades using artificial neural network. *Computers and Geotechnics, 161,* 105534.

Maila, G. B., Ntuli, F., & Lekgoba, T. (2020). Manufacture of fly ash based coagulants for use in water treatment, *Test Engineering and Management, 82*(2020), 5675–5681.

Medykowska, M., Wiśniewska, M., Szewczuk-Karpisz, K., Panek, R., & Franus, M. (2023). Simultaneous removal of inorganic and organic pollutants from multicomponent solutions by the use of zeolitic materials obtained from fly ash waste. *Clean Technologies and Environmental Policy, 25*(4), 1133–1148.

Miricioiu, M. G., Niculescu, V. C., Filote, C., Raboaca, M. S., & Nechifor, G. (2021). Coal fly ash derived silica nanomaterial for MMMs—Application in CO2/CH4 separation. *Membranes, 11*(2), 78.

Mohammadi, E. L., Najafi, E. K., Ranjbar, P. Z., Payan, M., Chenari, R. J., & Fatahi, B. (2023). Recycling industrial alkaline solutions for soil stabilization by low-concentrated fly ash-based alkali cements. *Construction and Building Materials, 393,* 132083.

Mohanty, S., Roy, N., Singh, S. P., & Sihag, P. (2021). Strength and durability of flyash, GGBS and cement clinker stabilized dispersive soil. *Cold Regions Science and Technology, 191,* 103358.

Munda, S., Nayak, A. K., Mishra, P. N., Bhattacharyya, P., Mohanty, S., Kumar, A.,..., & Thilagam, V. K. (2016). Combined application of rice husk biochar and fly ash improved the yield of lowland rice. *Soil Research, 54*(4), 451–459.

Mushtaq, F., Zahid, M., Bhatti, I. A., Nasir, S., & Hussain, T. (2019). Possible applications of coal fly ash in wastewater treatment. *Journal of Environmental Management, 240,* 27–46.

Mushtaq, S., Aslam, Z., Ali, R., Aslam, U., Naseem, S., Ashraf, M., & Bello, M. M. (2022). Surfactant modified waste ash for the removal of chloro and nitro group substituted benzene from wastewater. *Water Science & Technology, 86*(8), 1969–1980.

Mypati, V. N., & Saride, S. (2022). Feasibility of alkali-activated low-calcium fly ash as a binder for deep soil mixing. *Journal of Materials in Civil Engineering, 34*(1), 04021410.

Nayak, A. K., Raja, R., Rao, K. S., Shukla, A. K., Mohanty, S., Shahid, M.,..., & Swain, C. K. (2015). Effect of fly ash application on soil microbial response and heavy metal accumulation in soil and rice plant. *Ecotoxicology and Environmental Safety, 114*, 257–262.

Nguyen, N. T., Trinh, Q. H., Nguyen, T. T. H., & Ho, T. T. H. (2023). Study on the modification of fly ash as a coagulant. *Vietnam Journal of Science, Technology and Engineering, 65*(2), 20–25.

Noaman, M. F., Khan, M. A., Ali, K., & Hassan, A. (2022). A review on the effect of fly ash on the geotechnical properties and stability of soil. *Cleaner Materials, 6,* 100151.

Novais, R. M., Ascensao, G., Tobaldi, D. M., Seabra, M. P., & Labrincha, J. A. (2018). Biomass fly ash geopolymer monoliths for effective methylene blue removal from wastewaters. *Journal of Cleaner Production*, *171*, 783–794.

Oni, O., Fadairo, A. S., Gosnold, W., Egenhoff, S., Adeyemi, G., & Ling, K. (2023). Investigating the suitability of North Dakota fly ash as fluid loss reducing additive in densified water-based drilling fluid. *Powder Technology*, *430,* 118972.

Police, S., Maity, S., Chaudhary, D. K., Dusane, C. K., Sahu, S. K., & Kumar, A. V. (2020). Effect of coal fly ash's particle size on U adsorption in water samples and thermodynamic study on adsorption. *Environmental Chemistry and Ecotoxicology*, *2,* 32–38.

Ređenović, D., Kerkez, Đ., Pilipović, D. T., Dubovina, M., Šešlija, M., Tenodi, S., & Peško, I. (2024). The beneficial reuse of contaminated sediment: Long-term assessment of fly ash and lime-based mixtures. *Case Studies in Construction Materials*, *20,* e02711.

Rao, A., Kaushik, A., Kuppurangan, G., & Selvaraj, G. (2023). Characterization and application of novel fly ash blended ceramic membrane in MFC for low-cost and sustainable wastewater treatment and power generation. *Environmental Science and Pollution Research*, *30*(16), 45872–45887.

Renita, A. A., Amarnath, D. J., & Duraikannu, S. L. (2021). Synthesis of peanut-shell magnetized biocarbon for acid fuchsin dye removal. *Materials Today: Proceedings*, *43,* 3075–3078.

Rodwihok, C., Suwannakaew, M., Han, S. W., Lim, Y. J., Park, S. Y., Woo, S. W.,..., & Kim, H. S. (2023). Effective removal of hazardous organic contaminant using integrated photocatalytic adsorbents: Ternary zinc oxide/zeolite-coal fly ash/reduced graphene oxide nanocomposites. *Colloids and Surfaces A: Physicochemical and Engineering Aspects*, *662,* 131044.

Saride, S., & Mypati, V. N. (2024). Effect of area improvement ratio of geopolymer-based deep mixing columns on swell-shrink behavior of expansive soils. *Construction and Building Materials*, *417,* 135163.

Sharma, V., & Singh, S. (2020). Modeling for the use of waste materials (bottom ash and fly ash) in soil stabilization. *Materials Today: Proceedings*, *33,* 1610–1614.

Sheng, H., Wang, Z., Gao, L., Zhan, W., He, Z., & Zhang, J. (2023). Synergistic and competitive adsorption of NOx and SO2 using a microwave-assisted approach with industrial waste materials fly ash and carbide slag as sorbents. *Materials Today Sustainability*, *23,* 100475.

Silvani, C., Ibraim, E., Scheuermann Filho, H. C., Festugato, L., Diambra, A., & Consoli, N. C. (2022). Sand-fly ash-lime blends: Mechanical behavior under multiaxial stress condition. *Journal of Materials in Civil Engineering*, *34*(5), 04022059.

Singh, N. B., Agarwal, A., De, A., & Singh, P. (2022). Coal fly ash: An emerging material for water remediation. *International Journal of Coal Science & Technology*, *9*(1), 44.

Sivalingam, S., Kella, T., Maharana, M., & Sen, S. (2019). Efficient sono-sorptive elimination of methylene blue by fly ash-derived nano-zeolite X: Process optimization, isotherm and kinetic studies. *Journal of Cleaner Production*, *208*, 1241–1254.

Sun, G., Zhang, J., Hao, B., Li, X., Yan, M., & Liu, K. (2022). Feasible synthesis of coal fly ash based porous composites with multiscale pore structure and its application in Congo red adsorption. *Chemosphere*, *298,* 134136.

Tan, M., Li, X., Feng, Y., Wang, B., Han, L., Bao, W.,..., & Wang, J. (2023). Fly ash-derived mesoporous silica with large pore volume for augmented CO2 capture. *Fuel*, *351,* 128874.

Tan, T. H., Mo, K. H., Ling, T. C., & Lai, S. H. (2020). Current development of geopolymer as alternative adsorbent for heavy metal removal. *Environmental Technology & Innovation*, *18,* 100684.

Teng, X., Li, J., Wang, J., Liu, J., Ge, X., & Gu, T. (2021). Effective degradation of atrazine in wastewater by three-dimensional electrochemical system using fly ash-red mud particle electrode: Mechanism and pathway. *Separation and Purification Technology*, *267,* 118661.

Tian, Q., & Sasaki, K. (2019). Application of fly ash-based materials for stabilization/solidification of cesium and strontium. *Environmental Science and Pollution Research, 26,* 23542–23554.

Turan, C., Javadi, A. A., & Vinai, R. (2022). Effects of class C and class F fly ash on mechanical and microstructural behavior of clay soil—A comparative study. *Materials, 15*(5), 1845.

Umar, M., Ji, X., Kirikkaleli, D., & Alola, A. A. (2021). The imperativeness of environmental quality in the United States transportation sector amidst biomass-fossil energy consumption and growth. *Journal of Cleaner Production, 285,* 124863.

Varol, M., Sünbül, M., Aytop, H., & Yılmaz, C. (2019). Environmental, ecological and health risks of trace elements, and their sources in soils of Harran Plain, Turkey. *Chemosphere, 245,* 125592.

Wang, L., Tang, Y., Gong, Y., Shao, X., Lin, X., Xu, W.,…, & Kołodyńska, D. (2023). Remediation of micro-pollution in an alkaline washing solution of fly ash using simulated exhaust gas: Parameters and mechanism. *Sustainability, 15*(7), 5873.

Wang, N., Jin, L., Li, C., Liang, Y., & Wang, P. (2022). Preparation of coal fly ash-based Fenton-like catalyst and its application for the treatment of organic wastewater under microwave assistance. *Journal of Cleaner Production, 342,* 130926.

Wang, N., Sun, X., Zhao, Q., & Wang, P. (2021). Treatment of polymer-flooding wastewater by a modified coal fly ash-catalysed Fenton-like process with microwave pre-enhancement: System parameters, kinetics, and proposed mechanism. *Chemical Engineering Journal, 406,* 126734.

Wang, W., Qi, L., Zhang, P., Luo, J., & Li, J. (2022). Removal of COD in wastewater by magnetic coagulant prepared from modified fly ash. *Environmental Science and Pollution Research, 29*(34), 52175–52188.

Wang, X., Zhang, N., Zhang, Y., Liu, J., Xiao, X., Meng, K.,…, & Chu, P. K. (2020). Multiple flocculant prepared with dealkalized red mud and fly ash: Properties and characterization. *Journal of Water Process Engineering, 34,* 101173.

Xiao, H., Peng, H., Deng, S., Yang, X., Zhang, Y., & Li, Y. (2012). Preparation of activated carbon from edible fungi residue by microwave assisted K2CO3 activation—Application in reactive black 5 adsorption from aqueous solution. *Bioresource Technology, 111,* 127–133.

Xiao, R., Gao, T., Cui, X., Ji, Y., Zhang, Y., Chuai, X.,…, & Zhao, Y. (2022). Removal of elemental mercury from flue gas by recyclable CuCl2 modified magnetospheres catalyst from fly ash: Part 6. Commercial scale demonstration at a 1000MWth coal-fired power plant. *Fuel, 310,* 122219.

Xin, F., Xiao, R., Zhao, Y., & Zhang, J. (2022). Surface sulfidation modification of magnetospheres from fly ash for elemental mercury removal from coal combustion flue gas. *Chemical Engineering Journal, 436,* 135212.

Xu, D., Ji, P., Wang, L., Zhao, X., Hu, X., Huang, X., ... & Liu, F. (2021). Effect of modified fly ash on environmental safety of two soils contaminated with cadmium and lead. *Ecotoxicology and Environmental Safety, 215,* 112175.

Xu, G., Chen, F., Qian, Y., Lu, P., Hong, M., Zhang, Q., & Zhou, Q. (2023). Experimental investigation of removal of SO_3 from flue gas with modified fly ash adsorbents. *ACS Omega, 8*(19), 16656–16672.

Yadav, V. K., Amari, A., Gacem, A., Elboughdiri, N., Eltayeb, L. B., & Fulekar, M. H. (2023). Treatment of fly-ash-contaminated wastewater loaded with heavy metals by using fly-ash-synthesized iron oxide nanoparticles. *Water, 15*(5), 908.

Yao, Q., Peng, Y., Chen, M., Wang, Y., Ding, J., Ma, B.,…, & Lu, S. (2024). One-step high efficiency synthesis of zeolite from fly ash by mechanochemical method as a low-cost adsorbent for cadmium removal. *Journal of Environmental Chemical Engineering, 12*(1), 111877.

Yoobanpot, N., Jamsawang, P., Krairan, K., Jongpradist, P., & Likitlersuang, S. (2020). Laboratory investigation of the properties of cement fly ash gravel for use as a column-supported embankment. *Construction and Building Materials, 257,* 119493.

Zhang, J., Yan, M., Sun, G., & Liu, K. (2021). Simultaneous removal of Cu (II), Cd (II), Cr (VI), and rhodamine B in wastewater using TiO2 nanofibers membrane loaded on porous fly ash ceramic support. *Separation and Purification Technology, 272*, 118888.

5 Fly Ash-Based Sustainable Building Materials and Life Cycle Assessment of Fly Ash

Akash Anand, Naresh V. Patil, Amit Chauhan, Amit K. Jaiswal, Akash Saharan, Rahul Gautam and Harish Sharma

5.1 INTRODUCTION

The pressing concerns of natural resource scarcity, fossil fuel energy usage, and total greenhouse gas emissions demand urgent action. Fossil fuels, in particular, are in great demand due to the increasing world population and the expansion of the world economy. The observed increase in consumption can be attributed directly to the concerning escalation in greenhouse gas emissions. The relationship between agricultural nitrous oxide emissions and the scarcity of natural resources by considering the interplay between agriculture, forestry, fishery value addition, fossil fuels, and total greenhouse gas emissions in the leading countries from 1971 to 2020 was investigated. Several researchers found that there exists a positive correlation between limited availability of natural resources and various industrial sectors such as agriculture, forestry, fisheries, fossil fuel energy consumption, and overall greenhouse gas emissions (Mushtaq et al. 2019; Kalombe et al. 2020; Dwivedi and Jain 2014). In the present-day scenario, the energy demands revolve around the fossil fuels, coal is one of them (Gautam et al. 2021). The major portion of coal finds usage in the thermal power pants, which in turn produces fly ash (FA). Also, there has been a significantly increased dependency on sustainability within the construction industry (Dwivedi and Jain 2014). This has raised concerns on environmental degradation, deficient energy resources, and depletion of natural resources due to increase in pollution (Tian and Sasaki 2019). As a result, the construction industry is facing the heat to incorporate more sustainable practices. Hence, it is required to find an alternative material, capable of minimizing the environmental impact and also keeping the structural integrity and functional properties intact (Yuan et al. 2023). Now, the concrete manufacturing process compromises with the environment due to its predominant cement component, which is known to release substantial amounts of carbon dioxide. The environmental impact of concrete construction might potentially be

DOI: 10.1201/9781003498124-5

mitigated by minimizing cement usage by the incorporation of mineral admixtures, such as FA, while maintaining durability criteria (Lu et al. 2023; Vargas and Halog 2015). Finding a solution to the issue becomes significantly more challenging when there is a corresponding rise in the quantity of FA and a decrease in the landfill capacity (Hemalatha and Ramaswamy 2021). The findings by the International Energy Agency (IEI 2021) state that due to the increased use of industrial energy at the global level, there has been a considerable amount of increase in the emission of carbon dioxide (CO_2). The expected rise in the demand of coal for 2021 was 60%, which later led to an increase of 5% in CO_2 emissions on the global level. In similar lines of study conducted, it was found that the cement industry was ranked third in terms of utilization in industrial energy and emits about 7% of CO_2. Hence, the use of FA has become crucial in the field of civil engineering due to its significant economic and environmental implications (Radhika et al. 2022). The extent to which FA can substitute cement is constrained by the quantity of free lime present in the ash.

In the fast-developing world, coal is the largest contributor to the man-made increase of CO_2 in the atmosphere. Its environmental impact spreads far across many areas, from air pollution, water, and waste management, up to increased land uses. The face of greenhouse gas emissions, industries are seeking a new way to manage their environmental impact and reduce the carbon footprint of production. Bio-coal is also an alternative; it is carbon-neutral and aimed to replace the fossil fuels and fossil coal in industrial processes. Bio-coal is a product of bio-green pyrolysis and carbonization of raw biomass performed within controlled temperature and residence time conditions (Seow et al. 2022). Thermal conversion of biomass, which is done under the oxygen-free conditions, allows the removal of volatile organic compounds and cellulose components from the feedstock and creates a uniform, solid biofuel with characteristics similar to the ones in fossil coal. Such fuels are gaining increasing attention in the carbon-intensive industries and the launch of new bio-green production facilities dedicated to bio-coal production is heralding a new approach towards the use of fossil-free solid fuels. Bio-coal (biomass briquette) is a replacement for fossil coal, fire-wood. The demand and use of bio-coal briquettes is increasing constantly due to a number of reasons like hike in costs of other fuels and electricity; the need for energy resource security; and pollution control measures (Zhao et al. 2022).

As green building materials have been developed over the last few decades, the construction industry has adhered to very strict sustainability standards. With the increasing population, there is need for more power as well. The existence of different sources of carbon-neutral energy cannot ignore the fact that coal is inexpensive and readily available. Coal FA is a by-product which is produced by the combustion of coal, in large quantities worldwide, and this quantity continues to increase (Saraswathy et al. 2003; Hemalatha and Ramaswamy 2021). For finding sustainable alternatives for building materials, the substitution of non-renewable aggregates with recycled materials and the use of supplementary cementing materials (SCMs) such as, FA, blast furnace slag, etc., to partially or completely replace Portland cement are suggested. This substitution approach mitigates pollution in the production and manufacturing process by reducing the release of pollutants and mining waste. The researchers specifically highlighted recent developments in the characterization, compositional knowledge, hydration mechanism, activation methods, durability, and sustainability aspects of FA. The sustainable construction approach by emphasizing on the use of FA geopolymers is emerging (Gupta, Nagpal, and Gupta 2021). The synthetization of geopolymers in the absence of aggregates, sand, or cement resulted in the reduction of the CO_2 emissions. As a result, the geopolymers produced possessed high compressive strength due to the alkali activation of FA (Xu and Shi 2018).

Various researchers have proposed potential applications of FA in several civil engineering projects, due to the advantageous physical and chemical characteristics of FA. In Figure 5.1, it can be noted that utilization of FA in the manufacturing of brick and cement contributes to 5% and 10%, respectively, with a total of 15%. In a similar manner, FA could be used as an alternative material where it can replace cement in mortar and concrete, as shown in Figure 5.1. Subsequently, the same can be done in the construction of roads, highways, as a lightweight aggregate, structural infill and for the reclamation of low-lying areas (Kalombe et al. 2020; Mushtaq et al. 2019; Rastogi and Paul 2020).

In order to cover all the major areas involving the applications of FA, a specification for pulverized FA for use as pozzolana in cement, mortar, and concrete was also released in 2003 (IS: 3812, 2003), along with four additional sections. Similarly, FA was used to update the Indian Code of Practice for plain and reinforced concrete (IS: 456, 2000). Guidelines for the application of FA in road embankments and various pavement layers have also been provided by the Indian Road Congress (IRC) (IRC: 37, 2012). As a result, the use of FA in the cement and construction industries has significantly expanded. Also, the global production share of Portland Pozzolana Cement (PPC) increased from 21.8% in 1999 to 67% in 2018. However, it might become clear that using FA only in cement and concrete would not be sufficient to use larger volumes of FA in the future; instead, new channels must be opened to boost the use of FA in other applications (Ram and Mohanty 2022; Hower et al. 2017) (Figure 5.2).

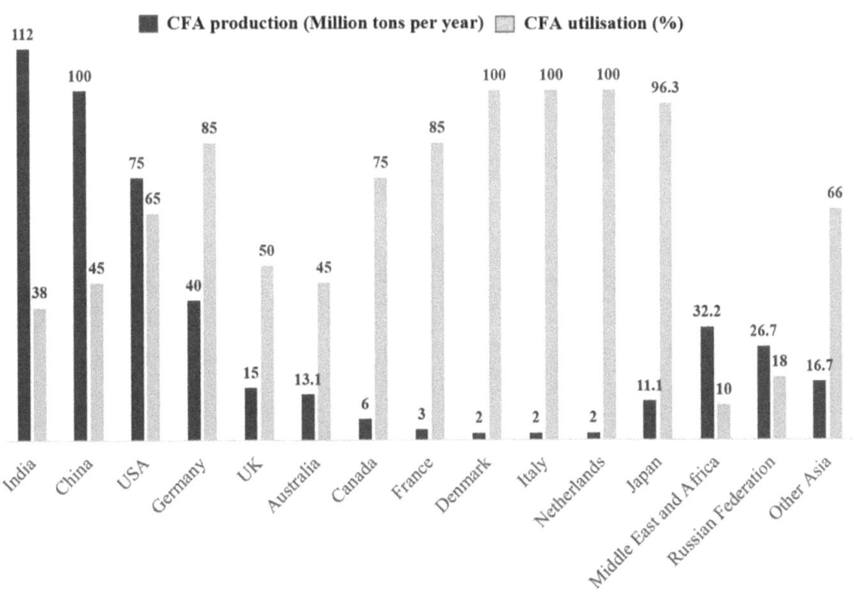

FIGURE 5.1 Production versus utilization of fly ash globally and source of origin.

5.2 FA: SOURCES, QUANTIFICATION, AND ISSUES

FA, a residual product derived from coal, has been widely employed as the predominant SCM in concrete due to its pozzolanic nature. The reaction degree of FA in cement-based materials is typically inadequate, and accurately quantifying its reaction degree is crucial for comprehending its pozzolanic hydration behaviour. Researchers identified certain components that were abundant in the unreacted FA particles but lacking in the surrounding paste. These elements were then utilized to separate the unreacted FA particles from the blended cement paste (Hower et al. 2017; Rastogi and Paul 2020). The sources of FA are illustrated in Figure 5.3 along with the percentage uses of FA shown in Figure 5.2.

5.2.1 Coal FA

It's a product of coal firing in coal power plants, at temperatures around 1,200°C–1,500°C (Yao et al. 2015). The composition of FA varies significantly depending on factors such as the type of coal, operating conditions, combustion conditions, cooling methods for flue gases, collectors, etc. This wide diversity in FA composition makes it challenging to characterize accurately.

5.2.2 Municipal Solid Waste FA

An increased generation of municipal solid waste (MSW) is observed, which is predominately incinerated or used in landfilling. MSW consists of organic materials and inorganic components are incinerated (Jianguo et al. 2004). It has varied compositions based on factors like location, lifestyle, education, awareness, recycling practices, etc. (Hasselriis and Licata 1996); it is a grey fine matrix with highly soluble salts, pH typically exceeding 12 (Lima et al. 2008).

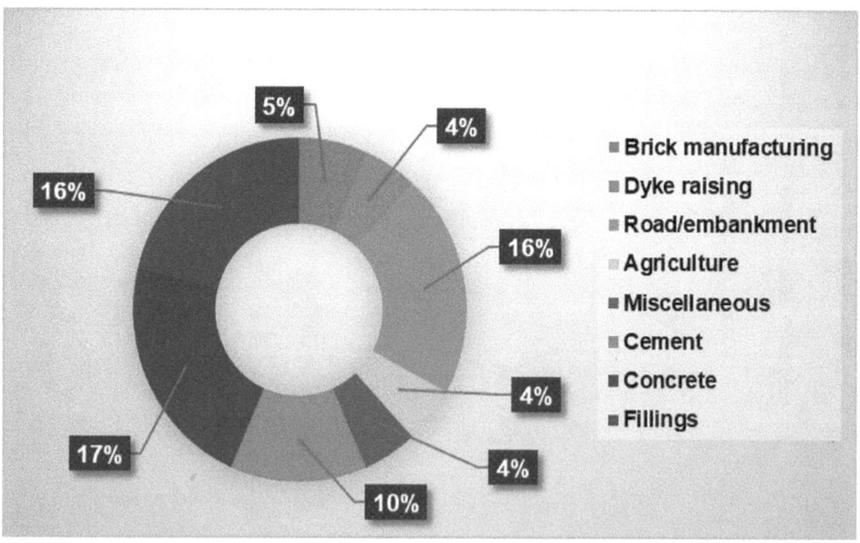

FIGURE 5.2 Percentage usage of fly ash.

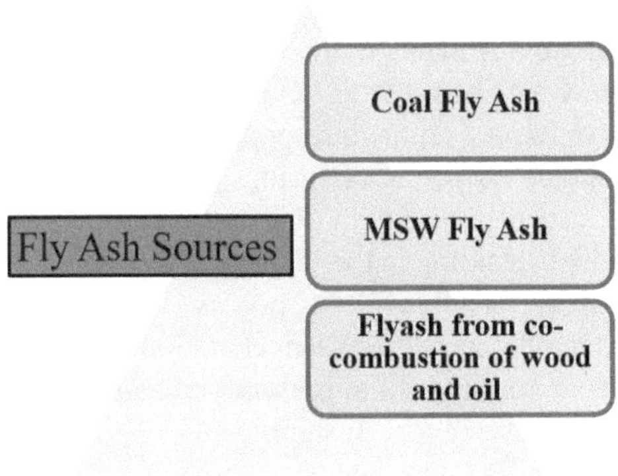

FIGURE 5.3 Sources of fly ash.

5.2.3 Ash from Co-Combustion of Wood and Oil

Combustion and co-combustion of wood and oil leads to thick grey ash. It has high pH, lesser Cl content but high iron content. It is a good source of cement production and verification processes (Lima et al. 2008).

5.2.4 Industrial Boilers and Bio-Coal

Industrial boilers from various industries contribute to FA. While bio-coal is a source of carbon-neutral energy, produced by the thermal conversion of biomass anaerobically and has characteristics similar to coal, bio-coal FA is lighter than FA from other sources.

Presently, there is a need for the novel approaches to precisely measure the quantity of unreacted FA particles in blended cement systems. Researchers suggested an approach for producing the total of distinctive components considering the inherent diversity of FA. Consequently, it is possible to separate the unreacted FA particles from the blended cement system by analysing the variations in chemical composition. This would help in accurately determining the proportion of unreacted FA and the degree of reaction of FA (Rashad 2015).

In India, the production of FA has consistently increased throughout the years, specifically from 1997 to 2015. FA generation was reported to be 70 MT in 1997, and it increased to 185 MT in 2015. The utilization of FA has experienced a significant growth, rising from 8% in 1997 to 70% in 2010. In addition, 40% of the FA is yet to be utilized (Singh 2017). Many researchers consider that FA hampers the environment significantly and can be judged as a hazardous material due to the presence of potentially harmful metals in FA that include lead (Pb), arsenic (As), mercury (Hg), aluminium (Al), cadmium (Cd), chromium (Cr), vanadium (V), boron (B), and selenium (Se). The proximity of these metals to the plants growing near ash dumps resulted in variations in enzyme activity. A significant reduction in microbiological action due to the presence of FA in soil has been observed (Jambhulkar, Shaikh, and Kumar 2018).

The soil enzymatic actions undergo fluctuations and the soil respiration is negatively impacted due to the presence of FA content in the soil. The accumulation of FA compromises with the groundwater quality in addition to that of the surface water. Further, if there are any plants in the vicinity of the FA dumping site, the soil and plants are adversely affected due to the presence of FA. The presence of coal FA basin had a negative impact on the bird population breeding in the area, leading to harmful impacts. Additionally, the offsprings of these birds led to an increase in the harmful substances such as strontium (Sr), cadmium (Cd), arsenic (As), and selenium (Se). The individuals residing in the vicinity of a thermal power plant (USEPA, 2007) experienced adverse health effects, such as cancer and respiratory diseases (Jambhulkar, Shaikh, and Kumar 2018; Chen et al. 2024).

Multiple experts have confirmed the presence of radionuclides in FA; however, the consequences are believed to be insignificant. The Bhabha Atomic Research Centre in Bombay reported that the detected radiation level in Indian coal is below the harmful limit and is very low. The presence of thorium and uranium radioactive elements was detected. However, it has been noted that their effects on the surrounding ecosystem are minimal. When FA is used for agricultural purposes, plants do not absorb any radioactive elements. The Central Fuel Research Institute in Dhanbad observed a minimal accumulation of radioactive elements by plants. FA harbours dangerous organic contaminants in addition to radionuclides. The dangerous organic pollutants include methyl sulphates, polychlorinated biphenyls, chlorinated benzofurans and dioxins, as well as polycyclic aromatic hydrocarbons.

The presence of trace elements and heavy metals in FA has the tendency to accumulate in living creatures and penetrate the food chain, resulting in hazardous consequences on both species and human health. Based on the hazardous nature of FA, especially towards the heavy metals, radionuclides, trace elements, and organic contaminants, which are of environmental importance, it can significantly affect humans, groundwater, soil, plants, birds, animals, and neighbouring water bodies. While FA is poisonous, it also contains essential minerals and trace elements that promote vegetation growth. The alkaline pH of the soil promotes plant development. Because of this characteristic, it can be utilized as both an amendment and a soil conditioner. Hence, it is essential to carry out an exploratory investigation regarding the toxicity of FA and its potential applications from an environmental standpoint (Rivera et al. 2015).

5.3 BUILDING MATERIALS

Building materials refer to substances or components used in the construction of structures such as buildings, bridges, roads, and other infrastructure. Here are some common types of building materials (Figure 5.4). Concrete is a mixture of cement, water, and aggregates (sand, gravel, or crushed stone) that solidifies into a strong and durable material. It is widely used in foundations, slabs, and structural elements. Other options are wood often used for framing, flooring, and finishing. Timber is a versatile and renewable resource. While bricks are made from clay or concrete, bricks are commonly used for constructing walls and facades.

These materials serve various purposes, including providing structural support, insulation, protection from the elements, and contributing to the overall aesthetics of the structure. Building materials can be natural or man-made and are chosen based

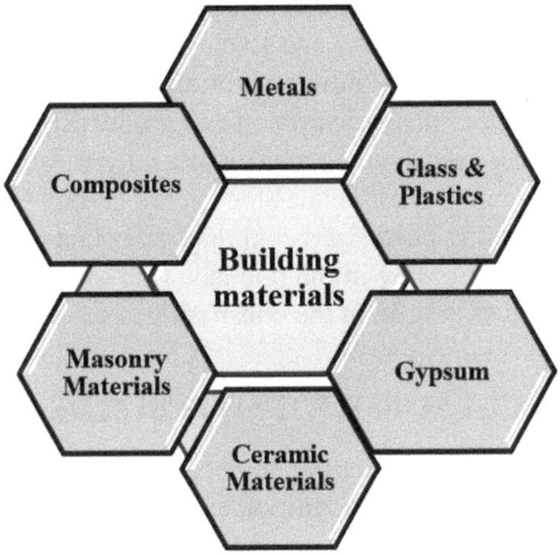

FIGURE 5.4 Different building materials.

on factors like durability, cost, availability, and the specific requirements of the construction project. Few other materials are metals like steel and aluminium that are commonly used for structural elements such as beams and columns. Copper and other metals are used for roofing and decorative purposes. Glass is used for windows, doors, and facades to allow natural light into buildings. It also contributes to the aesthetics of the structure. While plastics are used for various purposes, including pipes, insulation, and as a component in composite materials. Masonry materials are the materials besides bricks, materials like concrete blocks and natural stone are used for constructing walls. Ceramic materials are also used for tiles in flooring and walls, as well as other decorative elements. The gypsum products include drywall for interior walls and plaster for finishing.

Composite materials such as fibre-reinforced polymers (FRPs) and fibre cement, which combine different materials for improved properties.

5.4 SUSTAINABLE AND GREEN BUILDING MATERIALS

These materials are produced or taken from the environment and cause minimal impact on the environment and promote long-term ecological balance. They are often renewable, recycled, or are low in energy consumption, greenhouse gas emissions, etc. They are known for environmental friendliness, renewability, low toxicity, being durable, recyclable, and energy efficient. They offer lower carbon footprints, sustainability, and waste minimization. The key sustainable materials include bamboo, recycled materials, and eco-friendly options designed to minimize environmental impact (Kalombe et al. 2020; Yuan et al. 2023). The choice of building materials depends on various factors, including the type of construction, local climate conditions, building codes, and the specific requirements of the project. Advances in technology and "sustainable practices" continue to influence the development and use of new and innovative building materials. Apart from all the building materials listed above, it is very important to concentrate on sustainable or green building material such as recycled aggregate, plastic and steel, sustainable concrete, green roofing materials. Sustainable and green building materials are designed to minimize environmental impact throughout their life cycle, from production and use to disposal or recycling. These materials contribute to environmentally friendly construction practices, promote energy efficiency, and reduce the overall carbon footprint of buildings. FA is a by-product of burning pulverized coal in power plants, and it is composed of fine particles that are carried away in the flue gas. Due to its pozzolanic properties, FA has been widely used as a SCM in the construction industry.

5.5 FA AS A SUITABLE SUBSTITUTE FOR BUILDING MATERIALS

FA, a residual substance produced by the combustion of pulverized coal in electric power plants, has gained growing recognition as a significant construction material owing to its potential advantages and sustainability. There are several reasons why FA can be a viable alternative to conventional construction materials (Kalombe et al. 2020; Yuan et al. 2023) (Figure 5.5).

5.5.1 Utilization in Concrete Making

FA is frequently employed as a substitute for a portion of Portland cement in the making of concrete. These improvements include increased workability, less heat generation during hydration, enhanced long-term strength and durability, and reduced environmental impact of concrete manufacturing through decreased cement usage. Concrete that includes FA is commonly known as "fly ash concrete" or "fly ash blended cement concrete" (Rastogi and Paul 2020; Hemalatha and Ramaswamy 2017).

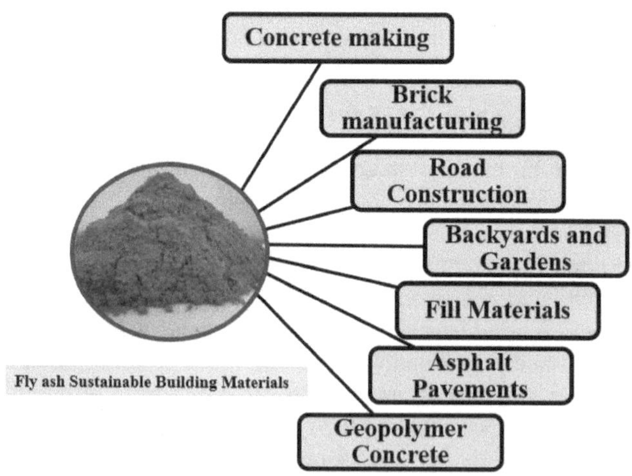

FIGURE 5.5 Fly ash-based building materials.

5.5.2 Brick Manufacturing

FA can be utilized as a substitute for clay in the production of bricks. FA bricks exhibit superior compressive strength, diminished efflorescence, and decreased water absorption in comparison to conventional clay bricks. Moreover, the utilization of FA in the manufacturing of bricks diminishes the need for clay and aids in the effective disposal of FA waste (Dwivedi and Jain 2014; Rastogi and Paul 2020).

5.5.3 Road Construction

FA can serve as a stabilizing agent in road building, namely in the construction of embankments and subgrades. When FA is combined with soil, it can promote compaction, decrease permeability, boost bearing capacity, and improve the overall performance of road infrastructure.

5.5.4 Fill Material

FA can serve as a fill material in building projects, including land reclamation, backfilling of trenches, and structural fills. Due to its lightweight composition, excellent compaction qualities, and strong load-bearing capability, it may effectively replace traditional fill materials such as dirt or gravel.

5.5.5 Asphalt Pavements

FA can be utilized in asphalt pavement construction either as a mineral filler or as a substitute for a portion of the fine aggregate. Implementing this approach can enhance the stability, longevity, and ability to withstand rutting and stripping of asphalt mixes, while simultaneously decreasing the overall expenses and environmental impact of pavement construction.

5.5.6 Geopolymer Concrete

FA can be utilized in the manufacturing process of geopolymer concrete, which involves activating aluminosilicate minerals using an alkaline solution. Geopolymer concrete exhibits superior mechanical qualities, elevated chemical resistance, and reduced greenhouse gas emissions in comparison to conventional Portland cement-based concrete (Xu and Shi 2018; Hemalatha and Ramaswamy 2021).

In construction, the use of FA provides several benefits such as waste reduction, preservation of resources, enhanced material functionality, and decreased environmental harm. Nevertheless, it is crucial to guarantee meticulous quality control and strict compliance with pertinent norms and laws when integrating FA into construction materials (Dwivedi and Jain 2014). Lu et al. (2023) provide insights into the importance of coal fly ash (CFA) as a secondary resource, and its significant role in large-scale utilization through building material production. The authors also discussed regarding the chemical and physical properties of CFA, classification, and the present status of its utilization in various building materials like cement, geopolymers, etc. (Rivera et al. 2015).

5.6 BENEFITS OF USING FA, ITS FUTURE, AND CHALLENGES

5.6.1 Benefits of Using FA

There are many benefits of using FA, it is a tool to serve for achieving a sustainable environment through utilization and waste management (Vargas and Halog 2015). The usage of FA is an important step in achieving the UN 7th, 9th, and 11th sustainable goals which emphasize affordable and clean energy; industry, innovation, and infrastructure; and sustainable cities and communities, respectively. It is noteworthy that based on the type of FA, its quality, the different applications are intended. Additionally, local regulations and standards should be considered when using FA in building materials.

The few benefits of using FA are discussed below.

a. Cement replacement in concrete
   FA has been extensively used as the partial replacement for cement in concrete. It enhances workability, increased strength, and durability of the concrete. It also reduces the heat of hydration in concrete, thereby preventing thermal cracking. It gives strength and reduces water demand (Kurda, de Brito, and Silvestre 2017; Li, Wu, and Wang 2022; Kameswara Rao, Achyutha Kumar Reddy, and Venkateswara Rao 2022).
b. Improved durability
   The use of FA in the concrete makes it denser and resistant to chemical attacks.
c. Economically viable
   FA is a product of coal combustion in thermal power plants. It is produced in huge quantities. By using it in construction activities or for energy derivation, the cost of manufacturing or operation can be reduced (Li, Wu, and Wang 2022).
d. Environmentally sustainable approach
   Being a waste product of thermal power plants, it poses a challenge of disposal. Land filling is an option, but land is also a treasure. By using it in construction activities, the greenhouse gas emissions associated with cement production can be reduced (Vargas and Halog 2015; Rivera et al. 2015).
e. FA-derived bricks and blocks offer enhanced strength and structural properties.
f. FA also contributes to soil stabilization, soil quality improvement, and land reclamation.
g. It offers an opportunity to be used in geopolymer construction reducing the carbon footprints (Ram and Mohanty 2022).
h. FA-based pavements are energy efficient, higher in strength, and stable.

5.7 FUTURE ASPECTS OF USING FA

As the future of the FA, scientific community is further aiming for practising green building projects for reduced carbon footprints. Also, it would contribute to reduction in the greenhouse gases emissions and to save climate. In future, to save environment and surroundings, the thermal power plants may be reduced, in that case the availability of FA will be reduced. In that case, the landfilled FA, bio-coal-based FA, or synthetic FA would be in use. Furthermore, the scientific fraternity will delve into exploring the further uses of FA and would come up with more sustainable approaches to use FA. It will continue to offer sustainability, waste minimization and efficient resource utilization (Li, Wu, and Wang 2022; Vargas and Halog 2015).

5.8 CHALLENGES WITH FA

As FA is a fine powder, it behaves like a fluid when the handling rate is too high. The fluidized FA causes discharge. Even its transportation is difficult, causing air pollution.

As the world is adopting sustainable and renewable energy, thermal power-based energy will not be used in the future, therefore it leads to decrease in the availability of FA. It will affect the construction industry. Also, they lead to low mechanical strength, which also causes longer setting periods and increased permeability. In case of inadequate curing, it will affect the performance of the FA cement.

FA pollutes the air, soil, and groundwater and when disposed openly, it affects aquatic life. There are also health hazards of FA, as it contains toxic elements like lead, arsenic, mercury, cadmium, and uranium posing health risks to humans and the environment.

It is potentially toxic leading to water pollution (particularly due to mercury).

5.9 LIFE CYCLE ASSESSMENT OF FA

Life cycle assessment (LCA) is a tool for comprehensive analysis to evaluate the impacts on environment by the product throughout its lifecycle. For FA, LCA helps to evaluate and sum up the benefits and drawbacks of each stage from production to disposal. The goal of LCA of FA is to evaluate the aptness and the sustainability of FA in different applications. It also aids the user in decision-making (Ersan et al. 2022; Fernando et al. 2021; Huber, Laner, and Fellner 2018).

FA is mainly used as construction materials; therefore here we are discussing the key components of LCA of FA (Huang and Chuieh 2015; Wang et al. 2017).

1. Goal – goal is to replace the construction material in original and to be replaced by FA for the environmental sustainability and also to compare its performance.
2. Scope – it describes the depth and detail of the study. It includes system boundaries which deals with the type of study, whether it is cradle to gate (extraction to production), cradle to grave (extraction to disposal), or cradle to cradle (extraction to disposal to recycling). The scope also covers for functional units, geographical boundaries considering unit of FA or concrete, and the location of the plant or the time frame for which LCA is practised.
3. Life cycle inventory – it includes data collection for the quantification of raw material acquisition; transportation of raw material to the plant; processing of FA (energy and material required, emissions out of the plant);

transportation of FA from plant to the construction sites. Then usage in the cement is evaluated to assess the reduction in cement usage, energy, and materials required to incorporate FA into concrete or other products. Finally, the end of life of the product is evaluated where disposal or recycling of FA-based material is considered.

4. Life cycle impact assessment

   It counts on the potential reduction of global warming due to emissions reduced by the replacement of cement by FA. It also deals with the reduction in the natural resource depletion by replacing the original material. FA is also a source of toxicity, therefore toxicity evaluation due to heavy metals present in FA, due to leachate on disposal, acidification potential, etc., are also assessed.

5. Interpretation

   It is the end step of LCA; it comprises the significant findings either positive or negative in terms of the environment, for example, key environmental benefits such as reduced CO^2 emissions, conservation of natural resources, waste minimization, toxicity, etc.

5.10 CONCLUSION

FA is a by-product of thermal power plants. Its massive production is a challenge. Therefore, it finds applicability in the different industries for sustainability. Its major contribution is in the construction industry, bio-energy, conserving natural resources, and lowering greenhouse gas emissions. It is getting popular for circular economy aspects and finds its concrete role in setting the sustainable standards. Therefore, this material needs to be explored more and more to reduce the environmental pollution. The scientific fraternity is making apt use of this and in future we will have more applications. For this, LCA is must. This chapter has discussed various possible applications, challenges, and steps in LCA of this useful material.

REFERENCES

Chen, Yi, Yingjie Fan, Yu Huang, Xiaoling Liao, Wenfeng Xu, and Tao Zhang. 2024. "A Comprehensive Review of Toxicity of Coal Fly Ash and Its Leachate in the Ecosystem." *Ecotoxicology and Environmental Safety* 269 (August 2023): 115905. https://doi.org/10.1016/j.ecoenv.2023.115905.

Dwivedi, Aakash, and Manish Kumar Jain. 2014. "Fly Ash – Waste Management and Overview : A Review." *Recent Research in Science and Technology*, 6 (1): 30–35.

Ersan, Yusuf Cagatay, Sedat Gulcimen, Tuba Nur Imis, Osman Saygin, and Nigmet Uzal. 2022. "Life Cycle Assessment of Lightweight Concrete Containing Recycled Plastics and Fly Ash." *European Journal of Environmental and Civil Engineering* 26 (7): 2722–2735. https://doi.org/10.1080/19648189.2020.1767216.

Fernando, Sarah, Chamila Gunasekara, David W. Law, M.C.M. Nasvi, Sujeeva Setunge, and Ranjith Dissanayake. 2021. "Life Cycle Assessment and Cost Analysis of Fly Ash–Rice Husk Ash Blended Alkali-Activated Concrete." *Journal of Environmental Management* 295 (December 2020): 113140. https://doi.org/10.1016/j.jenvman.2021.113140.

Gautam, R., Nayak, J. K., Daverey, A., & Ghosh, U. K. (2022). Emerging sustainable opportunities for waste to bioenergy: an overview. *Waste-to-energy approaches towards zero waste*, 1–55.

Gupta, Piyush, Garima Nagpal, and Namrata Gupta. 2021. "Fly Ash-Based Geopolymers: An Emerging Sustainable Solution for Heavy Metal Remediation from Aqueous Medium." *Beni-Suef University Journal of Basic and Applied Sciences* 10 (1). https://doi.org/10.1186/s43088-021-00179-8.

Hasselriis, Floyd, and Anthony Licata. 1996. "Analysis of Heavy Metal Emission Data from Municipal Waste Combustion." *Journal of Hazardous Materials* 47 (1–3): 77–102. https://doi.org/10.1016/0304-3894(95)00107-7.

Hemalatha, T., and Ananth Ramaswamy. 2017. "A Review on Fly Ash Characteristics – Towards Promoting High Volume Utilization in Developing Sustainable Concrete." *Journal of Cleaner Production* 147: 546–559. https://doi.org/10.1016/j.jclepro.2017.01.114.

Hemalatha, T., and Ananth Ramaswamy. 2021. "Fly Ash Cement." *Handbook of Fly Ash*, 547–563. https://doi.org/10.1016/B978-0-12-817686-3.00016-5.

Hower, James C., John G. Groppo, Uschi M. Graham, Colin R. Ward, Irena J. Kostova, Mercedes M. Maroto-Valer, and Shifeng Dai. 2017. "Coal-Derived Unburned Carbons in Fly Ash: A Review." *International Journal of Coal Geology* 179 (May): 11–27. https://doi.org/10.1016/j.coal.2017.05.007.

Huang, T. Y., and P. T. Chuieh. 2015. "Life Cycle Assessment of Reusing Fly Ash from Municipal Solid Waste Incineration." *Procedia Engineering* 118: 984–991. https://doi.org/10.1016/j.proeng.2015.08.539.

Huber, Florian, David Laner, and Johann Fellner. 2018. "Comparative Life Cycle Assessment of MSWI Fly Ash Treatment and Disposal." *Waste Management* 73: 392–403. https://doi.org/10.1016/j.wasman.2017.06.004.

Jambhulkar, Hemlata P., Siratun Montaha S. Shaikh, and M. Suresh Kumar. 2018. "Fly Ash Toxicity, Emerging Issues and Possible Implications for Its Exploitation in Agriculture; Indian Scenario: A Review." *Chemosphere* 213: 333–344. https://doi.org/10.1016/j.chemosphere.2018.09.045.

Jianguo, Jiang, Wang Jun, Xu Xin, Wang Wei, Deng Zhou, and Zhang Yan. 2004. "Heavy Metal Stabilization in Municipal Solid Waste Incineration Flyash Using Heavy Metal Chelating Agents." *Journal of Hazardous Materials* 113 (1–3): 141–146. https://doi.org/10.1016/j.jhazmat.2004.05.030.

Kalombe, Rosicky Methode, Victor Tunde Ojumu, Chuks Paul Eze, Sammy Mwasaha Nyale, John Kevern, and Leslie Felicia Petrik. 2020. "Fly Ash-Based Geopolymer Building Materials for Green and Sustainable Development." *Materials* 13 (24): 1–17. https://doi.org/10.3390/ma13245699.

Kameswara Rao, B., M. Achyutha Kumar Reddy, and A. Venkateswara Rao. 2022. "Effect of Flyash as Cement Replacement Material and Pore Filling Material in Concrete." *Materials Today: Proceedings* 52: 1775–1780. https://doi.org/10.1016/j.matpr.2021.11.444.

Kurda, Rawaz, Jorge de Brito, and José D. Silvestre. 2017. "Combined Influence of Recycled Concrete Aggregates and High Contents of Fly Ash on Concrete Properties." *Construction and Building Materials* 157: 554–572. https://doi.org/10.1016/j.conbuildmat.2017.09.128.

Li, Yang, Boren Wu, and Ruijun Wang. 2022. "Critical Review and Gap Analysis on the Use of High-Volume Fly Ash as a Substitute Constituent in Concrete." *Construction and Building Materials* 341 (March): 127889. https://doi.org/10.1016/j.conbuildmat.2022.127889.

Lima, A. T., L. M. Ottosen, A. J. Pedersen, and A. B. Ribeiro. 2008. "Characterization of Fly Ash from Bio and Municipal Waste." *Biomass and Bioenergy* 32 (3): 277–282. https://doi.org/10.1016/j.biombioe.2007.09.005.

Lu, Xuhang, Bo Liu, Qian Zhang, Quan Wen, Shuying Wang, Kui Xiao, and Shengen Zhang. 2023. "Recycling of Coal Fly Ash in Building Materials: A Review." *Minerals* 13 (1). https://doi.org/10.3390/min13010025.

Mushtaq, Farwa, Muhammad Zahid, Ijaz Ahmad Bhatti, Saqib Nasir, and Tajamal Hussain. 2019. "Possible Applications of Coal Fly Ash in Wastewater Treatment." *Journal of Environmental Management* 240 (February): 27–46. https://doi.org/10.1016/j.jenvman.2019.03.054.

Radhika, Devi, Archana Shivakumar, Deepak R. Kasai, Ravindranadh Koutavarapu, and Shaik Gouse Peera. 2022. "Microbial Electrolysis Cell as a Diverse Technology: Overview of Prospective Applications, Advancements, and Challenges." *Energies*. MDPI. https://doi.org/10.3390/en15072611.

Ram, Amit Kumar, and Supriya Mohanty. 2022. "State of the Art Review on Physiochemical and Engineering Characteristics of Fly Ash and Its Applications." *International Journal of Coal Science and Technology* 9 (1). https://doi.org/10.1007/s40789-022-00472-6.

Rashad, Alaa M. 2015. "A Brief on High-Volume Class F Fly Ash as Cement Replacement – A Guide for Civil Engineer." *International Journal of Sustainable Built Environment* 4 (2): 278–306. https://doi.org/10.1016/j.ijsbe.2015.10.002.

Rastogi, Abhijit, and Virendra Kumar Paul. 2020. "A Critical Review of the Potential for Fly Ash Utilisation in Construction-Specific Applications in India." *Environmental Research, Engineering and Management* 76 (2): 65–75. https://doi.org/10.5755/J01.EREM.76.2.25166.

Rivera, Felipe, Patricia Martínez, Javier Castro, and Mauricio López. 2015. "Massive Volume Fly-Ash Concrete: A More Sustainable Material with Fly Ash Replacing Cement and Aggregates." *Cement and Concrete Composites* 63: 104–112. https://doi.org/10.1016/j.cemconcomp.2015.08.001.

Saraswathy, V., S. Muralidharan, K. Thangavel, and S. Srinivasan. 2003. "Influence of Activated Fly Ash on Corrosion-Resistance and Strength of Concrete." *Cement and Concrete Composites* 25 (7): 673–680. https://doi.org/10.1016/S0958-9465(02)00068-9.

Seow, Yee Xuan, Yie Hua Tan, N. M. Mubarak, Jibrail Kansedo, Mohammad Khalid, Mohd Lokman Ibrahim, and Mostafa Ghasemi. 2022. "A Review on Biochar Production from Different Biomass Wastes by Recent Carbonization Technologies and Its Sustainable Applications." *Journal of Environmental Chemical Engineering* 10 (1): 107017. https://doi.org/10.1016/j.jece.2021.107017.

Singh, Surabhi. 2017. "Fly Ash in India: Generation Vis-à-Vis Utilization and Global Perspective." *International Journal of Applied Chemistry* 13 (1): 29–52.

Tian, Quanzhi, and Keiko Sasaki. 2019. "Application of Fly Ash-Based Materials for Stabilization/Solidification of Cesium and Strontium." *Environmental Science and Pollution Research* 26 (23): 23542–23554. https://doi.org/10.1007/s11356-019-05612-1.

Vargas, J., and A. Halog. 2015. "Effective Carbon Emission Reductions from Using Upgraded Fly Ash in the Cement Industry." *Journal of Cleaner Production* 103: 948–959. https://doi.org/10.1016/j.jclepro.2015.04.136.

Wang, Jing Jing, Yuan Feng Wang, Yi Wen Sun, Danielle Densley Tingley, and Yu Rong Zhang. 2017. "Life Cycle Sustainability Assessment of Fly Ash Concrete Structures." *Renewable and Sustainable Energy Reviews* 80 (May): 1162–1174. https://doi.org/10.1016/j.rser.2017.05.232.

Xu, Gang, and Xianming Shi. 2018. "Characteristics and Applications of Fly Ash as a Sustainable Construction Material: A State-of-the-Art Review." *Resources, Conservation and Recycling* 136 (April): 95–109. https://doi.org/10.1016/j.resconrec.2018.04.010.

Yao, Z. T., X. S. Ji, P. K. Sarker, J. H. Tang, L. Q. Ge, M. S. Xia, and Y. Q. Xi. 2015. "A Comprehensive Review on the Applications of Coal Fly Ash." *Earth-Science Reviews* 141: 105–121. https://doi.org/10.1016/j.earscirev.2014.11.016.

Yuan, Ning, Hao Xu, Yanjun Liu, Kaiqi Tan, and Yixiang Bao. 2023. "Synthesis and Environmental Applications of Nanoporous Materials Derived from Coal Fly Ash." *Sustainability* 15 (24): 16851. https://doi.org/10.3390/su152416851.

Zhao, Weizhong, Xianbo Su, Yifeng Zhang, Daping Xia, Shihui Hou, Yixuan Zhou, Haijiao Fu, Lufei Wang, and Xiangju Yin. 2022. "Microbial Electrolysis Enhanced Bioconversion of Coal to Methane Compared with Anaerobic Digestion: Insights into Differences in Metabolic Pathways." *Energy Conversion and Management* 259 (April): 115553. https://doi.org/10.1016/j.enconman.2022.115553.

6 Silica-Rich Zeolites from Fly Ash

Transforming Waste into Sustainable Resources for Materials Science

Bagia Maryam Ibrahim Yunus,
Alpeshkumar V. Patel and Amare Ayalew Abebe

6.1 INTRODUCTION

Sustainable development is an essential and powerful foundation for human existence and advancement. In this regard, rapid industrialization and urbanization cause heavy utilization of resources without giving much attention to environmental pollution, climatic change, and resource depletion. Further, waste management also emerged as an essential factor for sustainable development. Instead of seeing waste just as a burden on society and the environment, there is a growing recognition of its worth and potential for resource recovery. If appropriately channeled through technological advancement and creative approaches, waste could be an end-product with huge potential for reducing environmental impact while promoting resource efficiency (Belviso 2018).

Of the various waste streams from anthropogenic sources, fly ashes from coal and biomass combustion represent an abundant and underutilized resource with the potential for exploitation to generate high-value products (Liu 2022). The fly ash obtained from both sources has a similar mineralogical composition, including silica, alumina, iron oxide, and calcium oxide (Saceda, Leon, et al. 2011). The disposal of fly ash is burdensome to the environment, notably when considering landfilling and ash ponds. Biomass fly ash provides extra environmental advantages as it is a popular renewable and carbon-neutral energy source. Valorization of biomass residues results in bio fly ash, reducing dependence on fossil fuels and greenhouse gas emissions and positively contributing to sustainable energy production. Figure 6.1 depicts the chart summarizing the scenario of production versus utilization of fly ash in various countries. A rapidly developing country like India faces a massive amount of unutilized fly ash, which opens up a vast arena to address this alarming situation (Buzukashvili, Sommerville, et al. 2024).

DOI: 10.1201/9781003498124-6

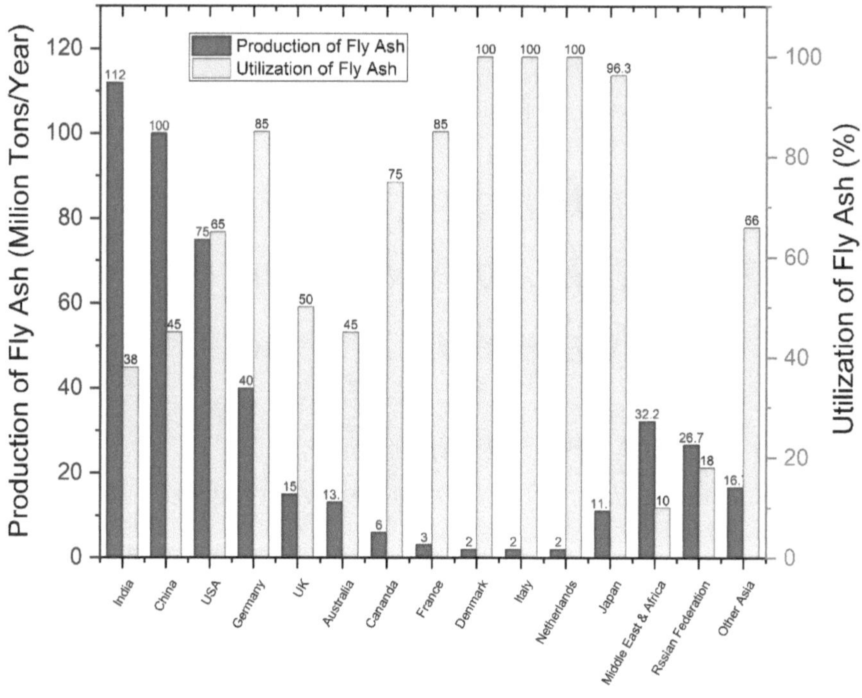

FIGURE 6.1 Production versus utilization of fly ash.

Recent material science and technology developments created potential scope for use in construction industries and transformation to high silica-rich zeolite, hence promising a pathway toward the circular economy and resource (Novais, Buruberri, et al. 2016). Fly ash exhibits pozzolanic properties, which makes it an excellent supplementary cementitious material. Adding fly ash to cement enhances concrete's durability and strength. This addition reduces the quantity of cement utilized in concrete mixes, lessening the environmental effect and reducing the cost (Nagao, Otake, et al. 2024). In road construction, fly ash provides better soil stability, improves bearing capacity, and reduces the settlement effect. It can also produce bricks and blocks to give them more strength and reduce energy consumption when fired in the kiln (Vidhya 2022).

Fly ash, in particular, comes from bio-coal containing high silica content, which is a good raw material for synthesizing silica-rich zeolites (Cao, Xuan, et al. 2023). Zeolites are three-dimensional crystal structures with molecular-sized holes and voids; some exist in nature, while others are produced in laboratories. Alkaline activation, hydrothermal treatment, and zeolite crystallization are used to manufacture zeolites of the desired quality from fly ash (Buzukashvili, Sommerville, et al. 2024; Guo, Hu, et al. 2024). A more flexible approach for pure silica to reach zeolite can be applied where silica from fly ash is extracted with alkali (Xing, Li, et al. 2024). Such extracted phase can then be transformed into zeolite with desired qualities (Shah, Patel, et al. 2019). The synthesized zeolite from fly ash reported excellent efficacy in molecular sieving, adsorption (Oluyinka, Patel, et al. 2020), and catalytic transformation (Shanti, Shanti and Meshram 2013), making them very valuable for material science and engineering applications. Using fly ash for zeolite synthesis reduces environmental impact and waste management costs. The production of zeolites from waste streams falls within the principles of the circular economy. It promotes resource efficiency, minimization of waste, and sustainable development toward a better and more evolved society.

This chapter presents revolutionary strategies for synthesizing zeolites from fly ash waste for the betterment and sustainability of resources in materials science applications (Shah, Patel, et al. 2017). This chapter provides a broad overview of the zeolite synthesis method, including alkali activation, hydrothermal conversion, solvothermal synthesis, and advanced techniques such as templete-aided and microwave-assisted synthesis. Advanced analytical methods for assessing the chemical as well as physical characteristics of fly ash and generated zeolites are briefly discussed (Shah, Abebe and Shah 2017). The applications of silica-rich zeolites in catalysis, adsorption, and environmental remediation processes are evaluated for their potential.

6.2 SYNTHESIS OF SILICA-RICH ZEOLITES FROM FLY ASH

Fly ash primarily consists of different metallic and non-metallic oxides, with the main components being silicon dioxide (SiO_2), aluminum oxide (Al_2O_3), iron oxide (Fe_2O_3), calcium oxide (CaO), magnesium oxide (MgO), and other minor minerals in crystalline and amorphous phases (Shah, Patel, et al. 2019). The mineral composition of coal fly ash or biomass fly ash depends upon the source of combustible material and additives included during the combustion process. The mineralogical composition of fly ash provides all the necessary elements for zeolite synthesis; hence, it is considered a cheap raw material. The zeolite synthesis methods, including alkali activation, fusion, hydrothermal crystallization, solvothermal synthesis, templet-aided synthesis, microwave-assisted synthesis, and combustion synthesis, are reported in the scientific literature and widely considered for large-scale production.

6.2.1 Alkali Activation

Synthetic zeolites are created by attempting to replicate natural zeolite synthesis. The procedures require alkali activation, followed by hydrothermal or non-hydrothermal crystallization. Alkali activation involves treating fly ash with a concentrated alkaline solution of NaOH or KOH. This powerful alkali dissolves the SiO_2 and Al_2O_3 species from fly ash, producing soluble silicate and aluminates. Identifying the crystalline and amorphous phases in fly ash is important because a greater quantity of amorphous phase causes more dissolution (Oluyinka, Patel, et al. 2020). The most common phases of fly ash include quartz, mullite, hematite, magnetite, and amorphous silica. (Grabias-Blicharz and Franus 2023). The mineralogy impacts its reactivity, strength, and other properties. During alkali activation, amorphous silica and some crystalline zeolite, like mullite, dissolve rapidly, while quartz and some other stable crystalline phases require harsh conditions to dissolve (Shah, Patel, et al. 2017). The possible reactions associated with fly ash dissolution are:

$$SiO_2 + 2NaOH \rightarrow Na_2SiO_3 + H_2O$$

$$Al_2O_3 + 2NaOH \rightarrow 2NaAlO_2 + H_2O$$

The soluble silicate and aluminate compounds formed through alkali activation amalgamate into a reactive gel phase within the fly ash. The gel phase acts as a precursor for zeolite formation, which leads to the subsequent crystallization process. The variety of physical and chemical parameters can dominate the alkali activation of fly ash. The effectiveness of alkali activation majorly depends on parameters such as the alkali solution concentration, reaction temperature, reaction time, and fly ash to alkali solution ratio. By carefully optimizing these factors, the silica and alumina breakdown can produce a gel, which ultimately crystallizes to produce the zeolite phase.

6.2.2 Alkali Fusion

The fusion procedure obtains the maximum quantity of silicate and aluminate from fly ash. Before synthesis, fly ash is pre-treated to eliminate impurities, and its composition needs to be adjusted for better reactivity. Pre-treatment procedures may include washing with acid or water to improve the silicone-to-aluminum ratio and sieving to obtain particles of the required size. The pre-treated fly ash is mixed with a strong alkali (sodium hydroxide, NaOH, or potassium hydroxide, KOH) or sodium carbonate (Na_2CO_3) in a predetermined ratio. After mixing thoroughly, the mixture was heated to a higher temperature for fusion, often between 500°C and 900°C. This fusion process breaks down the complex aluminosilicate structures in the fly ash, the amorphous silica, alumina, and some zeolite phases such as quartz, mullite, etc. This process helps dissolve the silica and alumina in the fly ash and form an alkali silicate or aluminate melt (Shah, Patel, et al. 2019). However, the ingot obtained by fusion must be crushed and pulverized to convert the material into a powdery substance, which helps dissolve these salts. The powder of this melt is then mixed with water and allowed to dissolve and crystallize for an extended period to obtain zeolite (Tang, Zhou, et al. 2019). The silicon-to-aluminum ratio, temperature, time, etc. are controlled to optimize zeolite synthesis (Sivalingam and Sen 2018; Liang, Liu, et al. 2024).

6.2.3 Hydrothermal Treatment

In zeolite synthesis, hydrothermal treatment is performed on the alkali-activated fly ash or alkali fusion material in an aqueous media. The therapy comprises an elevated temperature and pressure applied to the aqueous mixture for a prolonged time in a Teflon-lined autoclave. The temperature is generally between 100°C and 200°C for several hours to several days under autogenous pressure. During this step, the silica and alumina species in the solution rearrange and crystallize into zeolite structures. The mechanism comprises a complicated series of dissolution, nucleation, and crystallization stages dictated by the supersaturation of the solution to the zeolite phases. Hydrothermal conversion provides exact control over reaction parameters such as temperature, pressure, and reaction duration, creating well-defined zeolite crystals (Musyoka, Petrik, et al. 2012; Tauanov, Shah, et al. 2018). Following hydrothermal treatment, the zeolite-containing fly ash usually gets cooled to the ambient temperature and is aged appropriately. This cooling and aging step assists in stabilizing the zeolite crystals, improving their crystallinity and structural resilience. Filtration or centrifugation separates the liquid phase from the solid product, which includes produced zeolites. The zeolite crystals are then rinsed multiple times with distilled water to eliminate any leftover contaminants and unreacted alkalis. The washed zeolite crystals are dried at around 100°C to remove moisture.

A calcination step may follow to remove any organic templates or impurities, although this step majorly depends on the specific synthesis protocol and the desired zeolite properties. Several parameters impact the hydrothermal synthesis of zeolites from fly ash, including the Si/Al ratio in the fly ash, alkali content, and temperature, which determines the kind of zeolite that can be synthesized (Xing, Li, et al. 2024). The higher alkali concentrations typically promote faster dissolution of fly ash and may influence the crystallinity and type of zeolite formations. The temperature and pressure parameters during the hydrothermal treatment significantly affect the kinetics and extent of zeolite crystallization. Elevated temperatures and pressures generally accelerate crystallization and foster the development of larger zeolite crystals. One of the prominent parameters is the duration of the hydrothermal treatment for the crystallization process. Extended treatment durations permit more extensive zeolite crystal growth and maturation, forming well-defined crystalline structures (Boycheva, Zgureva and Shoumkova 2015). Maintaining optimal pH levels throughout hydrothermal treatment is imperative for steering the crystallization process and ensuring the formation of pure zeolite phases. Adjusting pH using acidic or alkaline solutions may be necessary to optimize zeolite crystal growth and minimize impurities (Längauer, Čablík, et al. 2021). Alkali activation and hydrothermal treatment are critical steps in synthesizing zeolites from fly ash. Alkali activation promotes the breakdown of silica and alumina to produce the reactive gel phase, whereas hydrothermal treatment induces the crystallinity in zeolite structures inside the fly ash matrix (Liu 2022). Optimizing reaction parameters and managing process conditions is crucial for producing high-quality zeolite products suitable for various applications.

6.2.4 Solvothermal Synthesis

Apart from water, solvothermal synthesis also involves using an organic solvent, ionic liquids, and supercritical fluids as the reaction medium. Alcohols like ethanol and isopropanol alter hydrolysis and condensation reactions during zeolite synthesis (Feng and Li 2017). Amines like ethylenediamine, diethylenetriamine, or hexamethylenetetramine act as structure-directing agents (Chen and Tang 2019), while ethylene glycol and other glycols are sometimes used to dissolve metal ions effectively.

Ionic liquids have been studied due to their special properties for synthesizing zeolites: close to zero vapor pressure, high thermal stability, and the dissolution power. Ionic liquids can act as solvents and templates, directing the assembly of zeolite structures (Meng and Xiao 2014). Supercritical CO_2 (carbon dioxide in its supercritical fluid state) has been explored as a solvent or co-solvent in zeolite synthesis. In its supercritical state, CO_2 possesses unique solvent properties that can be adjusted by changing the temperature and pressure, offering a way to fine-tune the synthesis process (Lopez-Periago and Domingo 2018).

In solvothermal synthesis, the solvent used is governed by the reaction's unique requirements, such as phase purity, crystal size, and zeolite shape (Feng and Li 2017). Researchers typically experiment with different solvents or combinations to optimize the synthesis conditions to produce zeolites with desired features (Buzukashvili, Sommerville, et al. 2024).

6.2.5 The Templete-Assisted Synthesis

Template-assisted synthesis is a crucial technique in materials research, enabling precise control over zeolites' pore size, structure, and framework. This methodology has been used in catalysis, ion exchange, gas separation, and adsorption applications because the resulting zeolites have well-defined pores with a high surface area. The synthesis is typically done by combining a silicon source, such as tetraethyl orthosilicate, and an aluminum source with a solution containing the template molecule. Such a mix is then subjected to hydrothermal treatment, which involves heating the mixture under pressure for an extended period. During this dissolution process, the silica and alumina sources dissolve in the solution, forming reactive species. These reactive species begin to nucleate and grow into a zeolitic framework, guided by the template molecules that dictate the formation of specific pore structures (White, Fischer, et al. 2014). Over time, a crystalline zeolite structure forms, encapsulating the template molecules within its pores.

Quaternary ammonium cation templating agents are most commonly used in zeolite synthesis. Their unique structure plays a pivotal role in determining the pore size and structure of the resulting zeolite (Liu 2022). The most widely used quaternary ammonium salt is cetyl trimethyl ammonium bromide (CTAB) (Shah, Patel, et al. 2017), used to synthesize Mobil Composition of Matter (MCM)-based zeolites. As discussed earlier, amines can act as both solvents and templates in the synthesis process, influencing the growth of the zeolite crystals and the formation of specific pore structures (Gao, Yang, et al. 2022). Polymers, such as polyethylene oxides, can also direct the formation of larger pore structures and even hierarchical zeolites with micro- and mesoporous structures (Tian, Liu, et al. 2016; Kan, Zhang, et al. 2020).

The method controls pore sizes and three-dimensional structures tailored to particular applications. The large pore obtained by the templates opens avenues for introducing organic functional groups into the zeolite structure, enhancing its selectivity and reactivity for various applications (Wu, Luan and Xiao 2022).

6.2.6 Microwave-Assisted Synthesis

Microwave-assisted zeolite synthesis is a relatively new, effective method to produce zeolites through the unique advantages that microwave irradiation offers in raising the reaction rate and increasing the product properties. The route has been attracting considerable attention for its dramatic reduction in synthesis times from days to hours or even minutes.

Microwave irradiation is responsible for material heating because of the interaction of microwave energy with the molecules of the substance, by which their rotation aligns with the rapidly changing electric field of the microwaves. The net outcome of this process is volumetric heating which is quite uniform and effective compared to conventional methods of heating by conduction and convection from outside inward into the interior of the material. In zeolite synthesis, microwave irradiation can directly interact with the reactant mixture and rapidly heat the synthesis mixture to the required reaction temperature, significantly speeding up the nucleation and crystal growth processes. The microwaves can also heat specific components within a mixture, potentially leading to more uniform nucleation sites and aiding in forming a homogeneous zeolite structure. The rapid and volumetric heating effect of microwaves can enhance mass transfer between the solid and liquid phases in the synthesis mixture, promoting more efficient use of reactants and faster crystal growth (Zhou, Jiang, et al. 2023).

Microwave-assisted synthesis can produce zeolites with higher crystallinity, greater purity, and an enhanced uniform particle size distribution compared to conventional hydrothermal synthesis. It allows rapid experimentation by altering parameters during synthesis, facilitating the optimization of zeolite properties for specific applications (Zeng, Hu, et al. 2021).

6.2.7 Combustion Synthesis

Synthesis of zeolite from fly ash using combustion route is also known as Self-propagating High-temperature Synthesis (SHS). The combustion route follows a chemical reaction that generates enough heat to sustain the reaction until completion without the need for external energy once initiated. In the synthesis, after adjusting the Si/Al ratio, a combustible reducing agent such as urea, glycine, or another compound is added (Boycheva, Zgureva and Shoumkova 2015). The mixture is then homogenized to produce a free-flowing powder, which is necessary for the combustion process to occur effectively. The ready mix is then placed in a furnace or on a hot plate at an appropriate temperature to initiate combustion. The ignition produces a self-sustaining combustion wave that propagates across the material, forming zeolite structures. The processed material may undergo subsequent steps like washing, drying, and calcination to eliminate unreacted precursors, organic matter, or other impurities (Belviso 2018). The physicochemical characteristics of synthesized zeolites, like pore size, surface area, and Si/Al ratio, can be tuned by the composition of the precursor mixture and combustion conditions.

The procedure delivers high temperatures and pressures, facilitating the development of zeolite phases inside the fly ash matrix. The method has several advantages over hydrothermal techniques, including simplicity, scalability, and less energy consumption. However, controlling the reaction parameters and product quality can be challenging (Pimraksa, Chindaprasirt and Setthaya 2010). The overall synthesis of zeolites includes chemical, physical, and thermal processes tailored to optimize the conversion of fly ash into crystalline zeolite phases. Each synthesis technique has distinct benefits and limitations, and the method chosen is determined by considerations such as desired zeolite characteristics, reaction conditions, and scalability for industrial applications. Recent studies and research aim to improve synthesis processes and investigate novel tactics for effectively transforming fly ash into valuable zeolite materials apart from alkali activation, fusion, and hydrothermal treatments. However, the extraction of silicate from fly ash has gained attention among researchers to synthesize pure zeolite.

6.3 EXTRACTION OF SILICATE FOR ZEOLITE SYNTHESIS

Silicates are the base material for the production of zeolites. Fly ash resulting from coal or biomass combustion usually contains quite a substantial amount of amorphous silica and some alkali-soluble zeolites. The extraction of silicates from fly ash is one of the crucial steps to synthesize zeolites. These silica phases can be dissolved in alkaline solutions and extracted as silicate-rich liquor. The typical process follows a basic step to remove impurity, involving sieving and washing. Sometimes, magnetic separation and acid washing are followed to remove magnetic particles and metal oxides. These pre-treatments increase the accessibility of silica within the fly ash matrix for subsequent extraction (Matlob, Kamarudin, et al. 2012). The pre-treated fly ash is mixed with a concentrated alkaline solution and then subjected to conventional hydrothermal heating, causing the dissolution of silica (SiO_2) and alumina (Al_2O_3) from the fly ash matrices. The dissolution can also be promoted by combining conventional and microwave heating (Shah, Patel, et al. 2019). The dissolution process causes the formation of soluble silicate and aluminates separated from the fly ash using the hot filtering process. The collected liquid may undergo additional purification procedures to eliminate contaminants and undesirable components, maintaining the quality of the extracted silicate solution (Liu 2022). It was then used to create pure zeolite using hydrothermal or template-assisted crystallization procedures. In both cases, the dissolved silicate and aluminate species react with the alkali to produce a reactive gel phase that essentially acts as a precursor for the formation of zeolite and aids in future crystallization processes.

The admixture underwent hydrothermal treatment during the crystallization process, in which dissolved silicate ions combine with alumina and structure-directing agents to induce the nucleation and development of zeolite crystals. Parameters such as temperature, pressure, treatment duration, and reaction mixture composition are carefully controlled to optimize zeolite crystal growth and ensure the formation of desired zeolite phases with specific properties (Moreno, Querol, et al. 2002). After hydrothermal treatment, the synthesized zeolite-containing material is washed to remove unreacted components and impurities. The washed material is then dried to obtain the final zeolite product in powder or crystalline form. Table 6.1 lists various types of silica-rich zeolites and their synthetic routes. Most researchers have relied on coal fly ash, where the silica content varies from 44% to 95%, making it an excellent raw material for zeolitic formation.

In summary, extracting silicate from fly ash involves alkali activation followed by selective leaching or extraction to obtain soluble silicate species. These extracted silicate ions serve as essential precursors for synthesizing zeolites through hydrothermal treatment, where controlled conditions promote the nucleation and growth of zeolite

TABLE 6.1
Type of silica-rich zeolites and their synthetic approach

Zeolite	Synthesis approach	Fly ash (%content)	Reference
ZSM-5, zeolite sodalite	Alkaline leaching, templet-assisted hydrothermal synthesis (tetrapropylammonium bromide)	Coal fly ash (SiO_2 = 54.4 ± 0.24, Al_2O_3 = 29.22 ± 0.10)	Ndlovu, Ameh, et al. (2023)
Zeolite faujasite (FAU)	Alkaline fusion method, hydrothermal method	Coal fly ash (SiO_2 = 44.24, Al_2O_3 = 33.20)	Hossi, Nkazi, et al. (2023)
Zeolite A	Fusion-assisted hydrothermal synthesis	Coal fly ash (SiO_2 = 54.43, Al_2O_3 = 24.22)	GV and Matli (2024)
Zeolite sodalite	Pre-acid washing, alkaline activation	Coal fly ash (SiO_2 = 60.8, Al_2O_3 = 29.9)	Abas and Fathy (2024)
Zeolite Y (Faujasite)	Microwave and ultrasound-assisted alkali activation followed by microwave synthesis	Coal fly ash (elemental composition is not mentioned	Chen, Song, et al. (2023)
Zeolite FAU	High-temperature alkali fusion, HCl, and zeolite seeds for crystallization	Coal fly ash (SiO_2 = 52.00, Al_2O_3 = 29.20)	Liu, Wu, et al. (2023)
Zeolite X	Alkali activation by conventional and microwave heating	Bituminous fly ash (SiO_2 = 53.60, Al_2O_3 = 25.4)	Truttim, Asavapisit and Piyaphanuwat (2023)
MCM-41	Templet-assisted hydrothermal synthesis (CTAB)	Bagasse fly ash (SiO_2=75.96, Al_2O_3=2.24)	Shah, Patel, et al. (2017)
Zeolite T	Alkaline hydrothermal synthesis	Rice husk ash (SiO_2 = 95.54, Al_2O_3 = 0.78)	Bohra, Dey, et al. (2013)
Zeolite Y	Acid washing, calcination, alkaline hydrothermal treatment	Rice husk ash (SiO_2 = 89.69, Al_2O_3 = 1.01)	Sivalingam and Sen (2019)

crystals (Saceda, Leon, et al. 2011). The synthesized zeolite product can be further characterized and utilized in various industrial applications.

6.4 CHARACTERIZATION TECHNIQUES

Characterization techniques are crucial for understanding the properties of fly ash and its derived zeolites. These techniques help assess the chemical composition, mineralogical phases, surface properties, and structural attributes, among other characteristics.

The elemental composition of the fly ash is derived using the X-ray fluorescence method (XRF) or inductively coupled plasma spectroscopy (ICP). These methods provide the amount of silicone and aluminum in the fly ash, which is crucial for zeolite synthesis. The X-ray diffraction (XRD) analysis offers information regarding amorphous and crystalline phases. From these data, the phase which is getting dissolved and converted into zeolite during the alkali hydrothermal method can be estimated. The fly ash's particle size, surface area, and thermal stability can also be determined to better understand dissolution kinetics (Alterary and Marei 2021).

After carefully understanding the analysis data of fly ash, a protocol for the zeolite synthesis can be designed. The synthesis can then be executed to obtain the zeolite with predetermined quality. The success of the synthetic protocol can be evaluated with a detailed characterization of the zeolite. Like fly ash, the elemental composition of synthesized zeolite is determined using XRF and ICP analysis. The crystal structure of the material is derived by XRD and small-angle XRD. The XRD patterns help in determining the various crystal phases present in the synthesized material, while small-angle XRD can determine the mesoporosity of the zeolite. The nitrogen adsorption-desorption isotherm is applied to obtain the surface area, porosity, and pore volume of the synthesized zeolite. Surface morphology and topologies can be observed using scanning electron microscopy (SEM), transmission electron microscopy (TEM), and atomic force microscopy (AFM) techniques. Surface chemistry and oxidation states of elements on zeolite can be derived from X-ray photoelectron spectroscopy (XPS), while functional groups are identified with Fourier transform infrared (FTIR) spectroscopy. Some other analyses, such as energy dispersive X-ray spectroscopy (EDS), nuclear magnetic resonance (NMR), and thermogravimetric analysis (TGA), can be done according to requirements and specific applications (Prodinger and Derewinski 2020). These analyses provide valuable insights into the composition and structural properties of fly ash, guiding the optimization of zeolite synthesis processes for desired product properties and applications (Alterary and Marei 2021).

6.5 APPLICATION OF SILICA-RICH ZEOLITES

Silica-rich zeolites, characterized by their porous crystalline structure and high silica content, find extensive applications across various industries owing to their unique functionalities and properties, such as high surface area, ion exchange capacity, shape selectivity, and catalytic activity. In this comprehensive overview, let us explore the diverse applications of silica-rich zeolites in detail.

6.6 ADSORPTION AND SEPARATION APPLICATIONS

The high surface area and customized pore volumes make silica-rich zeolites valuable adsorbents for gas separation and purification. Such zeolites' capacity to selectively adsorb gases based on size, shape, and polarity makes them ideal for natural gas purification, carbon dioxide capture, air separation, and other applications (Islam, Biswas and Naidu 2024). Zeolites are also widely utilized in liquid-phase adsorption processes, such as wastewater treatment, separation of isomeric compounds, and purification of solvents and chemicals.

6.7 CATALYSIS APPLICATIONS

In refinery applications, silica-rich zeolites act as catalysts in petroleum refining processes for the cracking and hydrocracking of hydrocarbons. They facilitate the conversion of heavy hydrocarbons into lighter fractions, improving yields of more valuable products such as gasoline, diesel, and jet fuel (Hamidi, Khoshbin and Karimzadeh 2021). Zeolites are efficient in the isomerization and alkylation of organic molecules and convert them into higher-value products. Silica-rich zeolites with tailored pore structures and acidity exhibit shape-selective catalytic properties, enabling precise control over product selectivity and distribution (Al-Zeer and MacKenzie 2019).

Silica-rich zeolites are used in catalytic converters in vehicle exhaust systems to selectively reduce nitrogen oxides (NOx) using reducing agents like ammonia (NH_3). They facilitate the conversion of harmful NOx emissions into nitrogen (N_2) and water vapor, reducing air pollution from vehicle exhaust (Duan, Dou, et al. 2020). Zeolites are used as membrane materials in membrane reactor systems for catalytic reactions. Silica-rich zeolite membranes selectively separate reactants and products, enhancing reaction efficiency and yield by continuously removing reaction byproducts and maintaining optimal reactant concentrations (Zhu, Ji, et al. 2018).

6.8 WATER PURIFICATION AND TREATMENT APPLICATIONS

Silica-rich zeolites have applications in various drinking water treatment processes, from removing heavy metals like arsenic and fluoride to organic pollutants. They act as adsorbents and ion exchangers, reducing the concentration of undesired substances to enhance the quality of water (Gao, Zhao, et al. 2020). They find applications in wastewater treatment by removing pollutants and nutrients from industrial effluents and municipal sewage. They are capable of adsorbing heavy metals, phosphates, and ammonia ions. This reduces environmental pollution and increases protection for aquatic ecosystems (Shah, Abebe and Shah 2017).

Zeolites are used in industrial water treatment applications for the purification of process water, boiler feedwater, and cooling water. They remove impurities such as hardness ions, silica, and organic contaminants, preventing scale formation, corrosion, and fouling in industrial equipment and systems (Shah, Oluyink, et al. 2018).

6.9 DRYING AND DEHYDRATION APPLICATIONS

Silica-rich zeolites are desiccants for drying gases, liquids, and solids in various industrial processes. They adsorb moisture from air, solvents, and gases, maintaining low humidity levels and preventing moisture-related degradation and corrosion in storage and transportation (Srimuk, Chirarattananon, et al. 2023). They are utilized as adsorbent materials in adsorption refrigeration and air conditioning systems. They adsorb and release water vapor during refrigeration, enabling energy-efficient cooling and temperature control without using traditional refrigerants (Costa and Paranhos 2020).

6.10 ENVIRONMENTAL REMEDIATION APPLICATIONS

Silica-rich zeolites are used for soil stabilization and remediation of contaminated landfills, brownfields, and mining sites. They immobilize heavy metals and radioactive ions in the soil, reducing their mobility and bioavailability and preventing their leaching into groundwater and surface water (Muhiddin and Tangkeallo 2020).

Groundwater remediation is one of the major areas of applicability as it encounters alarming levels of pollutants such as heavy metals, nitrates, and volatile organic compounds (VOCs). They adsorb contaminants from groundwater, reducing their concentration and protecting drinking water supplies and environmental health (Szerement, Szatanik-Kloc, et al. 2021).

Silica-rich zeolites are employed for the containment and immobilization of hazardous waste materials in landfills and disposal sites. They encapsulate toxic substances, preventing their release into the environment and mitigating the risks of soil and groundwater contamination (Al-Jubouri, Curry and Holmes 2016).

6.11 CONSTRUCTION MATERIALS APPLICATIONS

Zeolites are added to concrete mixtures as supplementary cementitious materials to improve concrete's strength, durability, and sustainability. They improve workability, reduce porosity, and strengthen resistance to chemical attack, contributing to the longevity of concrete infrastructure (Das, Adhikary and Rudzionis 2022).

They are used as additives in asphalt mixtures to improve pavement performance and durability. They enhance moisture resistance, reduce rutting and cracking, and increase the lifespan of asphalt pavements in road construction and maintenance applications (Amoni, Freitas, et al. 2019).

6.12 HYDROGEN PRODUCTION AND STORAGE APPLICATIONS

Silica-rich zeolites serve as adsorbents for hydrogen storage and purification in fuel cell applications. They adsorb and release hydrogen molecules at ambient temperatures and pressures, enabling the safe and efficient storage of hydrogen (Kurniawan, Romadiansyah, et al. 2018).

Zeolites are applied in catalysis for hydrogen production by water-gas shift, steam reforming, and methanol decomposition. The various conversions of hydrocarbons, alcohols, and water into hydrogen gas are facilitated by using suitable zeolites, providing a clean and renewable energy source both in industry and transport (Zhang, Kang, et al. 2013).

To sum it up, one can find numerous ways to utilize fly ash's raw and modified formats to obtain optimum output from this waste material. Silica-rich zeolites find a broad application domain across industries: petrochemicals, environment remediation, construction materials, water treatment, and energy production. Their versatile properties and different applications help to drive advancements in sustainability, resource conservation, and environmental protection; hence, they are core materials in modern technology and manufacturing. Continued research and innovation in zeolite science and engineering will be expected to expand the scope and impact of silica-rich zeolites in addressing global challenges and meeting societal needs.

REFERENCES

Abas, K. M. and N. A. Fathy (2024). “Sodalite zeolitic materials produced from coal fly ash for removal of congo red dye from aqueous solutions.” *International Journal of Environ mental Science and Technology* 21(5): 5165–5184.

Al-Jubouri, S. M., N. A. Curry and S. M. Holmes (2016). “Hierarchical porous structured zeolite composite for removal of ionic contaminants from waste streams and effective encapsulation of hazardous waste.” *Journal of Hazardous Materials* 320: 241–251.

Al-Zeer, M. I. and K. J. MacKenzie (2019). “Fly ash-based geopolymers as sustainable bifunctional heterogeneous catalysts and their reactivity in Friedel-Crafts acylation reactions.” *Catalysts* 9(4): 372.

Alterary, S. S. and N. H. Marei (2021). “Fly ash properties, characterization, and applications: A review.” *Journal of King Saud University-Science* 33(6): 101536.

Amoni, B. D. C., A. D. L. D. Freitas, A. R. Loiola, J. B. Soares and S. D. A. Soares (2019). “A method for NaA zeolite synthesis from coal fly ash and its application in warm mix asphalt.” *Road Materials and Pavement Design* 20(sup2): S558–S567.

Belviso, C. (2018). “State-of-the-art applications of fly ash from coal and biomass: A focus on zeolite synthesis processes and issues.” *Progress in Energy and Combustion Science* 65: 109–135.

Bohra, S., K. P. Dey, D. Kundu and M. K. Naskar (2013). “Synthesis of zeolite T powders by direct dissolution of rice husk ash: an agro-waste material.” *Journal of Materials Science* 48(22): 7893–7901.

Boycheva, S., D. Zgureva and A. Shoumkova (2015). “Recycling of lignite coal fly ash by its conversion into zeolites.” *Coal Combustion and Gasification Products* 7(1): 1–8.

Buzukashvili, S., R. Sommerville, N. A. Rowson and K. E. Waters (2024). “An overview of zeolites synthesised from coal fly ash and their potential for extracting heavy metals from industrial wastewater.” *Canadian Metallurgical Quarterly* 63(1): 130–152.

Cao, C., W. Xuan, S. Yan and Q. Wang (2023). “Zeolites synthesized from industrial and agricultural solid waste and their applications: A review.” *Journal of Environmental Chemical Engineering* 11(5): 110898.

Chen, Y. and S. Tang (2019). “Solvothermal synthesis of porous hydrangea-like zeolitic imidazole framework-8 (ZIF-8) crystals.” *Journal of Solid State Chemistry* 276: 68–74.

Chen, Z., G. Song, C. Li, W. Chen, Z. Li and S. Kawi (2023). “Coal fly ash to Y zeolite of great purity and crystallinity: A new and green activation method of combined in situ microwave and ultrasound.” *Solid State Sciences* 136: 107102.

Costa, J. A. S. and C. M. Paranhos (2020). “Mitigation of silica-rich wastes: An alternative to the synthesis eco-friendly silica-based mesoporous materials.” *Microporous and Mesoporous Materials* 309: 110570.

Das, M., S. K. Adhikary and Z. Rudzionis (2022). “Effectiveness of fly ash, zeolite, and unburnt rice husk as a substitute of cement in concrete.” *Materials Today:* Proceedings 61: 237–242.

Duan, X., J. Dou, Y. Zhao, S. Khoshk Rish and J. Yu (2020). “A study on Mn-Fe catalysts supported on coal fly ash for low-temperature selective catalytic reduction of NOx in flue gas.” *Catalysts* 10(12): 1399.

Feng, S.-H. and G.-H. Li (2017). Hydrothermal and solvothermal syntheses. In By Ruren Xu, Yan Xu (Eds.), *Modern inorganic synthetic chemistry*. Elsevier**:** 73–104.

Gao, J., J. Zhao, J. Zhang, Q. Li, J. Gao, M. Cai and J. Zhang (2020). “Preparation of a new low-cost substrate prepared from drinking water treatment sludge (DWTS)/bentonite/zeolite/fly ash for rapid phosphorus removal in constructed wetlands.” *Journal of Cleaner Production* 261: 121110.

Gao, M., L. Yang, S. Yang, T. Jiang, F. Wu and T. Nagasaka (2022). "Simple aminated modified zeolite 4A synthesized using fly ash and its remediation of mercury contamination: Characteristics and mechanism." *Sustainability* 14(23): 15924.

Grabias-Blicharz, E. and W. Franus (2023). "A critical review on mechanochemical processing of fly ash and fly ash-derived materials." *Science of the Total Environment* 860: 160529.

Guo, X., J. Hu and C. Wang (2024). "Microwave-assisted hydrothermal synthesis of autoclaved bricks from municipal solid waste incineration fly ash and coal fly ash (MSWIFA) and its carbon emission." *Journal of Sustainable Cement-Based Materials* 13(4): 609–621.

GV, S. K. and C. S. Matli (2024). "Synthesis and characterisation of pure zeolite A from fly ash for removal of lead and cadmium." *International Journal of Environmental Analytical Chemistry*: 1–23.

Hamidi, R., R. Khoshbin and R. Karimzadeh (2021). "A new approach for synthesis of well-crystallized Y zeolite from bentonite and rice husk ash used in Ni-Mo/Al2O3-Y hybrid nanocatalyst for hydrocracking of heavy oil." *Advanced Powder Technology* 32(2): 524–534.

Hossi, J. D. C., D. Nkazi, J. van Der Merwe and A. Joel (2023). "Fly ash derived FAU zeolite synthesis and characterization."

Islam, M. R., B. Biswas and R. Naidu (2024). "CO2 capture using zeolite synthesized from coal fly ash and its subsequent utilization for fire retardation and dye removal." *ACS Sustainable Resource Management* 1(4): 799–809.

Kan, L., L. Zhang, Y. Zhao and M. Wu (2020). "Properties of polyvinyl alcohol fiber reinforced fly ash based Engineered Geopolymer Composites with zeolite replacement." *Construction and Building Materials* 231: 117161.

Kurniawan, R. Y., Romadiansyah, T. Q., Tsamarah, A. D., & Widiastuti, N. (2018). Synthesis of zeolite-X from bottom ash for H2 adsorption. In *IOP Conference Series: Materials Science and Engineering* (Vol. 299, No. 1, p. 012083). IOP Publishing.

Längauer, D., V. Čablík, S. Hredzák, A. Zubrik, M. Matik and Z. Danková (2021). "Preparation of synthetic zeolites from coal fly ash by hydrothermal synthesis." *Materials* 14(5): 1267.

Liang, D., Y. Liu, R. Zhang, Q. Xie and L. Zhang (2024). "A review on the influence factors in the synthesis of zeolites and the transformation behavior of silicon and aluminum during the process." *Comments on Inorganic Chemistry*: *44*(6), 461–497.

Liu, H. (2022). "Conversion of harmful fly ash residue to zeolites: innovative processes focusing on maximum activation, extraction, and utilization of aluminosilicate." *ACS Omega* 7(23): 20347–20356.

Liu, P., Q. Wu, K. Yan, L. Wang and F.-S. Xiao (2023). "Solvent-free synthesis of FAU zeolite from coal fly ash." *Dalton Transactions* 52(1): 24–28.

Lopez-Periago, A. M. and C. Domingo (2018). "Features of supercritical CO2 in the delicate world of the nanopores." *The Journal of Supercritical Fluids* 134: 204–213.

Matlob, A. S., R. A. Kamarudin, Z. Jubri and Z. Ramli (2012). "Using the response surface methodology to optimize the extraction of silica and alumina from coal fly ash for the synthesis of zeolite Na-A." *Arabian Journal for Science and Engineering* 37: 27–40.

Meng, X. and F.-S. Xiao (2014). "Green routes for synthesis of zeolites." *Chemical Reviews* 114(2): 1521–1543.

Moreno, N., X. Querol, F. Plana, J. M. Andres, M. Janssen and H. Nugteren (2002). "Pure zeolite synthesis from silica extracted from coal fly ashes." *Journal of Chemical Technology & Biotechnology: International Research in Process, Environmental & Clean Technology* 77(3): 274–279.

Muhiddin, A. B., & Tangkeallo, M. M. (2020, July). Correlation of unconfined compressive strength and California bearing ratio in laterite soil stabilization using varied zeolite content activated by waterglass. In *Materials Science Forum* (Vol. 998, pp. 323–328). Trans Tech Publications Ltd.

Musyoka, N. M., L. F. Petrik, W. M. Gitari, G. Balfour and E. Hums (2012). "Optimization of hydrothermal synthesis of pure phase zeolite Na-P1 from South African coal fly ashes." *Journal of Environmental Science and Health, Part A* 47(3): 337–350.

Nagao, M., S. Otake, S. Kushibiki and T. Furusaki (2024). "Improving the activity of fly ash as a solid base catalyst through the pozzolanic reaction under mild conditions." *Journal of the Ceramic Society of Japan* 132(4): 184–188.

Ndlovu, N. Z. N., A. E. Ameh, L. F. Petrik and T. V. Ojumu (2023). "Synthesis and characterisation of pure phase ZSM-5 and sodalite zeolites from coal fly ash." *Materials Today Communications* 34: 105436.

Novais, R. M., L. Buruberri, M. Seabra and J. Labrincha (2016). "Novel porous fly-ash containing geopolymer monoliths for lead adsorption from wastewaters." *Journal of Hazardous Materials* 318: 631–640.

Oluyinka, O. A., A. V. Patel, B. A. Shah and M. I. Bagia (2020). "Microwave and fusion techniques for the synthesis of mesoporous zeolitic composite adsorbents from bagasse fly ash: Sorption of p-nitroaniline and nitrobenzene." *Applied Water Science* 10(12): 236.

Pimraksa, K., P. Chindaprasirt and N. Setthaya (2010). "Synthesis of zeolite phases from combustion by-products." *Waste Management & Research* 28(12): 1122–1132.

Prodinger, S. and M. A. Derewinski (2020). 3 – Synthetic zeolites and their characterization. In *Nanoporous materials for molecule separation and conversion*. J. Liu and F. Ding, Elsevier: 65–88.

Saceda, J.-J. F., R. L. d. Leon, K. Rintramee, S. Prayoonpokarach and J. Wittayakun (2011). "Properties of silica from rice husk and rice husk ash and their utilization for zeolite Y synthesis." *Quimica Nova* 34: 1394–1397.

Shah, B. A., A. A. Abebe and A. V. Shah (2017). "Microwave-synthesized barium-impregnated siliceous zeolitic material derived from bagasse fly ash for uptake of aniline." *Arabian Journal for Science and Engineering* 42: 139–152.

Shah, B. A., O. Oluyink, A. Pate and M. I. Bagia (2018). "Impacts of industrial and agricultural chemical wastes on freshwater resources." *JSM Chemistry* 6(1): 1052.

Shah, B. A., A. Patel, M. Bagia and A. Shah (2017). "Green approach towards the synthesis of MCM-41 from siliceous sugar industry waste." *International Journal of Applied Chemistry* 13(3): 497–514.

Shah, B. A., Patel, A. V., Bagia, M. I., & Oluyinka, O. A. (2019). Removal of Cr (VI) from aqueous solutions using EDCC-MCM-41: Isotherm, kinetics and thermodynamic evaluation. *Journal of Dispersion Science and Technology*, **40**(12), 1827–1841.

Shanti, M. D., K. Shanti and J. Meshram (2013). "Activated fly ash catalyzing the synthesis of novel potent 4-thiazolidinones of 4-amino antipyrine anils using CEM discover microwave methodology and their virtual screening." *Phosphorus, Sulfur, and Silicon and the Related Elements* 188(10): 1351–1360.

Sivalingam, S. and S. Sen (2018). "Optimization of synthesis parameters and characterization of coal fly ash derived microporous zeolite X." *Applied Surface Science* 455: 903–910.

Sivalingam, S. and S. Sen (2019). "Sono-assisted adsorption of As(V) from water by rice-husk-ash-derived iron-modified mesoporous zeolite Y: A cradle-to-cradle solution to a problematic solid waste material." *Industrial & Engineering Chemistry Research* 58(31): 14073–14087.

Srimuk, J., S. Chirarattananon, P. Chaiwiwatworakul, A. Nathakaranakule, P. Rakkwamsuk and S. Chiarakorn (2023). "Experimental investigation on the performance of a dehumidifier constructed from a water-to-air heat exchanger coated with composite desiccant of mesoporous silica gel and LiCl." *Applied Science and Engineering Progress* 16(4): 6821–6821.

Szerement, J., A. Szatanik-Kloc, R. Jarosz, T. Bajda and M. Mierzwa-Hersztek (2021). "Contemporary applications of natural and synthetic zeolites from fly ash in agriculture and environmental protection." *Journal of Cleaner Production* 311: 127461.

Tang, M., C. Zhou, N. Zhang, J. Pan, S. Cao, T. Hu, W. Ji, Z. Wen and T. Nie (2019). "Extraction of rare earth elements from coal fly ash by alkali fusion–acid leaching: Mechanism analysis." *International Journal of Coal Preparation and Utilization*, **42**(3), 536–555.

Tauanov, Z., D. Shah, V. Inglezakis and P. K. Jamwal (2018). "Hydrothermal synthesis of zeolite production from coal fly ash: A heuristic approach and its optimization for system identification of conversion." *Journal of Cleaner Production* 182: 616–623.

Tian, Q., Z. Liu, Y. Zhu, X. Dong, Y. Saih, J. M. Basset, M. Sun, W. Xu, L. Zhu and D. Zhang (2016). "Beyond creation of mesoporosity: The advantages of polymer-based dual-function templates for fabricating hierarchical zeolites." *Advanced Functional Materials* 26(12): 1881–1891.

Truttim, P., S. Asavapisit and R. Piyaphanuwat (2023). "Microwave synthesis of zeolite X from bituminous fly ash and its characterization." Materials Today*:* Proceedings.

Vidhya, K. (2022). "Eco-sustainable alkali activated brick using municipal incinerated ash." *International Journal of Coal Preparation and Utilization* 42(3): 331–348.

White, R. J., A. Fischer, C. Goebel and A. Thomas (2014). "A sustainable template for mesoporous zeolite synthesis." *Journal of the American Chemical Society* 136(7): 2715–2718.

Wu, Q., H. Luan and F.-S. Xiao (2022). "Targeted synthesis of zeolites from calculated interaction between zeolite structure and organic template." *National Science Review* 9(9): nwac023.

Xing, L., X. Li, P. Cao, J. Luo and H. Jiang (2024). "Stepwise extraction and utilization of silica and alumina from coal fly ash by mild hydrothermal process." *Process Safety and Environmental Protection* 182: 918–929.

Zeng, X., X. Hu, H. Song, G. Xia, Z.-Y. Shen, R. Yu and M. Moskovits (2021). "Microwave synthesis of zeolites and their related applications." *Microporous and Mesoporous Materials* 323: 111262.

Zhang, Y., L. Kang, J. Shang and H. Gao (2013). "A low cost synthesis of fly ash-based mesoporous nanocomposites for production of hydrogen by photocatalytic water-splitting." *Journal of Materials Science* 48: 5571–5578.

Zhou, Q., X. Jiang, Q. Qiu, Y. Zhao and L. Long (2023). "Synthesis of high-quality NaP1 zeolite from municipal solid waste incineration fly ash by microwave-assisted hydrothermal method and its adsorption capacity." *Science of the Total Environment* 855: 158741.

Zhu, L., J. Ji, S. Wang, C. Xu, K. Yang and M. Xu (2018). "Removal of Pb (II) from wastewater using Al2O3-NaA zeolite composite hollow fiber membranes synthesized from solid waste coal fly ash." *Chemosphere* 206: 278–284.

7 Aspects of Vermicomposting in Fly Ash Management

A Sustainable Approach

Nishat Fatima, Pankaj Kumar Singh, Anjali Singh, Keshav Singh and Marttin Paulraj Gundupalli

7.1 INTRODUCTION

Rapid urbanization and industrialization brought on by excessive population expansion raise the global demand for electricity. Power has been recognized as an essential component of progress in the majority of developing nations (Usmani et al., 2017). By 2030, electricity production in India is predicted to increase by up to 2–3 times due to coal, which has been considered to be the nation's primary fuel source. About 79% of the world's electricity is generated by coal-fired power stations every year (Satapathy et al., 2020). According to research by the Central Electricity Authority (CEA) of India, power plants used in the range of 730 MT of coal to produce 210 MT of fly ash (FA) on average in 2018–2019; by 2030, that amount might exceed 437 MT. India is one of the world's largest users of non-coking coal, which contains between 30% and 40% high ash (Senapati, 2011). Power plants produce fly ash, a byproduct with particles identical to glass, ranging in size from 0.01 to 100 mm (Davison et al., 1974). Fly ash is one of the world's most harmful solid wastes because it contains iron, silica, aluminum, arsenic, chromium, lead, calcium, magnesium, zinc, and other harmful oxides (Usmani & Kumar, 2017).

According to reports, fly ash (FA) can have a variety of negative health consequences and may persist in the atmosphere for an adequate duration of time (TERI, 1998). It also threatens the ecosystem. It blocks drainage, lowers the pH, makes the water less suitable for consumption, turbidites the water, affects plants, and has an additional negative impact on photosynthesis (Gupta et al., 2005). Soil, groundwater, and drinking water on the surface are all contaminated by FA that accumulates in landfills (Pandey et al., 2009a). Therefore, it is practically impossible to overstate how vital and important it is to find remedies to these challenges, including both safe disposal and profitable recycling of these materials (Chattopadhyaya & Bhattacharya, 2010). The severity of the issue has been acknowledged on a global scale, and several nations are researching alternative options for the efficient disposal and reuse of FA.

DOI: 10.1201/9781003498124-7

As FA is not classified as a toxic material, an immense amount of research is being done to find uses for it in different industries (Latifi et al., 2015). Because of this, FA pozzolanic features make it useful for a wide range of other applications in civil engineering, including building roads and constructing bricks, cement, ceramics, and other products. In addition to nitrogen (N), phosphorus (P), and potassium (K), FA provides all of the essential plant nutrients. Researchers are looking at utilizing it in agriculture. FA is a biofertilizer that helps crops grow and yield more by maintaining pH, improving soil aeration and percolation, and supplying nutrients. The agriculture industry in India utilizes about 2% of the FA produced (Kumar et al., 2005).

The soluble heavy metals present in FA have the potential to be utilized in agriculture; however, there are some drawbacks due to their slower rate of breakdown in the soil after application (Pandey & Singh, 2010; Pandey et al., 2009b). Utilizing conventional approaches to treat FA for heavy metal elimination is expensive as well as time-consuming (González & González-Chávez, 2006). Although direct FA application sustains pH and improves soil aeration and percolation, the biggest consequence for agricultural ecosystems is the delayed rate of heavy metal concentration breakdown. Among the most effective approaches for detoxifying and amending the soil while simultaneously getting rid of waste like FA is vermicomposting (Edwards & Bater, 1992). According to Arumugam et al. (2014), vermicomposting is a viable, cost-effective, and environmentally friendly approach to treating organic waste. Solid waste management is an issue that Indian cities need to address immediately. It has been observed that certain earthworm species can withstand heavy metals (Fatima & Singh, 2023a). Through the vermicomposting process, it is proved that earthworms effectively detoxify municipal solid waste (MSW), including industrial sludge (Fatima & Singh, 2023b; Singh & Singh, 2023). FA can be used in vermicomposting in very small to huge amounts (Muniraj et al., 2020).

FA vermicomposting supplies several benefits, including the ability to handle extremely low to massive quantities of FA, an effortless and economical process that works well for both small and large-scale applications, and easily accessible sectors for the vermicasts, which are used as soil enrichers (Usmani & Kumar, 2017). Earthworms enhance the fertility of soil supplemented with coal ash by improving the availability of mineral nutrients like NPK in the ash because they bioaccumulate heavy metals in their guts and minimize the level of harmful heavy metals present in the ash (Usmani & Kumar, 2017). This book chapter focuses on the physicochemical characteristics of FA, its function in plant growth, and its remediation through vermicomposting, an earthworm-mediated approach employed in agriculture. The main purpose of this book chapter consequently involves the biological management of FA.

7.2 BIO-COAL

The "bio-coal" product produced by thermally upgrading biomass is becoming increasingly popular worldwide as nations seek more renewable energy sources to lessen their dependency on coal and minimize greenhouse gas (GHG) emissions. Produced from biomass, bio-coal, often referred to as biochar or charcoal, has immense potential to replace fossil coal in the industrial sector. Bio-coal is usually produced via slow pyrolysis, which is thermally driven biomass breakdown at a low heating rate in an inert environment. Because bio-coal has a higher proportion of carbon, is more grindable, and has a better heating value than biomass, it is a desirable alternative to fossil coal (Mousa et al., 2016). Eco-friendly uses of bio-coal instead of fossil coal are very preferred in industries including metallurgical processes, co-firing power plants, and syngas generation (Molino et al., 2018; Dai et al., 2008; Suopajärvi et al., 2018). According to Mandova et al. (2018), the use of bio-based fuels and bio-reduction agents might lower GHG emissions by 42% from fossil fuels and reduce agents used in steel mills inside the European Union (EU). According to Norgate et al. (2012), adopting bio-coal instead of fossil fuels may cut the on-site carbon footprint of integrated steel facilities by 57%. In addition to its main benefit of reducing GHG emissions, bio-coal has lower ash, sulfur, and phosphorus contents than pulverized coal, which might result in a 50% reduction in the slag ratio (Feliciano-Bruzual, 2014). Compared to the initial raw materials, bio-coal is a solid fuel that has undergone thermal pretreatment and is meant to be more consistent. Low-quality fuel can undergo torrefaction to transform into a trade good with better qualities (Pawlak-Kruczek et al., 2020). Bio-coal is produced by converting biomass into a form that resembles conventional coal derived from fossil fuels, therefore providing a feasible alternative for coal consumers who intend to cut down on their emissions. Pyrolysis, the process of turning dry biomass into coal, is carried out at high temperatures in an inert atmosphere with no oxygen. The process is also referred to as torrefaction, depending on the final product's properties and the temperatures at which it happens. Similar to fossil coal, biomass has a high carbon content, but it also has a higher volatile matter content and a lower density. Bio-coal produced from biomass with pretreatment will have properties similar to those of fossil coal. There is a major difference between fossil coal and bio-coal, though, as the former influences its quality by raising its reactivity. As a result, bio-coal can only partially replace fossil coal in bio-coke. SiO_2 and Al_2O_3 constitute the majority of the ash in the bio-coal, but there are also some FeO, CaO, MgO, S compounds, P_2O_5, and alkalis (K_2O). Since bio-coal ash functions as an internal catalyst, these chemicals are anticipated to improve the reactivity of the coke with an increased quantity of ash present (Yu et al., 2014). To improve agricultural productivity, bio-coal ash can be used via liming (raising the pH of the soil in a specific area); however, this relies on the components of the ash waste (Mohamaed & KMuduli, 2021). Many approaches have been used to control ash, including adding forest soils and using them to make cement (Barbosa et al., 2013; Berra et al., 2015). Bio-coal is most commonly used in coal-fired power plants to boost co-firing rates and reduce carbon emissions. Additional possible uses include solid fuel, supercapacitor electrodes, adsorbent, soil amendment, and nanostructured carbon (Ahmad et al., 2023).

7.3 FA PROPERTIES

FA is made up of microscopic particles having a large surface area, a low to medium bulk density, a light structure, and a small average diameter (Yousuf et al., 2020). Particles with an angular shape comprise the carbonaceous material in FA. In general, fly ashes from the majority of bituminous coal have a particle size distribution that is identical to silt (smaller than 0.075 mm or number 200 sieve). Compared to bituminous coal fly ashes, sub-bituminous coal fly ashes are often somewhat coarser, although they are silt-sized as well. Depending on how much unburned carbon remains, FA can appear in a range of colors, from tan to gray to black. The carbon concentration decreases with color lightness (Yao et al., 2015). FA is chemically classified as an amorphous combination of ferro-aluminosilicate minerals. Although there are considerable variations in composition due to differences in coal sites, the principal elements are usually SiO_2, CaO, Al_2O_3, and Fe_2O_3, with minimal amounts of MgO, Na_2O, and so on. Fly ashes that come from bituminous coals are acidic, whereas those obtained from sub-bituminous coals are alkaline; this is significantly dependent on the ash's chemical components (Talbot et al., 1978; Roy & Griffin, 1984).

7.4 ENVIRONMENTAL IMPACT OF FA

When FA is released into the environment, plants are possibly able to absorb its minerals. Power plants emit substantial quantities of FA, which then accumulate on soil and vegetation surfaces, increasing the dispersion of numerous components into the surrounding environment. Even though these components are not highly concentrated in coal, their amounts in the ecosystem can nonetheless become substantial. Finer FA particles lead to higher quantities of volatile trace elements. These elements generally accumulate on particle surfaces at lower ambient temperatures after becoming volatilized during combustion (Davison et al., 1974; Quann et al., 1982). In comparison to typical farmland soils, FA samples often have a greater percentage of all necessary plant nutrients, except N (Adriano et al., 1980). Researchers have observed that plants' reactions to the macro and microelements in FA can range from advantageous benefits at low concentrations to detrimental ones at elevated levels (Chang et al., 1977). It has been shown that the nutritional balance of forest and agricultural soils can be disturbed by the persistent deposition of even Ca^{2+} and Mg^{2+} rich particulates (Dassler & Bortiz, 1988). Therefore, it is necessary to comprehend how FA affects soil structure as well as plant development and growth in the long term. In addition to contaminating the air, ash carried by the wind pollutes surrounding croplands by depositing coarse silica and other harmful substances. Occasionally, dissolved salts and non-dissolved ash suspensions get deposited into rivers and streams by intense rainfall. Groundwater becomes contaminated by ash pond runoff as well (Sett, 2017).

7.5 UTILIZATION OF FA IN AGRICULTURE

The majority of plant nutrients, except nitrogen and organic carbon, are present in large amounts in FA. Agronomic variables may benefit from FA due to its beneficial characteristics, as suggested by the numerous studies carried out in the forestry and agricultural sectors to investigate the impact of FA as a source of plant nutrients. In general, FA has been shown to improve crop growth or yield up to a specific threshold of application to soil, beyond which the favorable effects tend to wane. These impacts, however, are expected to vary by crop and soil condition (Chattopadhyay et al., 2017). Weinstein et al. (1989) reported that the application of FA ameliorant to the soil increased the yields of white clover, barley, Bermuda grass, and alfalfa. A slightly acidic soil (pH 6.0) treated with about 125 mt/ha of unweathered FA may increase the yield of sorghum, field corn, millet, potatoes, alfalfa, tomatoes, onions, carrots, beans, and cabbage (Furr et al., 1977). These plants have been shown to contain higher quantities of selenium, magnesium, boron, and arsenic. Applying 2%–4% FA to rice (*Oryza sativa*) crops significantly increased the amount of nitrogen, sulfur, calcium, sodium, and iron, based on a greenhouse experiment (Sikka & Kansal, 1994). Neelima et al. (1995) concluded that the application of several fly ash-soil combinations resulted in a considerable increase in plant nutrients and an increased production of aromatic grasses, specifically citronella (*Cymbopogon nardus*) and palmarosa (*Cymbopogon martini*). FA applied at a 25% dose produced great yields of brinjal, tomato, and cabbage along with positive responses from oil-seed crops like sunflower and peanut (Satapathy et al., 2020).

7.6 PLANTS NUTRIENTS PRESENT IN FA

A broad spectrum of agricultural plants has benefited from the widespread use of FA additives in soil. In general, FA improves plant development and nitrogen absorption when added to the soil (Aitken et al., 1984; Basu et al., 2009). Although FA contains plant nutrients in much higher amounts overall, the majority of these elements do not remain in bioavailable forms and require a considerable amount of time to be converted into forms that may be utilized. FA, when combined with organic materials and chemical fertilizers, increased grain yield in the context of wet-season rice, according to field studies on the use of coal ash in various combinations with organic manure and chemical fertilizers to enhance crop yield and soil productivity in rice-based cropping systems (Mitra et al., 2000). Additionally, the soils' physicochemical characteristics increased, showing improvements in pH, organic carbon, bulk density, and accessibility to nutrients (N, P, K, Ca, Mg, Zn, Cu, and Co). The activity of microorganisms was raised by the combination of fly ash and organic materials. A subsequent set of intricate field investigations revealed that applying fly ash and organic waste together was found to be more effective for growing rice, groundnut grain, and pods. These experiments assessed how various organic and industrial wastes might improve the productivity of acid-lateritic soil in a rice-groundnut cropping system (Iftikar et al., 2020). FA's nutrient availability must thus be significantly increased to raise FA's suitability as a source of plant nutrients. One effective approach in this regard might be to promote microbial activity to facilitate the chemical and biological breakdown of FA and the subsequent release of numerous nutrients in soluble form (Chattopadhyay et al., 2017). Bio-coal ash may contain some of the elements that are necessary for plant growth, such as potassium, calcium, magnesium, and so on, in a form that the plants can absorb and use for growth (Kowapradit, 2005).

7.7 FA APPLICATION IN SOLID WASTE MANAGEMENT

Solid waste management comprises all aspects of handling solid waste, from collecting it from its location to properly disposing it to prevent it from becoming an annoyance or negatively affecting the surrounding area (Dwivedi & Jain, 2014). For cities, disposing of MSW and sewage sludge from wastewater treatment facilities poses a constant threat to the environment (Mikavica, 2023). Mainly, the sludge contains enhanced heavy metals and microorganisms that, if not controlled appropriately, could pose potential threats to humans (Fatima & Singh, 2023b). Therefore, it is essential to use these resources as efficiently as possible in a well-thought-out way to avoid adverse effects on the environment, plant growth, or soil health.

Combining FA, sewage sludge waste along with MSW for the production of compost is an efficient approach to making the best possible use of these waste materials.

Furthermore, there is an increase in activity from inorganic to organic farming as a result of awareness of the harmful effects of overusing chemical fertilizers. A significant amount of compost or organic manure would be needed for organic farming. It is viable to effectively fulfill the requirement for organic manure. FA is a soil supplement that may enhance the soil's texture, porosity, and nutritional value (Mikavica, 2023). Comprehensive nutrient management techniques are essential to maintain soil fertility and sustainably produce crops. FA from coal has been effectively converted into a nutrient-rich product by co-composting with sewage sludge (Mao-Fa et al., 1999).

7.8 FA RECYCLING THROUGH VERMICOMPOSTING

The possibility of recycling vast amounts of FA and various organic wastes has gained interest due to the benefits of utilizing FA in conjunction with organic substances. Therefore, it was essential to implement an effective composting technology to recycle FA and organic materials at the same time. Vermicomposting biotechnology is one of the best techniques for composting organic wastes; it has become increasingly common recently since it is a simple and highly efficient method of making compost of higher quality than conventional composting processes (Fatima & Singh, 2023a). According to Singh and Fatima (2019), the earthworm species most frequently utilized for vermicomposting comprise *Eisenia fetida*, redworm, or brandling worm. According to Haimi and Huhta (1990), vermicomposting additionally facilitates the utilization of other species, such as *Eisenia andrei*, *Perionyx excavatus*, *Eudrilus eugeniae*, and *Lumbricus rubellus*. These optimum species have traits such as being resistant, having a healthy diet, being prolific breeders, and being good species for ecological evaluation. They are also excellent at vermicomposting (Römbke et al., 2005). FA, combined with other organic substances, can be broken down by earthworms during gastrointestinal transit. During the vermicomposting process, earthworms release millions of "decomposer microbes" and "nitrogen-fixing" microorganisms into their stomachs. Vermicomposted FA also contains considerable amounts of Ca, Mg, Zn, and Mn. Earthworms' biological activities transform heavy metal levels in FA into enhanced soil nutrients by amending with organic material on a small scale before expanding to large-scale production (Satapathy et al., 2020). Compared to other wastes, vermicomposts produced from different FA wastes have significantly greater concentrations of the main nutrients like N, P, K, Ca, and Mg (Table 7.1) (Usmani & Kumar, 2016).

7.9 EARTHWORMS' FUNCTION IN MANAGING FA

Class Oligochaeta earthworms, which are considered among the most significant members of soil communities, have an essential ecological role in building and maintaining soil structure (Singh & Fatima, 2022). They have the potential to be significant agents in the monitoring of both organic and inorganic toxicants since they reside in an area with polluted soil (Cooke et al., 1992). According to Lee (1985), they have been suggested to be an intriguing biological indicator of several metals found in soils. They can break down "fly ash" from coal-fired power plants, which causes a significant dumping issue because of its high heavy metal concentration. Earthworms accumulate heavy metals in their body tissues during the process of transforming FA

TABLE 7.1
Nutrient content analysis before and after vermicomposting with several earthworm species

Location (site)	Substrate	Earthworm species	Exposure duration	Combination used	Nutrient composition								References
					OC (mg/kg)		*N* (mg/kg)		*P* (mg/kg)		*K* (mg/kg)		
					BV	AV	BV	AV	BV	AV	BV	AV	
BHEL, Tiruchirapalli, Tamil Nadu	Fly ash (FA)	*Eudrilus eugeniae*	60	T_0 = FA	7.98	7.90	0.19	0.2	0.03	0.04	1.15	1.18	Venkatesh and Eevera (2007)
				T_1 = CD	10.49	7.05	0.68	1.40	0.16	0.72	0.98	1.69	
				T_2 = FA + CD (1:1)	8.98	6.98	0.19	0.91	0.04	0.22	1.20	1.64	
				T_3 = FA + CD (1:3)	9.19	7.50	0.22	0.98	0.06	0.26	1.18	1.86	
				T_4 = FA + CD (3:1)	8.79	7.25	0.19	0.63	0.03	0.18	1.22	1.74	
Durgapur TPP	Fly ash (FA)	*Eisenia fetida*	45–75	FA: OM									Bhattacharya et al. (2012)
				1:0	–	–	470	518	18.3	24.2	51	62	
				0:1	–	–	640	120	29.0	33.0	82	118	
				1:1	–	–	517	1,078	26.7	38.5	102	138	
				1:3	–	–	628	924	24.5	19.8	80	138	
				3:1	–	–	544	742	23.1	18.5	50	110	
Neyveli Lignite Corporation, Tamil Nadu	Fly ash (FA)	*Lampito mauritii*	50	CD: FA									Ananthakrishnasamy et al. (2009)
				1:1	15.7	13.2	0.41	0.69	0.5	0.8	0.10	0.15	
				2:1	31.8	25.7	0.72	1.01	1.0	1.5	0.19	0.27	
				3:1	44.8	27.65	1.12	1.65	1.2	1.8	0.24	0.38	
–	Fly ash (FA)	*Lampito mauritii*	–	FA: PM: CD									Manivannan et al. (2012)
				1:1:8	437.7	351.6	11.5	21.3	5.1	9.6	14.5	13.3	
				2:2:6	437.7	351.6	11.8	20.1	4.7	8.8	14.1	12.6	
				3:3:4	422.3	353.3	11.6	20.1	4.5	8.5	13.5	12.2	
				4:3:3	416.4	381.5	11.3	15.4	4.2	7.1	12.4	12.2	
				3:4:3	420.4	379.3	11.9	16.6	4.6	7.2	12.9	11.3	

(Continued)

TABLE 7.1
(Continued)

Location (site)	Substrate	Earthworm species	Exposure duration	Combination used	Nutrient composition								References
					OC (mg/kg)		*N* (mg/kg)		*P* (mg/kg)		*K* (mg/kg)		
					BV	AV	BV	AV	BV	AV	BV	AV	
Raichur TPP, Karnataka	Fly ash (FA)	*Eudrilus eugeniae*	28	FA:PM:CD:CO2									Senappa (2012)
				2:2:1	–	–	1.87	–	0.89	-	-	-	
				2:2:0:1	–	–	1.08	–	0.15	-	-	-	
				2:0:2:1	–	–	0.38	–	0.36	-	-	-	
				2:2:0:0	–	–	1.42	–	0.15	-	-	-	
				2:0:2:0	–	–	0.96	–	0.58	-	-	-	
Panki TPP	Fly ash (FA)	*Eisenia fetida*	60	FA:PM									Niyazi and Chaurasia (2014)
				1:0	1.26	0.94	0.13	0.16	0.28	0.34	0.06	0.05	
				3:2	6.34	3.56	0.31	0.42	0.51	0.63	0.09	0.11	
				2:3	13.84	7.04	0.54	0.77	0.79	1.07	0.12	0.14	
				1:4	20.72	1.19	0.81	1.19	1.09	1.56	0.16	0.21	
				4:1	27.44	14.08	1.09	1.53	1.40	1.93	0.20	0.24	
				0:1	53.23	15.48	1.28	1.81	1.60	2.37	0.23	0.29	

Abbreviations: CD = Cow dung, OC = Organic carbon, N = Nitrogen, P = Phosphorous, K = Potassium, BV = Before vermicomposting, AV = After vermicomposting, TPP = Thermal power plant, OM = Organic matter, PM = Press mud, T = Treatment, BHEL = Bharat Heavy Electricals Limited.

into beneficial vermicompost. FA is an efficient source of plant nutrients in agriculture; its microbiological activities are further enhanced when combined with organic compounds (Mohapatra & Rao, 2001; Thomas et al., 1998). Earthworms raise the levels of N, P, and K in FA, thereby increasing its availability.

7.10 HEAVY METALS PRESENT IN FA: ACCUMULATION BY EARTHWORM SPECIES

This is the mechanism by which an earthworm's tissues accumulate more chemicals than FA or soil because they have been exposed to chemicals via all possible routes (such as nutritional absorption). *Perionyx excavatus, Eudrilus eugeniae, Eisenia fetida*, and *Perionyx sansibaricus* are among the earthworm species that are significant bio-accumulators and bioindicators of environmental contamination with persistent pollutants, including heavy metals (Suthar et al., 2008; Wang et al., 2009; Fatima & Singh, 2023a). They have also been shown to effectively minimize FA's toxicity and raise its level of nutrients (N, P, and K) (Bhattacharya & Chattopadhyay, 2006).

The amount of metal found in FA at any particular relative concentration determines how much metal accumulates in the earthworm tissues. Earthworm-based elimination of metal from FA is gaining a lot of curiosity. Because they may accumulate metals in their guts, earthworms have demonstrated that they are valuable organisms in the handling of FA, assisting in reducing the metal concentration of the waste product (Figure 7.1). Consequently, several investigations have demonstrated a considerable reduction in the concentration of heavy metals in FA before and after vermicomposting, as described in Table 7.2. The feed combination, including FA+PM (press mud) at an approximate ratio of 1:4, was determined to have the maximum

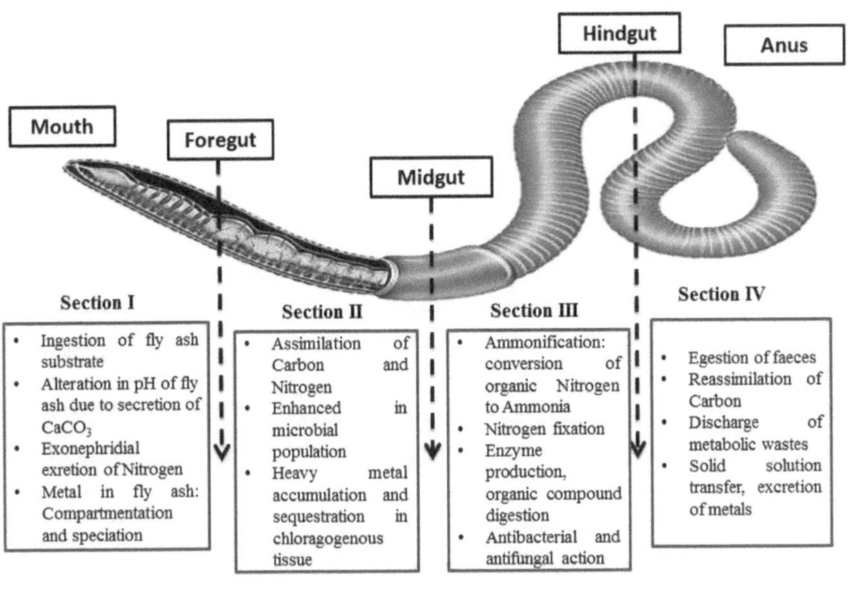

FIGURE 7.1 Physicochemical changes occur in earthworm's body due to ingestion of FA.

TABLE 7.2
Summarized data of heavy metal remediated through vermicomposting from fly ash

Location (site)	Sub strate used	Earthworm species	Exposure duration (days)	Combi nations used	Metal concentration (mg/kg) (FA)						
					Before vermicomposting						
					Cr	Cd	Pb	Ni	Zn	Cu	Mn
Panki thermal power plant, Kanpur	Fly ash	*Eisenia fetida*	–	FA: PM							
				1:0	124	–	52	–	84	79	–
				4:1	101	–	41	–	107	86	–
				3:2	83	–	34	–	128	91	–
				2:3	61	–	27	–	147	66	–
				1:4	41	–	21	–	169	66	–
				0:1	28	–	20	–	186	56	–
Durgapur, India	Fly ash	*Eisenia fetida*	45–75	FA: OM							
				1:0	0.86	0.36	1.8	–	–	–	–
				0:1	0.38	0.92	0.9	–	–	–	–
				1:1	0.22	0.14	1.0	–	–	–	–
				1:3	0.35	0.23	0.7	–	–	–	–
				3:1	0.29	0.28	1.3	–	–	–	–
Thermal power plant	Fly ash	*Eisenia fetida*	180	FA: CD							
				1:4	23.7	–	–	–	–	22.1	–
				4:1	81.3	–	–	–	–	57.9	–
Thermal power station, Tamil Nadu, India	Fly ash, horse dung (HD)	*Eudrilus eugeniae*	60	FA: HD in different ratios							
				–	High	High	High	–	–	–	–
Australia	Fly ash	*Megascolecid* sp. and *Aporrectodea trapezoids*	60	Combinations for composting that include two species of earthworms and three rates of fly ash (Where a = *Megascolecid* sp, b = *Aporrectodea trapezoids*) Copper – a. 23.3, b. 20.9 Chromium – a. 23.5, b. 15.7 Zinc – a. 221, b. 530							
–	Fly ash	*Eudrilus eugeniae*	60	MSW + MW + FW							
					Cr	Cd	Pb	Ni	Zn	Cu	Mn
				Estimation of metals	–	0.10	0.61	–	4.85	0.74	7.98
National TPP, Noida, India	Fly ash	*Eisenia fetida*	90	Vermicompost reactor							
				VC1: BP: F (1:1)	50.55	0.90	21.60	–	26.55	21.72	–
				VC2: BP: F (2:1)	43.66	0.90	15.76	–	25.60	23.26	–
				VC3: BP: F (3:1)	36.23	0.79	10.74	–	25.50	23.12	–

Metal concentration (mg/kg) (FA)							
After vermicomposting							
Cr	**Cd**	**Pb**	**Ni**	**Zn**	**Cu**	**Mn**	**References**
							Niyazi and Chaurasia (2014)
110	–	52	-	74	73	-	
84	–	41	-	85	79	-	
63	–	33.5	-	92	74	-	
43	–	26	-	92	41	-	
25	–	21	-	95	40	-	
15	–	20	-	94	52	-	
							Bhattacharya et al. (2012)
0.84	0.3	2.48	-	-	-	-	
0.35	0.14	0.45	-	-	-	-	
0.19	0.92	0.67	-	-	-	-	
0.27	0.12	0.60	-	-	-	-	
0.24	0.19	0.92	-	-	-	-	
							Gupta et al. (2005)
12.8	–	–	–	–	13.9	–	
60.9	–	–	–	–	51.1	–	
							Dharani et al. (2010)
Low	Low	Low	–	–	–	–	
							Muir et al. (2007)
							Pattnaik and Reddy (2010)
Cr	Cd	Pb	Ni	Zn	Cu	Mn	
–	0.05	0.31	–	1.49	0.47	5.16	
							Karwal and Kaushik (2020)
33.06	0.67	16.39	–	18.37	16.57	–	
29.79	0.68	12.03	–	17.76	17.39	–	
19.17	0.44	7.67	–	17.62	15.25	–	

(Continued)

TABLE 7.2
(Continued)

Location (site)	Sub strate used	Earthworm species	Exposure duration (days)	Combi nations used	Metal concentration (mg/kg) (FA)						
					Before vermicomposting						
					Cr	Cd	Pb	Ni	Zn	Cu	Mn
Chandra pura TPP, Jhark hand, India	Fly ash	*Eisenia fetida*, *Eudrilus eugeniae*, and *Lumbricus rubellus*	90	*Eisenia fetida*							
				Ef1	23.78	4.46	34.38	23.51	98.56	66.54	–
				Ef2	28.42	3.38	32.57	17.90	86.43	56.45	–
				Ef3	18.60	2.55	24.98	12.69	66.85	35.75	–
				Ef4	33.99	4.87	35.23	28.7	98.09	67.93	–
				Ef5	5.66	6.47	1.15	0.84	12.67	10.50	–
				Eudrilus eugeniae							
				Ee1	23.86	4.58	34.36	23.87	98.93	66.16	–
				Ee2	28.35	3.33	32.71	17.74	85.36	56.59	–
				Ee3	18.65	2.59	24.23	12.65	69.77	34.83	–
				Ee4	23.77	6.92	33.25	28.5	88.58	54.77	–
				Ee5	9.57	6.80	2.67	1.58	12.83	10.73	–
				Lumbricus rubellus							
				Lr1	23.83	4.55	34.25	23.49	97.45	64.96	–
				Lr2	28.09	3.31	32.62	17.52	84.30	56.50	–
				Lr3	18.63	4.60	24.16	12.63	67.15	34.72	–
				Lr4	33.61	2.54	43.67	24.30	98.35	68.35	–
				Lr5	5.68	7.21	1.14	2.85	12.72	10.74	–

Abbreviations: FA = Fly Ash, PM = Press mud, OM = Organic matter, CD = Cow dung, MSW = Municipal solid waste, MW = Municipal waste, FW = Flower waste, Buffalo dung + Press mud = BP, VC = Vermicompost, TPP = Thermal power plant, Ef = *Eisenia fetida*, Ee = *Eudrilus eugeniae*, Lr = *Lumbricus rubellus*.

Metal concentration (mg/kg) (FA)							
After vermicomposting							
Cr	Cd	Pb	Ni	Zn	Cu	Mn	References
							Usmani et al. (2017)
20.51	2.51	30.13	20.06	95.41	58.22	–	
17.54	1.65	23.40	11.31	71.78	46.05	–	
7.66	1.25	14.38	9.57	50.79	15.43	–	
17.24	2.43	22.81	16.2	54.45	38.90	–	
2.31	3.29	0.57	0.55	8.34	2.31	–	
17.94	2.59	31.07	18.28	90.71	20.47	–	
15.43	2.10	16.39	10.52	70.42	45.96	–	
7.60	1.76	14.05	3.55	32.70	14.36	–	
12.54	3.54	13.11	21.58	56.74	27.33	–	
7.06	2.54	1.48	1.21	4.47	5.99	–	
22.13	2.49	31.13	20.59	94.92	58.24	–	
17.93	1.45	22.29	13.82	72.62	50.42	–	
15.56	2.84	15.39	7.52	52.47	16.78	–	
22.50	1.24	32.23	11.47	64.19	45.58	–	
2.54	2.7	0.82	2.54	8.31	6.44	–	

amount of metal accumulation by earthworms. Similar findings have been reported by Morgan and Morgan (1992) and Das et al. (2012), revealing that inter-specific variations in nutritional metal intake and physiological consumption may have an impact on patterns of metal deposition in earthworm tissues.

Physical and chemical alterations in the earthworm's numerous sections, such as variations in the concentration of nutrients and the accumulation of heavy metals after FA consumption indicate potential physiological stress and bioaccumulation risks, highlighting the need for careful evaluation of FA's environmental impact on soil biota.

7.11 MECHANISM OF HEAVY METAL ACCUMULATION

Earthworms' chloragogenous tissue around the posterior alimentary canal is the main region of metal deposition (Morgan & Morris, 1982). Metallothioneins are produced by earthworms in response to heavy metal exposure. These proteins are rich in cysteines, low in molecular weight, and highly responsive to specific trace metal ions, such as zinc, copper, and cadmium (Dallinger, 1994). These proteins are essential for shielding organisms and cells against harmful stress caused by metals. These proteins possess an identifier designating them as pollution markers. One of the most important aspects influencing the mobility of heavy elements in soil or FA and the absorption of metals by soil organisms, including isopods, amphipods, and earthworms, is the fractionation of heavy elements among the components (Becquer et al., 2005; Gál et al., 2008). A significant portion of the metals found in an earthworm's body are found in its posterior alimentary canal (PAC), which is where the worms tend to deposit a significant amount of cadmium (Cd), lead (Pb), zinc (Zn), and calcium (Ca). *Aporrectodea caliginosa* has a different calcium distribution mechanism than *Lumbricus rubellus* and *Dendrodrilus rubidus*. The alimentary canal in both of these two species had the greatest calcium concentrations, accounting for around 70% of the overall calcium weight in this particular tissue component (Ireland, 1975; Morgan & Morgan, 1990).

7.12 CONCLUSION

Thermal power plants emit enormous amounts of FA, which is one of the principal causes of environmental pollution. The quantity won't be reduced soon due to the rise in power demand. In the last few years, FA treatment has become more complex. Under FA management, researchers are investigating how to convert FA into appropriate productivity through the implementation of an efficient vermicomposting process. The addition of NPK levels to substrates rose as a result of the earthworms' ingestion of FA, which reduced the heavy metals in it. The changes in physicochemical properties, the capacity of earthworms to survive, and the transformation of FA's heavy metal concentration into organic substances that are sustainable are all covered in this book chapter. The result of recycling FA into a resource for high-yield organic fertilizers may tackle important environmental protection concerns, enhance crop yields, boost economic growth, and improve soil fertility. This chapter explains how frequently FA could be used in agriculture to ensure safe disposal and effective management.

REFERENCES

Adriano, D. C., Page, A. L., Elseewi, A. A., Chang, A. C., & Straughan, I. R. (1980). Utilization and disposal of fly ash and other coal residues in terrestrial ecosystems: A review. *Journal of Environmental Quality*, *9*(3), 333–344.

Ahmad, K. A., Ahmad, E., Mesfer, M. K. A., & Nigam, K. (2023). Bio-coal and bio-coke production from agro residues. *Chemical Engineering Journal*, *473*, 145340.

Aitken, R. L., Campbell, D. J., & Bell, L. C. (1984). Properties of Australian fly ashes relevant to their agronomic utilization. *Soil Research*, *22*(4), 443.

Ananthakrishnasamy, S., Sarojini, S., Gunasekaran, G., & Manimegala, G. (2009). Fly ash – A lignite waste management through vermicomposting by Indigenous earthworms *Lampito mauritii*. *American-Eurasian Journal of Agricultural & Environmental Sciences*, *5*(6),720–724.

Arumugam, K., Ganesan, S., Muthunarayanan, V., Vivek, S., Sugumar, S., & Munusamy, V. (2014). Potentiality of *Eisenia fetida* to degrade disposable paper cups—An eco-friendly solution to solid waste pollution. *Environmental Science and Pollution Research*, *22*(4), 2868–2876.

Barbosa, R., Lapa, N., Dias, D., & Mendes, B. (2013). Concretes containing biomass ashes: Mechanical, chemical, and ecotoxic performances. *Construction & Building Materials*, *48*, 457–463.

Basu, M., Pande, M., Bhadoria, P. B. S., & Mahapatra, S. K. (2009). Potential fly-ash utilization in agriculture: A global review. *Progress in Natural Science: Materials International*, *19*(10), 1173–1186.

Becquer, T., Dai, J., Quantin, C., & Lavelle, P. (2005). Sources of bioavailable trace metals for earthworms from a Zn-, Pb- and Cd-contaminated soil. *Soil Biology & Biochemistry*, *37*(8), 1564–1568.

Berra, M., Mangialardi, T., & Paolini, A. E. (2015). Reuse of woody biomass fly ash in cement-based materials. *Construction & Building Materials*, *76*, 286–296.

Bhattacharya, S., & Chattopadhyay, G. N. (2006). Effect of vermicomposting on the transformation of some trace elements in fly ash. *Nutrient Cycling in Agroecosystems*, *75*(1–3), 223–231.

Bhattacharya, S., Iftikar, W., Sahariah, B., & Chattopadhyay, G. N. (2012). Vermicomposting converts fly ash to enrich soil fertility and sustain crop growth in red and lateritic soils. *Resources, Conservation and Recycling*, *65*, 100–106.

Chang, A. C., Lund, L. J., Page, A. L., & Warneke, J. E. (1977). Physical properties of fly ash-amended soils. *Journal of Environmental Quality*, *6*(3), 267–270.

Chattopadhyay, G. N., & Bhattacharya, S. S. (2010). *Use of coal ash in agriculture*. Bull.1. Coal Ash Institute of India, Kolkata.

Chattopadhyay, G. N., Roy, G., & Iftikar, W. (2017). Use of vermicomposting biotechnology for recycling fly ash as a source of plant nutrients. *Linnaeus Eco-Tech*, 517–529.

Cooke, A. S., Greig-Smith, P. W., & Jones, S. A. (1992). Consequences for vertebrate wildlife of toxic residues in earthworm prey. *Ecotoxicology of Earthworms*, 109–115.

Dai, J., Sokhansanj, S., Grace, J. R., Bi, X., Lim, C. J., & Melin, S. (2008). Overview and some issues related to co-firing biomass and coal. *Canadian Journal of Chemical Engineering*, *86*(3), 367–386.

Dallinger, R. (1994). Invertebrate organisms as biological indicators of heavy metal pollution. *Applied Biochemistry and Biotechnology*, *48*(1), 27–31.

Das, D., Bhattacharyya, P., Ghosh, B. N., & Banik, P. (2012). Effect of vermicomposting on calcium, sulphur and some heavy metal content of different biodegradable organic wastes under liming and microbial inoculation. *Journal of Environmental Science and Health, Part B, Pesticides, Food Contaminants, and Agricultural Wastes*, *47*(3), 205–211.

Dassler, H. C., & Bortiz, S. (1988). *Air pollution and its influence on vegetation causes/effects/prophylaxis and therapy*. Dr. W Junk Publishers, Dordrecht, pp. 169–188.

Davison, R. L., Natusch, D. F. S., Wallace, J. R., & Evans, C. A. (1974). Trace elements in fly ash. Dependence of concentration on particle size. *Environmental Science & Technology*, *8*(13), 1107–1113.

Dharani, S., Manimegala, G., Gunasekaran, G., Ananthakrishnaswamy, S., & Sarojini, S. (2010). Vermicomposting of flyash: A potential hazardous waste converted into nutritive compost by earthworm (*Eudrilus eugeniae*). *International Journal of Global Environmental Issues*, *10*(3/4), 384.

Dwivedi, A. K., & Jain, M. K. (2014). Fly ash – Waste management and overview: A Review. *Recent Research in Science and Technology*, *6*(1), 30–35.

Edwards, C. A., & Bater, J. (1992). The use of earthworms in environmental management. *Soil Biology & Biochemistry*, *24*(12), 1683–1689.

Fatima, N., & Singh, K. (2023a). Accumulation of heavy metals in soil: Sources, toxicity, health impacts, and remediation by earthworms. *European Journal of Biological Research*, *13*(3), 129–143.

Fatima, N., & Singh, K. (2023b). Accumulation of heavy metals from the combination of different biological wastes by earthworm *Lampito mauritii* Kinberg. *International Journal of Zoological Investigations*, *9*(1), 257–271.

Feliciano-Bruzual, C. (2014). Charcoal injection in blast furnaces (Bio-PCI): CO2 reduction potential and economic prospects. *Journal of Materials Research and Technology*, *3*(3), 233–243.

Furr, A. K., Parkinson, T., Hinrichs, R., Van Campen, D. R., Bache, C., Gutenmann, W. H., St John, L. E., Pakkala, I. S., & Lisk, D. J. (1977). National survey of elements and radioactivity in fly ashes. Absorption of elements by cabbage grown in fly ash-soil mixtures. *Environmental Science & Technology*, *11*(13), 1194–1201.

Gál, J., Markiewicz-Patkowska, J., Hursthouse, A., & Tatner, P. (2008). Metal uptake by woodlice in urban soils. *Ecotoxicology and Environmental Safety*, *69*(1), 139–149.

González, R., & González-Chávez, M. D. (2006). Metal accumulation in wild plants surrounding mining wastes. *Environmental Pollution*, *144*(1), 84–92.

Gupta, S., Tewari, A., Srivastavá, R. C., Murthy, R., & Chandra, S. V. (2005). Potential of *Eisenia foetida* for sustainable and efficient vermicomposting of fly ash. *Water, Air, & Soil Pollution*, *163*(1–4), 293–302.

Haimi, J., & Huhta, V. (1990). Effect of earthworms on decomposition processes in raw humus forest soil: A microcosm study. *Biology and Fertility of Soils*, *10*(3), 178–183.

Iftikar, W., Roy, G., & Chattopadhyay, G. N. (2020). Use of vermicomposted fly ash as an important component for integrated nutrient management of potato. *Open Access Journal of Waste Management & Xenobiotics*, *3*(3), 1–7.

Ireland, M. (1975). Distribution of lead, zinc and calcium in *Dendrobaena rubida* (oligochaeta) living in soil contaminated by base metal mining in wales. *Comparative Biochemistry and Physiology. B. Comparative Biochemistry*, *52*(4), 551–555.

Karwal, M., & Kaushik, A. (2020). Co-composting and vermicomposting of coal fly ash with press mud: Changes in nutrients, micro-nutrients and enzyme activities. *Environmental Technology and Innovation*, *18*, 100708.

Kowapradit, S. (2005). Study of bottom ash in increase the growth of Cassava. *Reports of Agricultural Extension Department*, *3*(1), 129–136.

Kumar, V., Singh, G., & Rai, R. (2005). *Fly ash: A material for another green revolution*. Fly Ash India, New Delhi.

Latifi, N., Marto, A., Rashid, A. S. A., & Yii, J. L. J. (2015). Strength and physico-chemical characteristics of Fly Ash–Bottom Ash mixture. *Arabian Journal for Science and Engineering*, *40*(9), 2447–2455.

Lee, K. E. (1985). *Earthworms: their ecology and relationships with soils and land use* (p. 411). Sydney; Orlando : Academic Press.

Mandova, H., Leduc, S., Wang, C., Wetterlund, E., Patrizio, P., Gale, W., & Kraxner, F. (2018). Possibilities for CO2 emission reduction using biomass in European integrated steel plants. *Biomass & Bioenergy, 115*, 231–243.

Manivannan, S., Anbalagan, M., & Prakasam, B. A. (2012). Influence of earthworm *Lampitto mauritii* (Kinberg) and organic additives for efficient vermicomposting of fly ash. *American-Eurasian Journal of Scientific Research, 7*(2), 58–63.

Mao-Fa, F., Wong, J. W. C., K, K., & Wong, M. H. (1999). Co-composting of sewage sludge and coal fly ash: nutrient transformations. *Bioresource Technology, 67*(1), 19–24.

Mikavica, G. (2023). Composition of fly ash in solid waste management and their disposal methods. *International Journal of Waste Resources, 13*, 524.

Mittra, B. N., Ghosh, B. C., Karmakar, S., George, J., Rautaray, S. K., & Basu, S. (2000). Utilization of industrial and solid wastes in augmenting crop production and soil productivity. *Waste Recycling and Resource Management in the Developing World*. pp. 185Á/195. University of Kalyani and International Ecological Engineering Society, Switzerland.

Mohamaed, A., & KMuduli, K. (2021). Utilization of biomass flyash in cementitious application. *Journal of Physics. Conference Series, 2129*(1), 012005.

Mohapatra, R., & Rao, J. (2001). Some aspects of characterisation, utilisation and environmental effects of fly ash. *Journal of Chemical Technology & Biotechnology, 76*, 9–26.

Molino, A., Larocca, V., Chianese, S., & Musmarra, D. (2018). Biofuels production by biomass gasification: A review. *Energies, 11*(4), 811.

Morgan, A., & Morris, B. (1982). The accumulation and intracellular compartmentation of cadmium, lead, zinc and calcium in two earthworm species (*Dendrobaena rubida* and *Lumbricus rubellus*) living in highly contaminated soil. *Histochemistry (Berlin), 75*(2), 269–285.

Morgan, J. E., & Morgan, A. (1990). The distribution of cadmium, copper, lead, zinc and calcium in the tissues of the earthworm *Lumbricus rubellus* sampled from one uncontaminated and four polluted soils. *Oecologia, 84*(4), 559–566.

Morgan, J. E., & Morgan, A. (1992). Heavy metal concentrations in the tissues, ingesta and faeces of ecophysiologically different earthworm species. *Soil Biology & Biochemistry, 24*(12), 1691–1697.

Mousa, E., Wang, C., Riesbeck, J., & Larsson, M. (2016). Biomass applications in iron and steel industry: An overview of challenges and opportunities. *Renewable & Sustainable Energy Reviews, 65*, 1247–1266.

Muir, M., Yunusa, I., Burchett, M., Lawrie, R., Chan, K. Y., & Manoharan, V. (2007). Short-term responses of two contrasting species of earthworms in an agricultural soil amended with coal fly ash. *Soil Biology & Biochemistry, 39*(5), 987–992.

Muniraj, S., Ravi, L., Ganesh, H., Sethuraman, M., & Muthunarayanan, V. (2020). Effective utilization of fly ash for vermicompost production by employing *Eisenia fetida. Nature Environment and Pollution Technology, 19*(02), 687–694.

Neelima, M. R., Khandual, S., Tripathy, A., & Sahu, R. K. (1995). Fly ash management in the state of Orissa. In *Conference Proceeding, April*. RRL, Bhubhaneshwar, pp.76–89.

Niyazi, R., & Chaurasia, S. (2014). Vermistabilization of fly ash amended with pressmud by employing *Eisenia foetida. International Journal of Pharmaceutical, Chemical, and Biological Sciences, 4*(1), 85–95.

Norgate, T., Haque, N., Somerville, M., & Jahanshahi, S. (2012). Biomass as a source of renewable carbon for iron and steelmaking. *ISIJ International, 52*(8), 1472–1481.

Pandey, V. C., Abhilash, P., & Singh, N. (2009a). The Indian perspective of utilizing fly ash in phytoremediation, phytomanagement and biomass production. *Journal of Environmental Management, 90*(10), 2943–2958.

Pandey, V. C., Abhilash, P., Upadhyay, R. N., & Tewari, D. D. (2009b). Application of fly ash on the growth performance and translocation of toxic heavy metals within *Cajanus cajan* L.: Implication for safe utilization of fly ash for agricultural production. *Journal of Hazardous Materials*, *166*(1), 255–259.

Pandey, V. C., & Singh, N. (2010). Impact of fly ash incorporation in soil systems. *Agriculture, Ecosystems & Environment*, *136*(1–2), 16–27.

Pattnaik, S., & Reddy, M. V. (2010). Assessment of municipal solid waste management in Puducherry (Pondicherry), India. *Resources, Conservation and Recycling*, *54*(8), 512–520.

Pawlak-Kruczek, H., Arora, A., Gupta, A., Saeed, M. A., Niedzwiecki, L., Andrews, G., Phylaktou, H., Gibbs, B., Newlaczyl, A., & Livesey, P. M. (2020). Biocoal – Quality control and assurance. *Biomass & Bioenergy*, *135*, 105509.

Quann, R. J., Neville, M., Janghorbani, M., Mims, C. A., & Sarofim, A. F. (1982). Mineral matter and trace-element vaporization in a laboratory-pulverized coal combustion system. *Environmental Science & Technology*, *16*(11), 776–781.

Römbke, J., Jänsch, S., & Didden, W. (2005). The use of earthworms in ecological soil classification and assessment concepts. *Ecotoxicology and Environmental Safety*, *62*(2), 249–265.

Roy, W. R., & Griffin, R. A. (1984). Illinois Basin coal fly ashes. 2. Equilibria relationships and qualitative modeling of ash-water reactions. *Environmental Science & Technology*, *18*(10), 739–742.

Satapathy, S., Nayak, Y., & Satapathy, K. B. (2021). Soil Amendment with Flyash through Vermicomposting by the Action of Epigeic Earthworm. *Annals of the Romanian Society for Cell Biology*, *25*(2), 1990–1999.

Senapati, M. R. (2011). Fly ash from thermal power plants–waste management and overview. *Current Science*, *100*(12), 1791–1794.

Senappa, S. N. (2012). Larm as a necessity during composting and/or vermicomposting of kitchen refuses. *Universal Journal of Environmental Research and Technology*, *2*(1), 97–100.

Sett, R. (2017). Flyash: Characteristics, problems and possible utilization. *Advances in Applied Science Research*, *8*(3), 32–50.

Sikka, R., & Kansal, B. (1994). Characterization of thermal power-plant fly ash for agronomic purposes and to identify pollution hazards. *Bioresource Technology*, *50*(3), 269–273.

Singh, A., & Singh, K. (2023). Estimation of physico-chemical parameters of different combinations of buffalo dung with cane sugar bagasse by vermic activity of earthworm *Lampito mauritii*. *International Journal of Zoology and Applied Biosciences*, *8*(2), 40–48.

Singh, K., & Fatima, N. (2019). Role of earthworms in heavy metal accumulation. *Biospectra*, *14*(2), 21–36.

Singh, K., & Fatima, N. (2022). The efficiency of earthworms as a biomarker for environmental pollution. *International Journal of Biological Innovations*, *04*(01), 104–112.

Suopajärvi, H., Umeki, K., Mousa, E., Hedayati, A., Romar, H., Kemppainen, A., Wang, C., Phounglamcheik, A., Tuomikoski, S., Norberg, N., Andefors, A., Öhman, M., Lassi, U., & Fabritius, T. (2018). Use of biomass in integrated steelmaking – Status quo, future needs and comparison to other low-CO_2 steel production technologies. *Applied Energy*, *213*, 384–407.

Suthar, S., Singh, S., & Dhawan, S. (2008). Earthworms as bioindicator of metals (Zn, Fe, Mn, Cu, Pb and Cd) in soils: Is metal bioaccumulation affected by their ecological category? *Ecological Engineering*, *32*(2), 99–107.

Talbot, R. W., Anderson, M. A., & Andrén, A. (1978). Qualitative model of heterogeneous equilibriums in a fly ash pond. *Environmental Science & Technology*, *12*(9), 1056–1062.

TERI. (1998). *Reclaiming ash pond by means of mycorrhizal-organobiofertilizer: Quarterly reports submitted to fly ash mission*. TIFAC, DST, Tata Energy Research Institute, New Delhi.

Thomas, G. V., Prabhu, S. R., Reeny, M. Z., & Bopaiah, B. M. (1998). Evaluation of lignocellulosic biomass from coconut palm as substrate for cultivation of Pleurotus sajor-caju (Fr.) Singer. *World Journal of Microbiology and Biotechnology*, *14*, 879–882.

Usmani, Z., & Kumar, V. (2016). Management of fly ash through Vermicomposting: A rational approach. *Environmental Quality Management*, *25*(3), 53–66.

Usmani, Z., & Kumar, V. (2017). The implications of fly ash remediation through vermicomposting: A review. *Nature Environment and Pollution Technology*, *16*(2), 363–374.

Usmani, Z., Kumar, V., & Mritunjay, S. K. (2017). Vermicomposting of coal fly ash using epigeic and epi-endogeic earthworm species: nutrient dynamics and metal remediation. *RSC Advances*, *7*(9), 4876–4890.

Venkatesh, R. M., & Eevera, T. (2007). Mass reduction and recovery of nutrients through vermicomposting of fly ash. *Applied Ecology and Environmental Research*, *6*(1), 77–84.

Wang, Q., Zhou, D., Cang, L., Li, L., & Zhu, H. (2009). Indication of soil heavy metal pollution with earthworms and soil microbial biomass carbon in the vicinity of an abandoned copper mine in Eastern Nanjing, China. *European Journal of Soil Biology*, *45*(3), 229–234.

Weinstein, L. H., Osmeloski, J. F., Rutzke, M., Beers, A., Mc Cahan, J. B., Bache, C., & Lisk, D. J. (1989). Elemental analysis of grasses and legumes growing on soil covering coal fly ash landfill sites. *Journal of Food Safety*, *9*(4), 291–300.

Yao, Z., Ji, X., Sarker, P. K., Tang, J. H., Ge, L., Xia, M., & Xi, Y. (2015). A comprehensive review on the applications of coal fly ash. *Earth-Science Reviews*, *141*, 105–121.

Yousuf, A., Manzoor, S. O., Youssouf, M., Malik, Z. A., & Khawaja, K. S. (2020). Fly ash: Production and utilization in India – An overview. *Journal of Materials and Environmental Science*, *11*(6), 911–921.

Yu, C., Thy, P., Wang, L., Anderson, S., VanderGheynst, J., Upadhyaya, S., & Jenkins, B. (2014). Influence of leaching pretreatment on fuel properties of biomass. *Fuel Processing Technology*, *128*, 43–53.

8 Fly Ash as a Low-Cost Adsorbent for the Removal of Industrial Dyes from Aqueous Media

Angela Martins and Nelson Nunes

8.1 INTRODUCTION

Water is among the planet's most valuable resources, yet only around 1.0% is accessible for human use. However, freshwater sources are increasingly contaminated by a mix of organic and inorganic pollutants (Adeleye et al., 2016; Aigbe et al., 2021). Dye waste is a particularly challenging group of contaminants because these substances are toxic to freshwater and marine life due to their oxidation and light stability and, as such, are not easily biodegradable. They are also noticeable to the naked eye and detract from the aesthetic quality of the environment. Approximately 7×10^6 tons are produced annually, and more than 10,000 commercial dyes are available today (Dutta et al., 2021). The major industries that generate dye wastes are textile, plastics, leather, cosmetics, pharmaceuticals, food preparation, and paints. However, it is the textile industry that generates the highest volume of wastewater pollutants (D'Antoni et al., 2017). The multi-stage wet processing involved in turning fibers into fabrics generates large quantities of colored wastewater, particularly during scouring, bleaching, and dyeing processes.

Dye molecules are composed of a chromophore component that is responsible for the color. Approximately 12 distinct categories of chromogenic groups exist. Among these, the azo class is the most prevalent, followed by anthraquinone. Other less important chromogenic groups include indigoid, nitroso, nitro, and triarylmethane dyes, among others (Ali, 2010). The classification of dyes is shown in Table 8.1, according to their chemical structure.

Azo dyes cover 60%–70% of all dyes used in the food and textile industries (Benkhaya et al., 2020). Azo bonds, chemically denoted as – R-N = N-R', R and R', are responsible for imparting color to a compound, and they are typically aryl or substituted aryl groups. Certain azo compounds, including Direct Red 2, Direct Blue 15, and Congo Red, carry a substantial threat to human health due to their toxic

DOI: 10.1201/9781003498124-8

TABLE 8.1
Categorization of dyes based on the chemical structure

Class	Chromophore	Example
Nitro dyes	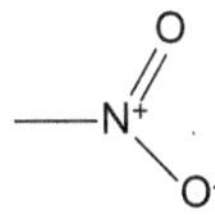	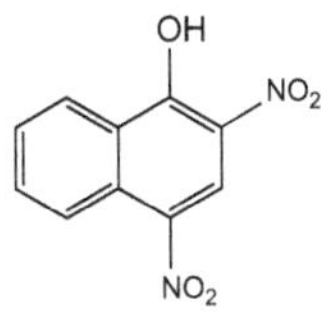 Acid Yellow 24
Nitroso dyes	$O = N-$	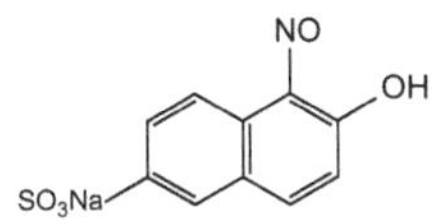 Acid Green 1
Azo dyes	$-N = N-$	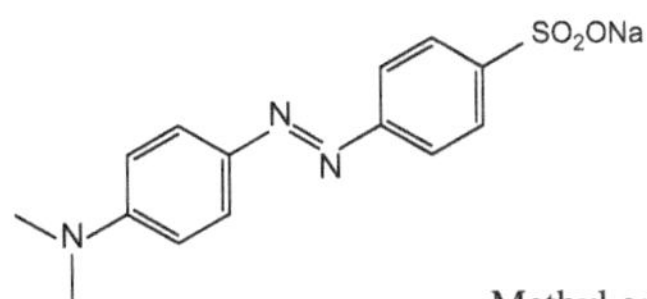 Methyl orange
Triarylmethane dyes	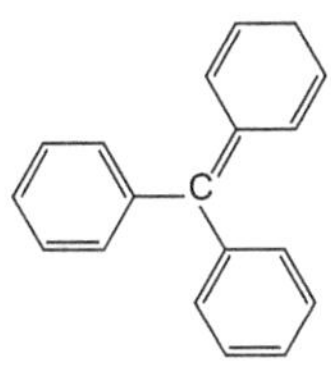	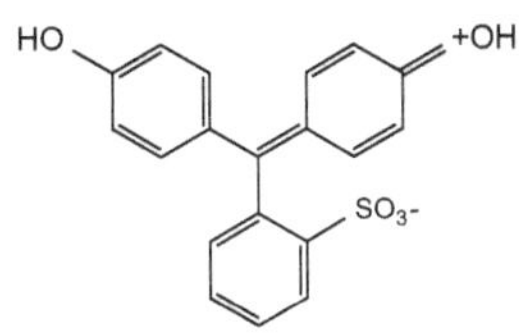 Phenol red
Phthalein dyes	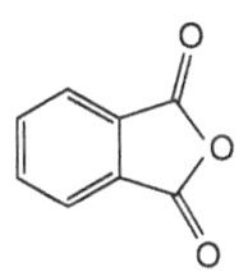	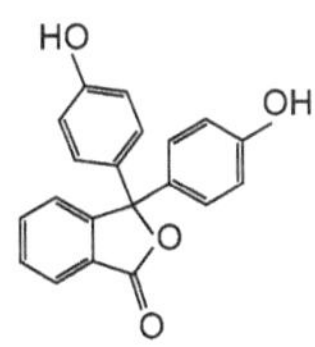 Phenolphthalein
Indigo dyes	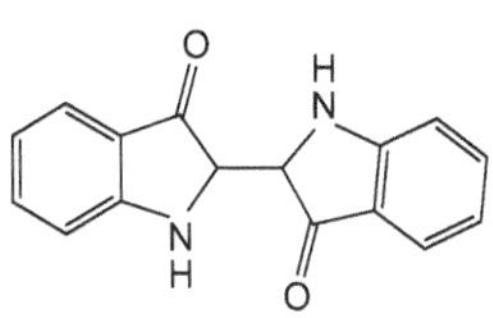	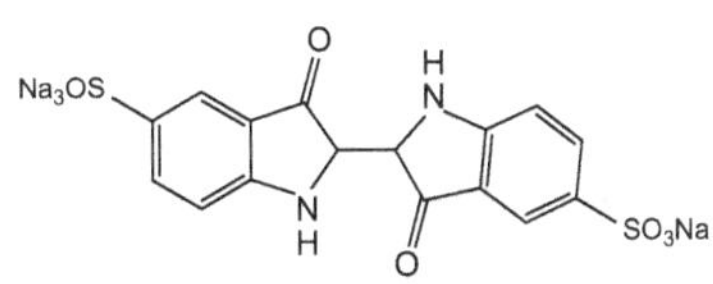 Indigo Carmine
Anthraquinone dyes	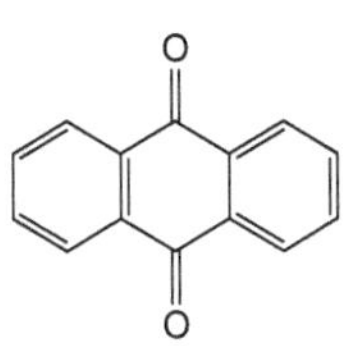	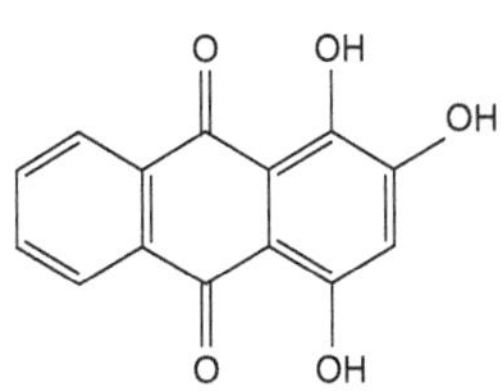 Purpurin

nature, ability to cause genetic mutations, and potential for inducing cancer (Wekoye et al., 2020). Anthraquinone dyes occupy the second position in terms of importance within the dye industry, with fewer commercial applications compared to azo dyes due to their higher production costs. Unlike azo dyes, which are fully synthetic, the anthraquinone chromogen is responsible for all major natural red dyes. The chemical structure of anthraquinone is based on anthracene, which has three interconnected benzene rings and features two carbonyl functionalities on its central ring. Although colorless, this structure produces color upon substitution of the aromatic rings, which becomes deeper as the basicity of the substituent increases (Routoula & Patwardhan, 2020). There are several groups of dyes, including classifications as anionic, cationic, and non-ionic. Anionic dyes include acid, reactive, direct, and azo dyes, while cationic dyes are mostly alkaline (Dawood & Sen, 2014).

8.2 DYE REMOVAL THROUGH PHYSICAL ADSORPTION

The methods of color removal can be classified into three main categories: biological, chemical, and physical (Bal & Thakur, 2021; D'Antoni et al., 2017), which are briefly presented in Table 8.2.

One of the most common approaches for eliminating dyes from wastewaters involves physical adsorption or simply adsorption. The mechanism of adsorption involves the attachment of dye molecules onto an adsorbent's surface through various intermolecular forces such as Van der Waals forces, hydrophobic interactions, hydrogen bonds, and dipole-dipole interactions. The degree of adsorption phenomena is influenced by the characteristics of the adsorbed molecules, including their molecular weight, structure, size, polarity, and concentration. The particle size, surface area, and surface charge of the adsorbent materials also significantly impact the adsorption process (Dawood & Sen, 2014). Physical adsorption has the advantage of removing high concentrations of dyes without byproducts, using small amounts of adsorbent materials. In addition, regeneration of the adsorbent materials is usually feasible, with the possibility of multiple reuses. However, this method is not effective for all dyes and the production of the required adsorbent materials can be costly.

TABLE 8.2
Methods for wastewater color removal

Strategies for de-colorization

Physical	Chemical	Biological
• Physical adsorption	• Ozone treatment	• Aerobic/anaerobic degradation
• Flocculation/ coagulation	• Oxidative degradation	• Fungi
• Membranes	• Fenton reactions	• Algae
• Irradiation	• Electrocoagulation	• Bioremediation
• Ion exchange	• Indirect oxidation	• Activated sludge tanks

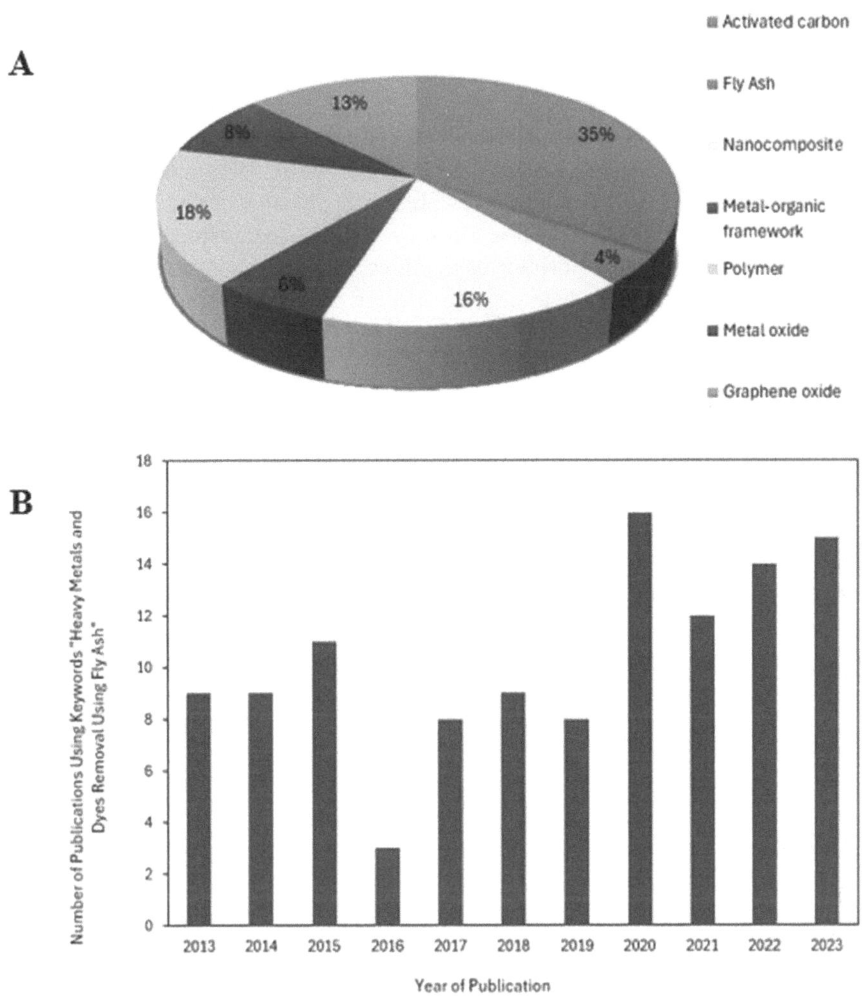

FIGURE 8.1 (A) Materials employed in the elimination of dyes from wastewaters. (B) Analysis of articles published in indexed journals over the past ten years.

A variety of adsorbent materials are available for tackling the elimination of dyes from wastewater. The pattern of studies concerning the application of various adsorbents for dye elimination is displayed in Figure 8.1A. The data, expressed as a percentage of total published works, were compiled from the "Web of Science" database. As can be observed, the most common materials are polymers and activated carbons.

Recently materials like orange peel, rice husk, lemongrass leaf, and peanut shells (among others) have gained recognition as valuable carbon sources to derive biochars and activated carbons, that can be used as effective adsorbent materials in dye remediation (Khalfaoui et al., 2022; Khan et al., 2018; Nunes et al., 2023). However, despite their availability, the number of steps required, and the low yields obtained in converting these resources into effective adsorbents make them expensive and may prevent their large-scale use.

Figure 8.1A illustrates fly ash as a material that has received relatively less attention in research. An analysis of articles published in indexed journals over the past ten years, obtained from "Web of Science" (Figure 8.1B), reveals a noticeable rise in the number of publications during the last four years. This upward trend suggests a burgeoning research interest and increased application of fly ash as a valuable industrial byproduct. Since fly ash is a costless and abundant waste, it is important to investigate its potential adsorption capabilities. The main concern in using fly ash is the possibility of releasing heavy metals that are part of its composition, but fortunately, the concentration of potentially released heavy metals has been always below the legal standards (Janoš et al., 2003).

8.3 FLY ASH AS AN AFFORDABLE ADSORBENT MATERIAL FOR ELIMINATING DYES

Coal power generation leads to the creation of fly ash, making it a significant waste product within the industry. Coal combustion for electricity generation results in the generation of approximately 500 million tons of fly ash on a yearly basis, but only an average of 25%–35% is reused (Mathapati et al., 2021; Xu & Shi, 2018). This is due to its complex composition, containing anthropogenic and toxic components, which makes this material difficult to manage. Landfills serve as the primary destination for most fly ash, given the high economic burden of employing agricultural land for its disposal. On the other hand, it can cause groundwater contamination through leakage and leaching. Given these circumstances, it is essential to pursue research on maximizing the value and utility of fly ash, thereby transforming an industrial waste into a raw material for multiple uses in different fields, as shown schematically in Figure 8.2.

In recent times, biomass-based coal substitutes like bio-coal briquettes derived from agricultural waste have shown remarkable promise in comparison to traditional coal sources (Rawat & Kumar, 2022). Preliminary experiments have shown that fly ash generated from bio-coal mimics the properties of conventional material (Naik et al., 2023), thereby opening a wide range of applications for these materials.

The composition of fly ash includes finely ground, predominantly spherical particles, which can manifest as either solid or hollow structures. The carbonaceous component is composed of irregular particles. Their properties depend on the type of coal they are made from, and the combustion techniques used. In general, fly ash

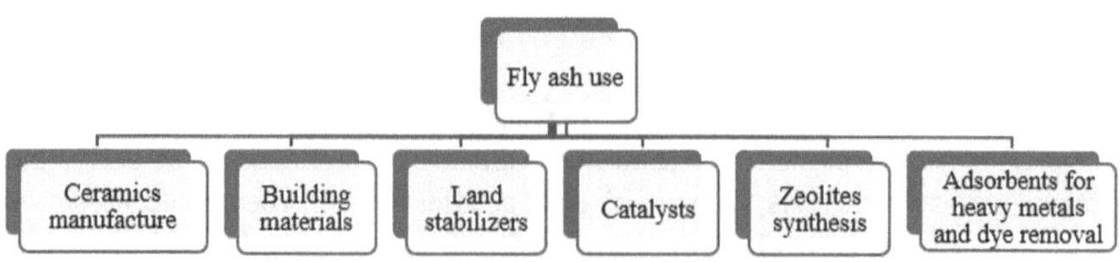

FIGURE 8.2 Fly ash utilization in several fields.

is rich in aluminosilicate compounds SiO_2 and Al_2O_3, making it suitable to produce zeolites, which can be further used as catalysts and adsorbents (Querol et al., 2002). Other oxides such as CaO, MgO, Na_2O, and TiO_2 are also present. Furthermore, it holds various elements like phosphorus (P), potassium (K), magnesium (Mg), zinc (Zn), iron (Fe), manganese (Mn), and others.

Given the analogous chemical makeup of clay and fly ash, with a good lime binding capacity, it can be used to produce cement, clay bricks, and other building materials (Xu & Shi, 2018). Because of their reduced weight per unit volume, substantial water retention ability, and suitable pH levels, these materials are well-suited for agricultural purposes as fertilizers and soil enhancers (Jala & Goyal, 2006). They can be applied in the removal of atmospheric pollutants, SO_2 (Izquierdo & Rubio, 2008), NO_x (Lu & Do, 1991), and even an important capacity for carbon capture and storage (Wee, 2013) has been reported. In wastewater treatment, the range of applications is much broader and includes adsorption of heavy metals like mercury (Kapoor & Viraraghavan, 1992), lead (Zhu et al., 2018), nickel (Banerjee et al., 2003), cadmium (Huang et al., 2020), to name a few. Other chemical species like fluorides (Chatuverdi et al., 1990), or phenol (Viraraghavam & Alfaro, 1998) have also been reported. However, the removal of colored matter from wastewaters is the most explored application of fly ash as adsorbents.

8.3.1 Factors Influencing the Dye Adsorption on to Fly Ash

A wide body of research on dye adsorption points to the textural and chemical attributes of fly ash as critical determinants in effective dye removal (Astuti et al., 2019) as well as the wastewater properties like pH and temperature (Kisku et al., 2015).

The pH value of the effluent plays a vital role in the efficiency of adsorption, as it influences the chemical environment that governs the adsorption process between pollutants and the adsorbent. At different pH, the ionization degree of adsorbate molecules shifts along with modifications to the materials' surface characteristics. The interaction between dyes and fly ash can be assessed by determining the pH value at which the overall charged surface of the particles is neutral (point of zero charge, pH_{PZC}). The relationship between the solution pH and the pH_{PZC} of an absorptive material's surface can forecast the nature of their adsorption interaction. When the solution pH surpasses the pH_{PZC} value, the material acquires a negative charge, fostering the attachment of cationic dye molecules. In contrast, if the solution's pH falls below the pH_{PZC}, the adsorbent surface assumes a positive charge, drawing anionic dye species toward it (Gregg & Sing, 1982). For example, the adsorption of Mordant Orange 1, a dye characterized by anionic properties, onto fly ash with pH_{PZC} equal to 11. This process is optimally facilitated under acidic conditions, and the adsorption efficiency substantially decreases as pH rises. In contrast, Methylene Blue (MB), a cationic dye, demonstrates enhanced adsorption under basic conditions (Rosa et al., 2021).

The impact of adsorbent dosage, a crucial factor in determining a material's adsorption capacity, was investigated for Reactive Blue 25 in the presence of a fly ash sample from India (Deshannavar et al., 2017). In this study, the fly ash amount changed from 1 to 5 g/l, while maintaining the other operation parameters constant. By increasing the fly ash dosage from 1 to 2.5 g/l, the removal efficiency increased from 81.65% to 88.56%, while further increase in fly ash amount did not affect the removal efficiency.

TABLE 8.3
Temperature effect of several dyes' adsorption by fly ash

Dye	Temp. range (K)	Process type	ΔH^o (kJ/mol)	Reference
Methylene Blue	298–308	Exothermic	−13.44	Karaca et al. (2018)
Rhodamine B	293–313	Endothermic	2.81	Rosa et al. (2021)
Acid Red 1	303–333	Endothermic	9.83	Hsu (2008)
Mordant Orange 1	293–313	Exothermic	−15.07	Rosa et al. (2021)
Methyl Orange	293–313	Endothermic	13.36	Potgieter et al. (2021)
Congo Red	298–323	Endothermic	5.32	Harja et al. (2022)
Acid Black 1	293–313	Endothermic	35.70	Sun et al. (2010)
Reactive Blue 171	293–313	Endothermic	13.70	Sun et al. (2010)

Temperature is another important parameter to be considered in dyeing wastewater treatment since several industrial dying processes involve elevated temperatures. The impact of temperature on the adsorptive capabilities of fly ash must be considered. Temperature can affect the adsorption process in two ways: (i) endothermic processes ($\Delta H^o > 0$), as the temperature rises, there is an enhancement in the quantity of adsorbed dye by the adsorbent due to enhanced mobility, which facilitates surface interaction and (ii) exothermic processes ($\Delta H^o < 0$), the rise in temperature results in a reduction of the adsorption capacity because the binding affinity between the adsorbate and adsorption sites lessens as temperatures increase (Yagub et al., 2014). Several dye-fly ash adsorption cases, depicted in Table 8.3, highlight the role of temperature as a critical factor.

8.3.2 Kinetics and Adsorption Models

Understanding the dynamics of dye adsorption on adsorbents plays a crucial role in determining optimal operating conditions. Identifying the most suitable kinetic model for an adsorption system enables the determination of the rate constant – a vital factor in setting the appropriate contact duration between the adsorbent and adsorbate, leading to efficient dye elimination. The kinetic models generally applied to fly ash adsorption are: pseudo-first order (Özacar & Şengil, 2006) and pseudo-second order (Ho & McKay, 1999), Eqs. 8.1 and 8.2, respectively.

$$\ln\left(q_e - q_t\right) = \ln\ q_c - k_{pl} \times t \tag{8.1}$$

$$\frac{t}{q_t} = \frac{1}{k_{p2} \times q_e^2} + \frac{1}{q_e} \times t \tag{8.2}$$

In this context, q_t and q_e denote the quantities of the adsorbed solute (mg/g) at given time t and equilibrium, respectively. The pseudo-first-order rate constant (min^{-1}) is represented by k_{p1}, while the pseudo-second-order rate constant (g/(mg min)) is denoted by k_{p2}. In most investigations, it has been observed that the pseudo-second-order kinetic model outperforms other models in characterizing the adsorption of dyes onto fly ash. As a case in point, the pseudo-first and pseudo-second order kinetic models were employed to study the adsorption behavior of Mordant Orange 1 dye onto fly ash (Rosa et al., 2021). The regression analyses yielded correlation coefficients of 0.0628 and 0.9997 for the pseudo-first and pseudo-second-order models, respectively, indicating that the latter provides a superior fit to the experimental data, as evidenced by previous research on dye adsorption onto fly ash, like Acid Red 1 (Hsu, 2008), Reactive Blue 25 (Deshannavar et al., 2017), or MB (Karaca et al., 2018).

Adsorption isotherms are essential to understand the relationship between the adsorbent and adsorbate, as they enable the quantification of the adsorption capacity of the adsorbent. In the context of fly ash, the amount of dye adsorbed at equilibrium can be represented using various adsorption models. Several isotherm models are available in the literature (Gregg & Sing, 1982) but the most widely used are the Freundlich and Langmuir models. The Langmuir model has proven effective in numerous liquid adsorption processes and assumes the formation of a monolayer. It can be described by Eq. 8.3.

$$q_e = \frac{q_m \times K_L \times C_e}{1 + K_L \times C_e} \tag{8.3}$$

In this context, q_m signifies the monolayer's adsorption capacity (mg/g), while K_L represents the Langmuir constant (L/mg) inherent to the interaction between the adsorbate and the adsorbent, these parameters are characteristic of each system. The Freundlich model considers an irregular adsorption surface comprising varying sites with dissimilar adsorption energies, it is described in Eq. 8.4.

$$q_e = K_F C_e^{1/n} \tag{8.4}$$

K_F symbolizes the Freundlich constant (mg$^{1-1/n}$ L$^{1/n}$/g), conveying the adsorption capacity, while $1/n$ reflects the adsorption intensity or surface heterogeneity, with values favorable for the range of 0 to 1.

As previously stated, employing isotherm models allows the quantification of the adsorption capacity for a given adsorbent-adsorbate system. However, it is sometimes problematic to predict the real sorption capacity of dyes, especially at high concentrations, due to the following reasons (Atun et al., 2019): (i) in concentrated dyestuff solutions, the dye molecules might exist as aggregates rather than monomeric forms capable of forming a monolayer on an adsorbent's surface, (ii) interactions between dye molecules, (iii) different adsorption mechanisms may occur for dilute and concentrated solutions, (iv) interactions that occur between adsorbed molecules and those free in solution. Accordingly, the use of more sophisticated models is required. One instance is the Liu isotherm model, which integrates elements of both the Freundlich and Langmuir equations. This model, expressed in Eq. 8.5, involves three parameters that allow the mathematical flexibility to explain solute-solute interactions that are not considered on classic model isotherms.

$$q_e = q_{m1}\frac{\left(K_{g1}a_e\right)^{nL}}{1+\left(K_{g1}a_e\right)^{nL}} + \left(q_{m2} - q_{e,sp}\right)\frac{\left[K_{g2}\left(a_e - a_{e,sp}\right)\right]^{nL2}}{1+\left[K_{g2}\left(a_e - a_{e,sp}\right)\right]^{nL2}} \tag{8.5}$$

In the Liu model, $a_{e,sp}$ denotes the activity of the solute at equilibrium during the commencement of the second adsorption plateau, while $q_{e,sp}$ represents the corresponding sorbed quantity. K_g signifies the Liu sorption constant, and n_L indicates the Liu exponent, with distinct values for the first (1) and second (2) plateaus in the isotherm. The Liu isotherms become Langmuir equation when their exponents are unitary, this model describes L-shaped curves for values less than 1 and S-shaped curves when n_L exceeds 1.

8.3.3 Dye Removal with Raw and Modified Fly Ash. Assessment of Selected Case Studies

Table 8.4 lists data from various examples involving the adsorption of MB, a prevalently used cationic dye for imparting color to textiles like cotton, wood, and silk. The presence of this dye in the environment, even at trace levels, causes toxicological effects and extensive research efforts have been devoted toward eliminating MB from wastewater due to its widespread application as a water-soluble dye for various materials. This research includes a significant number of applications using fly ash as an adsorbent material (Dutta et al., 2021).

Employing untreated fly ash as a resource represents a strategy to capitalize on an industrial byproduct, aiming at cutting down processing expenses without prior treatment. The adsorption capacity of fly ash not only varies depending on coal source, combustion conditions but also the place where the fly ash sample is collected. An example is fly ash obtained from an electrostatic precipitator at a Western Australian power plant, fueled by sub-bituminous black coal, with a specific surface area (S_{BET}) of 15.6 m^2/g and a pH of 4.4. This sample exhibits an adsorption capacity for MB dye of 2.96 mg/g (Wang et al., 2005). In contrast, another fly ash specimen from a Portuguese power plant that utilizes bituminous coal from Colombia, the S_{BET} value of this sample is 13.2 m^2/g and its pH level is 11, and this sample is capable of retaining 4.0 mg/g of MB dye (Rosa et al., 2021). The potential of fly ash as an adsorbent can be significantly enhanced through various pre-treatment methods, as evidenced by the data presented in Table 8.4.

For example, a fly ash sample from a Vietnamese power plant was exposed to an alkaline treatment with solid NaOH at a weight ratio of 1.5:1, homogenized, and maintained at a temperature of 90°C for a duration of 6 hours. After cooling, the sample was grounded, mixed with distilled water using a volume/weight ratio of 10, stirred at a rate of 200 rotations per minute for three hours while maintaining a temperature of 90°C, and finally washed to neutral pH.

Employing this simple and cost-effective treatment method led to an increase in the specific surface area of fly ash from an initial value of 3.55 m^2/g to 66.39 m^2/g, enabling a maximum adsorption capacity for MB dye of 28.65 mg/g, which is substantially higher than any known value for raw fly ash (Dinh et al., 2021). A more sophisticated treatment was performed on a fly ash sample obtained in a Portuguese

TABLE 8.4
Adsorption capacity (qm) of MB dye quantified by Langmuir adsorption model at 20°C–25°C using raw or modified fly ash samples produced in different power stations

Power station location	Modification	q_m (mg/g)	Reference
Portugal	Raw	4.0	Rosa et al. (2021)
Turkey	Raw	0.12	Karaca et al. (2018)
Australia	Raw	2.96	Wang et al. (2005)
Vietnam	Alkaline thermal hydrolysis treatment	28.65	Dinh et al. (2021)
Portugal	Geopolymer monoliths from biomass fly ash	15.4	Novais et al. (2018)
Undetermined	Mechanochemical treatment	7.97	Li et al. (2018)
India	Granulated (prepared by extrusion)	3.62	Padmashri et al. (2016)

co-generation plant. The primary goal of this research was to explore the generation of geopolymeric monoliths. In pursuit of this aim, an approach was devised, encompassing the combination of fly ash with alkaline activators, sodium silicate, and a 10 M NaOH solution. Subsequently, H_2O_2 served as a blowing agent using a timeframe of 2 minutes. Following the completion of the mixing process, the resulting slurry underwent a curing phase of 24 hours at a temperature of 40°C and a relative humidity level of 65%. Post-demolding, the specimens were left to cure at ambient temperature for a duration of 28 days. This procedure allows to obtain the modified fly ash as monolithic bodies (22 mm × 3 mm cylindrical discs), with handling advantages over the typical powder form. The fly ash geopolymers show a significant increase of 5.4–15.4 mg/g, attributed to the fact that the porosity of the modified sample was doubled when compared to the original fly ash. Furthermore, regeneration and reuse were investigated and up to five cycles were achieved, without compromising the adsorption performance (Novais et al., 2018).

Although most of the published work deals with the adsorption of one isolated dye, as in the previous examples on MB, real industrial effluents usually contain more than one pollutant simultaneously.

To exemplify the potential of fly ash as an adsorbent for concurrently removing two distinct types of dyes – anthraquinone-based Acid Blue 127 (AB-127) and azo-based Acid Yellow 17 (AY-17) – separate experiments were conducted utilizing varying AY-17/AB-127 molar ratios of 0.1:1, 1:1, and 5:1 in conjunction with a Turkish fly ash sample (Atun et al., 2019). Two-step isotherms were evident in both individual and mixed systems, signifying that multilayer adsorption took place due to interactions between the dyes in highly concentrated solutions. At 25°C, the Liu isotherm model provided the optimal fit to the individual adsorption data sets of AB-127 and AY-17 dyes, resulting in adsorption capacities of 7.70×10^{-5} and 23×10^{-5} mol/g, respectively.

For binary systems, two-step isotherms were also found, and the Liu isotherm model was found to be a suitable fit for the experimental data, but in this case the occurrence of adsorption competition between the two dye molecules was observed in concentrated solutions. However, since the adsorption capacities are, in some cases, comparable to activated carbons, fly ash adsorbents can be considered as potential candidates for effective on-site applications in various plant treatment processes.

A fly ash sample sourced from a Romanian thermal power plant underwent investigation for its potential in decolorizing a genuine textile wastewater stream in a one-step treatment using the wastewater stream from a private company in North Romania after dyeing and rinsing steps of cotton fabrics (Zaharia & Suteu, 2013). The dyeing bath has two azo dyes Remazol Arancio 3R and Remazol Rose RB as well as Na_2CO_3, NaOH, and other elements such as electrolytes, anti-crust, and reducing components. The results obtained showed a high discoloration degree, ranging from 55% to 83% and organic matter elimination, conveyed as chemical oxygen demand (COD) of 44%–61%. The adsorption was most effective when carried out at a pH below 2, driven by the electrostatic attraction between the fly ash's positive surface and the negatively charged dye molecules. Moreover, temperatures above 25°C facilitated the process due to its endothermic nature. Post-treatment, the textile effluent conforms to quality standards and can either be reused within the fabric production or discharged following a neutralization phase. The acquired data present a promising basis for employing fly ash as an efficient adsorbent in a one-step treatment process, potentially replacing conventional multi-step processes involving flocculation, chlorination, and ozonation.

8.4 FINAL REMARKS

This work offers a review of the application of fly ash as a cost-effective adsorbent to eliminate industrial dyes from wastewaters. Fly ash is produced in vast quantities on a worldwide basis as an industrial byproduct when coal is burned for energy generation. Regardless, only an average of 25%–35% has some kind of reuse. The application of this material as adsorbent material still has a low profile among adsorbent materials for dye removing, but interest in using this industrial waste has been growing steadily as a response to the call to valorize waste in the context of a more sustainable economy. Fly ash's adsorption capacity relies heavily on the specific textural properties and surface chemistry of the solid material, along with the unique features of the wastewater, such as its pH and temperature.

Most studies available on the literature deal with the removal of dyes under optimized conditions and, in most cases, using dilute dye solutions. In these cases, simple adsorption models like Langmuir isotherm model can be applied. When more concentrated solutions or more than one dye species are present, a more sophisticated isotherm model is required to quantify the adsorption capacity. Conversely, the modification of fly ash by a simple and low-cost process can result in a notable increase in its ability to absorb dyes. In conclusion, existing bibliography demonstrates that fly ashes possess considerable potential to be used in the removal of dyes from wastewater and, although large-scale applications are still unknown, some preliminary results show that they can positively contribute to replace multi-step processes of wastewater treatment in the near future.

ACKNOWLEDGMENTS

This work was funded by the Fundação para a Ciência e Tecnologia (FCT) through UIDB/00100/2020, (https://doi.org/10.54499/UIDB/00100/2020), UIDP/00100/2020, (https://doi.org/10.54499/UIDP/00100/2020), and LA/P/0056/2020, (https://doi.org/10.54499/LA/P/0056/2020), and Instituto Politécnico de Lisboa (IPL) through Project IPL/IDI&CA2024/Ash2ZeoCat_ISEL.

REFERENCES

Adeleye, A. S., Conway, J. R., Garner, K., Huang, Y., Su, Y., & Keller, A. A. (2016). Engineered nanomaterials for water treatment and remediation: Costs, benefits, and applicability. *Chemical Engineering Journal, 286*, 640–662. https://doi.org/10.1016/j.cej.2015.10.105

Aigbe, U. O., Ukhurebor, K. E., Onyancha, R. B., Osibote, O. A., Darmokoesoemo, H., & Kusuma, H. S. (2021). Fly ash-based adsorbent for adsorption of heavy metals and dyes from aqueous solution: A review. *Journal of Materials Research and Technology, 14*, 2751–2774. https://doi.org/10.1016/j.jmrt.2021.07.140

Ali, H. (2010). Biodegradation of synthetic dyes—A review. *Water, Air, and Soil Pollution, 213*(1–4), 251–273. https://doi.org/10.1007/s11270-010-0382-4

Astuti, W., Chafidz, A., Wahyuni, E. T., Prasetya, A., Bendiyasa, I. M., & Abasaeed, A. E. (2019). Methyl violet dye removal using coal fly ash (CFA) as a dual sites adsorbent. *Journal of Environmental Chemical Engineering, 7*(5), 1–10. https://doi.org/10.1016/j.jece.2019.103262

Atun, G., Ayar, N., Kurtoğlu, A. E., & Ortaboy, S. (2019). A comparison of sorptive removal of anthraquinone and azo dyes using fly ash from single and binary solutions. *Journal of Hazardous Materials, 371*(January), 94–107. https://doi.org/10.1016/j.jhazmat.2019.03.006

Bal, G., & Thakur, A. (2021). Distinct approaches of removal of dyes from wastewater: A review. *Materials Today: Proceedings, 50*, 1575–1579. https://doi.org/10.1016/j.matpr.2021.09.119

Banerjee, S. S., Jayaram, R. V., & Joshi, M. V. (2003). Removal of nickel(II) and zinc(II) from wastewater using fly ash and impregnated fly ash. *Separation Science and Technology, 38*(5), 1015–1032. https://doi.org/10.1081/SS-120018121

Benkhaya, S., M'rabet, S., & El Harfi, A. (2020). Classifications, properties, recent synthesis and applications of azo dyes. *Heliyon, 6*(1). https://doi.org/10.1016/j.heliyon.2020.e03271

Chatuverdi, A. K., Yadova, K. P., & Singh, V. N. (1990). Deflourination of water by adsorption of fly-ash. *Water Air Soil Pollution, 9*, 51–61.

D'Antoni, B. M., Iracà, F., & Romero, M. (2017). Abstract review on: Current treatment technologies and practical approaches on textile wastewater dyes removal. *Panta Rei Srl-Water Solutions, April 2017*, 1–10. https://doi.org/10.13140/RG.2.2.11472.71689

Dawood, S., & Sen, T. K. (2014). Review on dye removal from its aqueous solution into alternative cost effective and non-conventional ads. *Journal Chemical and Process Engineering Research, 1*, 104.

Deshannavar, U. B., Katageri, B. G., El-Harbawi, M., Parab, A., & Acharya, K. (2017). Fly ash as an adsorbent for the removal of reactive blue 25 dye from aqueous solutions: Optimization, kinetic and isotherm investigations. *Proceedings of the Estonian Academy of Sciences, 66*(3), 300. https://doi.org/10.3176/proc.2017.3.10

Dinh, N. T., Vo, L. N. H., Tran, N. T. T., Phan, T. D., & Nguyen, D. B. (2021). Enhancing the removal efficiency of methylene blue in water by fly ashviaa modified adsorbent with alkaline thermal hydrolysis treatment. *RSC Advances, 11*(33), 20292–20302. https://doi.org/10.1039/d1ra02637b

Dutta, S., Gupta, B., Srivastava, S. K., & Gupta, A. K. (2021). Recent advances on the removal of dyes from wastewater using various adsorbents: A critical review. *Materials Advances, 2*(14), 4497–4531. https://doi.org/10.1039/d1ma00354b

Gregg, S. J., & Sing, K. S. W. (1982). *Adsorption, Surface Area and Porosity* (Second Ed.). Academic Press.

Harja, M., Buema, G., & Bucur, D. (2022). Recent advances in removal of Congo Red dye by adsorption using an industrial waste. *Scientific Reports*, *12*(1), 1–18. https://doi.org/10.1038/s41598-022-10093-3

Ho, Y. S., & McKay, G. (1999). Pseudo-second order model for sorption process. *Process Biochemistry*, *34*(6), 451–465. https://doi.org/10.1021/acs.oprd.7b00090

Hsu, T. C. (2008). Adsorption of an acid dye onto coal fly ash. *Fuel*, *87*(13–14), 3040–3045. https://doi.org/10.1016/j.fuel.2008.03.026

Huang, X., Zhao, H., Hu, X., Liu, F., Wang, L., Zhao, X., Gao, P., & Ji, P. (2020). Optimization of preparation technology for modified coal fly ash and its adsorption properties for Cd2+. *Journal of Hazardous Materials*, *392*(March), 122461. https://doi.org/10.1016/j.jhazmat.2020.122461

Izquierdo, M. T., & Rubio, B. (2008). Carbon-enriched coal fly ash as a precursor of activated carbons for SO2 removal. *Journal of Hazardous Materials*, *155*(1–2), 199–205. https://doi.org/10.1016/j.jhazmat.2007.11.047

Jala, S., & Goyal, D. (2006). Fly ash as a soil ameliorant for improving crop production—A review. *Bioresource Technology*, *97*(9), 1136–1147. https://doi.org/10.1016/j.biortech.2004.09.004

Janoš, P., Buchtová, H., & Rýznarová, M. (2003). Sorption of dyes from aqueous solutions onto fly ash. *Water Research*, *37*(20), 4938–4944. https://doi.org/10.1016/j.watres.2003.08.011

Kapoor, A., & Viraraghavan, T. (1992). Adsorption of mercury from wastewater fly-ash. *Adsorption, Science and Technology*, *9*, 130–147.

Karaca, H., Altıntığ, E., Türker, D., & Teker, M. (2018). An evaluation of coal fly ash as an adsorbent for the removal of methylene blue from aqueous solutions: Kinetic and thermodynamic studies. *Journal of Dispersion Science and Technology*, *39*(12), 1800–1807. https://doi.org/10.1080/01932691.2018.1462191

Khalfaoui, A., Bouchareb, E. M., Derbal, K., Boukhaloua, S., Chahbouni, B., & Bouchareb, R. (2022). Uptake of Methyl Red dye from aqueous solution using activated carbons prepared from Moringa Oleifera shells. *Cleaner Chemical Engineering*, *4*(July), 100069. https://doi.org/10.1016/j.clce.2022.100069

Khan, A., Wang, X., Gul, K., Khuda, F., Aly, Z., & Elseman, A. M. (2018). Microwave-assisted spent black tea leaves as cost-effective and powerful green adsorbent for the efficient removal of Eriochrome black T from aqueous solutions. *Egyptian Journal of Basic and Applied Sciences*, *5*(2), 171–182. https://doi.org/10.1016/j.ejbas.2018.04.002

Kisku, G. C., Markandeya, Shukla, S. P., Singh, D. S., & Murthy, R. C. (2015). Characterization and adsorptive capacity of coal fly ash from aqueous solutions of disperse blue and disperse orange dyes. *Environmental Earth Sciences*, *74*(2), 1125–1135. https://doi.org/10.1007/s12665-015-4098-z

Li, H., Dai, M., Dai, S., Dong, X., & Li, F. (2018). Methylene blue adsorption properties of mechanochemistry modified coal fly ash. *Human and Ecological Risk Assessment*, *24*(8), 2133–2141. https://doi.org/10.1080/10807039.2018.1440527

Lu, G. Q., & Do, D. D. (1991). Adsorption properties of fly-ash particles for NOx removal from flue gases. *Fuel Process Technology*, *27*, 95–107.

Mathapati, M., Amate, K., Prasad, C. D., Jayavardhana, M. L., & Raju, T. H. (2021). A review on fly ash utilization. *Materials Today: Proceedings*, *50*, 1535–1540. https://doi.org/10.1016/j.matpr.2021.09.106

Naik, S. S., Pandey, S., Pawar, S. N., Shinde, B. H., & Prakash, C. (2023). Innovative and interactive methodology for development of geopolymer mortar using fly ash of agricultural waste briquettes. *International Journal on Interactive Design and Manufacturing (IJIDeM)*. https://doi.org/10.1007/s12008-023-01280-1

Novais, R. M., Ascensão, G., Tobaldi, D. M., Seabra, M. P., & Labrincha, J. A. (2018). Biomass fly ash geopolymer monoliths for effective methylene blue removal from wastewaters. *Journal of Cleaner Production*, *171*, 783–794. https://doi.org/10.1016/j.jclepro.2017.10.078

Nunes, N., Santos, M. T., & Martins, A. (2023). Low cost modified biochars from peanut shells for the removal of textile dyes. *Journal of the Brazilian Chemical Society*, *34*(6), 885–893. https://doi.org/10.21577/0103-5053.20220156

Özacar, M., & Şengil, I. A. (2006). A two stage batch adsorber design for methylene blue removal to minimize contact time. *Journal of Environmental Management*, *80*(4), 372–379. https://doi.org/10.1016/j.jenvman.2005.10.004

Pethkar, P. A., Pethkar, A. V., & Nalawade, P. M. (2016). Removal of methylene blue dye from solutions using granulated fly ash. *IIOABJ*, *7*(October), 49–54.

Potgieter, J. H., Pardesi, C., & Pearson, S. (2021). A kinetic and thermodynamic investigation into the removal of methyl orange from wastewater utilizing fly ash in different process configurations. *Environmental Geochemistry and Health*, *43*(7), 2539–2550. https://doi.org/10.1007/s10653-020-00567-6

Querol, X., Moreno, N., Umaa, J. C., Alastuey, A., Hernández, E., López-Soler, A., & Plana, F. (2002). Synthesis of zeolites from coal fly ash: An overview. *International Journal of Coal Geology*, *50*(1–4), 413–423. https://doi.org/10.1016/S0166-5162(02)00124-6

Rawat, S., & Kumar, S. (2022). Critical review on processing technologies and economic aspect of bio-coal briquette production. *Preparative Biochemistry & Biotechnology*. https://www.tandfonline.com/doi/abs/10.1080/10826068.2021.2001754

Rosa, T., Martins, A., Santos, M., Trindade, T., & Nunes, N. (2021). Coal fly ash waste, a low-cost adsorbent for the removal of mordant orange dye from aqueous media. *Journal of the Brazilian Chemical Society*, *00*(00), 1–12. https://doi.org/10.21577/0103-5053.20210116

Routoula, E., & Patwardhan, S. V. (2020). Degradation of anthraquinone dyes from effluents: A review focusing on enzymatic dye degradation with industrial potential. *Environmental Science and Technology*, *54*(2), 647–664. https://doi.org/10.1021/acs.est.9b03737

Sun, D., Zhang, X., Wu, Y., & Liu, X. (2010). Adsorption of anionic dyes from aqueous solution on fly ash. *Journal of Hazardous Materials*, *181*(1–3), 335–342. https://doi.org/10.1016/j.jhazmat.2010.05.015

Viraraghavam, T., & Alfaro, F. M. (1998). Adsorption of phenol from wastewater by peat flys-ash and bentonite. *Journal of Hazardous Materials*, *57*, 59–70.

Wang, S., Boyjoo, Y., & Choueib, A. (2005). A comparative study of dye removal using fly ash treated by different methods. *Chemosphere*, *60*(10), 1401–1407. https://doi.org/10.1016/j.chemosphere.2005.01.091

Wee, J. H. (2013). A review on carbon dioxide capture and storage technology using coal fly ash. *Applied Energy*, *106*, 143–151. https://doi.org/10.1016/j.apenergy.2013.01.062

Wekoye, J. N., Wanyonyi, W. C., Wangila, P. T., & Tonui, M. K. (2020). Kinetic and equilibrium studies of Congo red dye adsorption on cabbage waste powder. *Environmental Chemistry and Ecotoxicology*, *2*, 24–31. https://doi.org/10.1016/j.enceco.2020.01.004

Xu, G., & Shi, X. (2018). Characteristics and applications of fly ash as a sustainable construction material: A state-of-the-art review. *Resources, Conservation and Recycling*, *136*(August 2017), 95–109. https://doi.org/10.1016/j.resconrec.2018.04.010

Yagub, M. T., Sen, T. K., Afroze, S., & Ang, H. M. (2014). Dye and its removal from aqueous solution by adsorption: A review. *Advances in Colloid and Interface Science*, *209*, 172–184. https://doi.org/10.1016/j.cis.2014.04.002

Zaharia, C., & Suteu, D. (2013). Coal fly ash as adsorptive material for treatment of a real textile effluent: Operating parameters and treatment efficiency. *Environmental Science and Pollution Research*, *20*(4), 2226–2235. https://doi.org/10.1007/s11356-012-1065-z

Zhu, L., Ji, J., Wang, S., Xu, C., Yang, K., & Xu, M. (2018). Removal of Pb(II) from wastewater using Al2O3-NaA zeolite composite hollow fiber membranes synthesized from solid waste coal fly ash. *Chemosphere*, *206*, 278–284. https://doi.org/10.1016/j.chemosphere.2018.05.001

9 Valorization of Fly Ash Waste in the Preparation of Advanced Ceramic Materials

Brahim Achiou, Dounia Beqqour,
Majda Breida, Alain Tèebwaoga Sina,
Jamal Ait Brahim, Redouane Beniazza,
Mouhamed Ouammou and Saad Alami Younssi

9.1 INTRODUCTION

The global demand for energy is exponentially increasing due to the population growth. By 2050, an increase of over 15% is expected compared to 2021 (Chanthakett et al. 2024). A major part of energy production worldwide is derived from fossil energy resources including coal. During electricity production, coal-firing generates large amounts of by-products namely fly ash (FA) (Chafiq Elidrissi et al. 2023). The high demand for electricity has led to the construction of several coal-fired power plants around the world, generating 800 million tons of FA worldwide in 2019 (Rafieizonooz, Khankhaje, and Rezania 2022).

The utilization of FA in ceramics is intended to be an important means of valorization given their chemical composition which is rich in Si, Al, Fe, Ca, and O. Additionally, the development of high-added value and high-consuming applications of FA could promote the circular economy in coal industries.

This chapter unveils a pioneering approach to the synthesis of ceramics-based FA. It marks a significant advancement in both sustainable waste management and material science, particularly in the realm of ceramics. In fact, the synthesis of ceramic FA-based phases such as mullite, cordierite, and feldspar minimizes waste and enhances the ceramics' structural and functional features (Figure 9.1). In parallel, this chapter explores the various uses of FA in fabrication ceramic membranes, building materials, and innovative glass-ceramic products. Thus, emphasizing FA's versatility and potential to revolutionize the ceramics industry. Through detailed examination of FA's application and effectiveness as a material for high-temperature and electrical insulation, this chapter underscores the material's adaptability and importance in meeting contemporary industrial needs. Besides the technical and environmental

 DOI: 10.1201/9781003498124-9

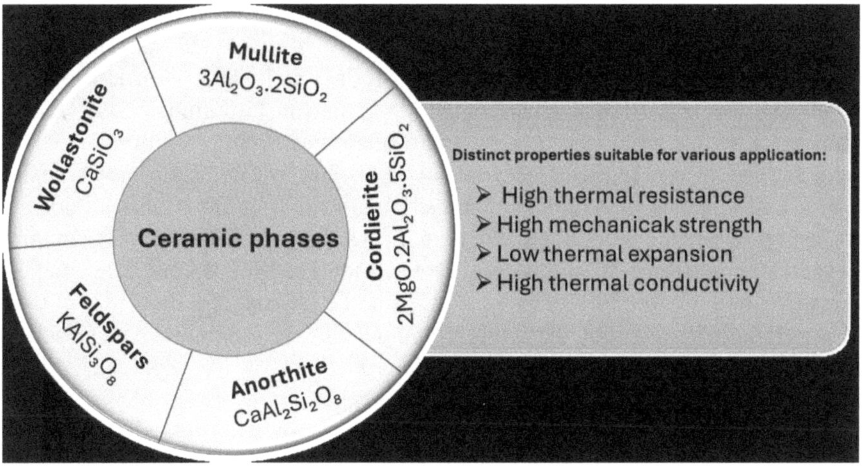

FIGURE 9.1 Synthesized ceramic phases derived from FA-based materials.

challenges of incorporating FA into ceramic production, this chapter optimistically concludes with a call to action for a shift toward sustainable material practices in the ceramics sector, aiming to inspire further research, innovation, and policy changes toward a more sustainable future.

9.2 PROPERTIES OF FA RELEVANT TO CERAMICS

9.2.1 Physical Properties

FA has a loose powder appearance consisting of fine particles in the form of dust trapped in the combustion furnace gases, with a color that varies from brown to black depending on the unburned carbon content in the ashes. The predominant particles are composed of irregular debris, solid spheres, hollow microspheres, and are predominantly glassy (amorphous) in nature. Overall, combustion temperature and cooling time affect the morphology of FA particles (Wang et al. 2021). For instance, the spherical shape of FA is obtained at high combustion temperatures ranging from 1,300°C to 1,700°C and using pulverized coal combustion technology. The angular shape is obtained at low combustion temperatures ranging from 700°C to 900°C and by using fluidized bed combustion technology (Zimar et al. 2022). The ability of FA to be compacted into a solid layer with low unit weight makes it effective in filling applications, such as ceramics, due to its particle size distribution and the spherical shape of its particles (Mathapati et al. 2022).

9.2.2 Chemical Properties

The chemical properties of FA are primarily dependent on the origin of coal. There are mainly four types of coal namely anthracite, bituminous, sub-bituminous, and lignite. Each varies in terms of its calorific value, chemical composition, ash content, and geological origin (Mathapati et al. 2022). FA from bituminous coal is predominantly composed of SiO_2, Al_2O_3, Fe_2O_3, and CaO, with variable amounts of carbon as measured by loss on ignition (LOI). Lignite and sub-bituminous coal FA are characterized by high content of CaO and MgO, with low content of SiO_2, Fe_2O_3, and a low carbon content compared to bituminous CFA. As for anthracite FA, it is produced in very small quantities because anthracite coal is rarely burned in power plants (Mathapati et al. 2022). The classification of FA generally depends on its chemical composition according to the American Society for Testing Materials (ASTM C618) standard. Calcareous FA (Class C) contains between 50% and 70% of SiO_2, Al_2O_3, Fe_2O_3, and over 18% of CaO. Siliceous FA (Class F) contains over 70% of SiO_2, Al_2O_3, and Fe_2O_3, and less than 18% of CaO (Wang et al. 2021). Besides these major components, FA contains oxides such as CaO, MgO, K_2O, Na_2O, TiO_2, and various trace elements such as Hg, Cr, Ni, Pb, V, As, and Se (Mathapati et al. 2022).

The use of FA in the ceramics sector is explained by its chemical composition which is rich in SiO_2 and Al_2O_3, similar to that of clays. Additionally, FA contains fluxing elements such as K_2O and Na_2O, and a significant LOI likely due to incomplete coal combustion.

9.2.3 Mineralogical Properties

The mineralogical composition of FA is influenced by geological factors associated with the genesis and deposition of coal, as well as the conditions of combustion. FA is mainly composed of three phases namely quartz, mullite, and hematite (Luo et al. 2021). Furthermore, the content of the vitreous phase in FA could play a crucial role in the vitrification of ceramic materials, promoting good compactness and interesting mechanical performance of the final product.

9.3 ADVANCED CERAMIC MATERIALS FROM FA

9.3.1 Classification of Ceramics

9.3.1.1 Porous Ceramics

Porous ceramics can be classified into three main categories based on pore size, namely macropore (more than 50 nm), mesopore (between 2 and 50 nm), and micropore (less than 2 nm). Also, porous ceramics could be classified as materials with open and closed porosity. This feature is fundamental in determining the functionalities of the various materials. For instance, porous ceramics with open porosity (interconnected pores) is advantageous for fluid transport applications due to its permeability (Wang et al. 2021). Porous ceramics essentially offer chemical and thermal stability, mechanical strength, and abrasion resistance. Additionally, they exhibit poor thermal conductivity, low mass, and relatively low density. These unique characteristics make them suitable for a variety of applications such as thermal insulation, bone substitutes, hot gas collectors, filters, bioreactors, dust collectors, and engine components. Porous ceramics have been designed for the removal of contaminants ranging in size from several micrometers down to the nanometer range from various fluids (Yang et al. 2019).

9.3.1.2 Porous Glasses and Glass-Ceramics

The interest in porous glasses and ceramics is mainly related to the properties of cellular ceramics, which include high permeability, high surface area, high thermal and acoustic insulation, low specific heat, low density, and high chemical resistance. Different strategies, including but not limited to glass foams, can be applied due to the special nature of glass ceramics. Generally, most glass foams are made by sintering glass materials and foaming agents that release gases within the pyroplastic mass, leading to a highly porous glass structure. The release of gas is associated with either oxidation or decomposition reactions. Oxidizing reactions involve the release of COx gases (carbon monoxide or carbon dioxide) from C-containing compounds including carbon black, graphite, SiC, and organic materials that react with atmospheric oxygen (Krasnyi et al. 2021).

9.3.1.3 Foamed Ceramics

One of the most common methods for preparing foamed ceramics is direct foaming. The resulting pores are circular in shape. Recently, direct foaming has been employed to prepare a wide range of ceramic foams, including those made from hydroxyapatite, alumina, zirconia, silica, magnesia, and other materials. The principal manufacturing processes for glass-ceramic foams include the powder sintering method, the organic slurry impregnation method, the sol-gel method, etc. For the synthesis of glass-ceramic foams from industrial solid waste as a raw material, the powder sintering method is the most widely used. Furthermore, spherical pore foams can be prepared using CFA cenospheres as well as direct foaming with a foaming agent.

9.3.2 FA-Derived Ceramic Phases

9.3.2.1 Mullite

Mullite is an oxide ceramic consisting of 72 wt% alumina and 28 wt% silica ($3Al_2O_3.2SiO_2$). It has a low density (3.19), high refractoriness (>1,700°C), high resistance to oxidants, a low coefficient of thermal expansion ($4.5–5.6 \times 10–^{6}$ $K–^{1}$), and excellent thermal stability with a melting point higher than 1800°C. Due to its unique properties namely high-temperature strength, thermal shock resistance, and low dielectric constant, it is one of the most widely used ceramic materials. The mullite phase is typically synthesized through a thermal reaction of natural or synthetic materials containing Al_2O_3 and SiO_2, including clay, kaolinite, alumina, aluminum isopropoxide, and pyrophyllite. A variety of waste materials can be used as sources of mullite ceramics. CFA is an abundant waste material with a high content of silica and alumina as well as metal oxides. In combination with aluminum-rich components such as aluminum powder, Al_2O_3, $Al(OH)_3$, and AlF_3, CFA was used for the synthesis of mullite ceramic. It can be used as a promising candidate for substrates in electronic packaging due to its low dielectric constant (ε¼ 6.7, compared to 9.8 for Al_2O_3) (Wang et al. 2021).

9.3.2.2 Cordierite

Cordierite ($2MgO{\cdot}2Al_2O_3{\cdot}5SiO_2$) is an important ceramic structure with distinctive unique chemical and physical properties such as high thermal and chemical stability and excellent insulating properties. Cordierite is a promising material to be used as a ceramic support for a variety of applications such as membranes, filters, and catalysts. Due to the high cost of the raw material, the synthesis of cordierite phase is expensive. Dealing with this challenge, researchers have explored the potential of involving solid waste for the synthesis of low-cost cordierite (Senthil Kumar et al. 2019).

9.3.2.3 Anorthite

Anorthite ($CaAl_2Si_2O_8$) is a common natural raw material with a density of 2.74–2.76 that contains $[SiO_4]^{4-}$ and $[AlO_4]^{5-}$ tetrahedral linked by Ca^{2+}. It offers high abrasion and corrosion resistance, good thermal shock resistance, high load-bearing capacity, and low thermal expansion coefficient (4.82×10^{-6} K^{-1}). These attributes make anorthite suitable for high-temperature applications such as ceramic substrates. Furthermore, cost-effective anorthite-based ceramic membranes have been successfully synthesized from various solid waste materials, including lime sludge, FA, paper processing residues, and construction and demolition wastes (Wang et al. 2021).

9.3.2.4 Feldspars

Feldspar ($KAlSi_3O_8$) is a key material in the ceramic industry that has been employed as a flux element to produce extensive ceramics including ceramic wares, stoneware materials, porcelain stoneware, porcelains, glass-ceramics, and bricks. Alkali feldspar, which is primarily composed of Na and K, plays a crucial role in lowering the melting temperature of the glass batch and favoring the melting of other minerals present in the formulation. Conversely, the contribution of feldspars with Al and Ca enhances both the chemical and physical resistance of the product. Feldspars can be obtained from natural quarries: granitic pegmatites, aplites, granitoid, metamorphic, feldspathic arenites, albitic or epithermal alterations (Luo et al. 2019). The presence of feldspars enhances the features of the obtained ceramics in terms of durability, whiteness, strength, toughness, color stability, and chemical resistance. Additionally, feldspars facilitate the solidification of crystalline phases and the melting, softening, and wetting of other batch components. All these properties allow feldspars to be used in various utilizations, with variable contents. These utilizations include sanitary ware, tableware (17%–20%), wall tiles (0%–11%), floor tiles (55%–60% feldspar content), dental porcelains (60%–80%), high-voltage electrical porcelains (25%–35%), kitchen and oven ware (10%), refractory products, etc. Feldspars can be regarded as one of the main "backbones of the ceramics industry". Their purity is highly determined by the type of final product (Klima et al. 2022).

9.3.2.5 Wollastonite

Wollastonite ($CaSiO_3$) is a white mineral, that is chemically composed of calcium silicate. The abundance of silicon and calcium in the earth's crust facilitates the production of Wollastonite, which is distinguished by low sintering temperature and high mechanical strength. Wollastonite has emerged as a potential competitor to porous ceramics due to its thermal stability, low dielectric loss, and good biodegradability. Additionally, it exhibits low thermal expansion, thermal conductivity, and dielectric constant as well as good biocompatibility. These properties make Wollastonite a promising material for bone tissue regeneration (Klima et al. 2022).

9.4 APPLICATIONS OF FA-BASED CERAMICS

9.4.1 Ceramic Membranes

9.4.1.1 Membrane Prepared Via Pressing Method

One approach for CFA membranes preparation is the application of the pressing method in both dry and wet routes. Yan et al. investigated the addition of α-Al_2O_3 into FA for flat membranes preparation using a dry pressing method at 12 MPa and a sintering temperature of 1,250°C for 0.5 hours (Yang et al. 2019). The porous membranes exhibit a porosity ≥28% with the addition of 25 wt% alumina. It was proven that Al_2O_3 reacts with SiO_2 in FA leading to mullite formation, while the increase of alumina induces quartz decrease. On the other hand, Al_2O_3 aids in reducing membrane shrinkage during the sintering process caused by increasing the distance between the ceramic particles. In the fabrication process of FA circular ceramics (with glycerol and polyvinyl alcohol (PVA) as binders) via the dry pressing method under 20 MPa, it was proven that sintering above 800°C leads to the melting of FA particles and the fusion of necks with each other. Higher than the critical temperature, the densification process becomes dominant and dense membranes are obtained. Therefore, sintering temperature of FA membranes is critical for achieving the uniform morphological structure of pores.

9.4.1.2 Membrane Prepared Via Extrusion Method

Extrusion method was used to prepare CFA tubular ceramic membranes (length 80 cm) with high efficiency for water and heat recovery (Huang et al. 2022). In this method, the feeding raw materials in the form of paste is forced through a die opening using an endless screw or piston. The membranes were prepared using CFA (as the main raw material at 91 wt%), dextrin and carboxymethyl cellulose (CMC) as the pore forming agent and organic binder at 3 and 6 wt%, respectively, while water and glycerin (7.70 wt%) were respectively used as solvent and plasticizer. The mass ratio of powdery materials and glycerin solution was fixed at 4:1. The extruded tubular membranes were firstly dried at room temperature for 24 hours and then sintered at different temperatures from 1,050°C to 1,200°C for 2 hours. It was revealed that the increase in sintering temperature enhances particle merging and connection and reduces membrane bulk density. This leads to a decrease in the apparent porosity. It was demonstrated that the CFA calcinated at high temperatures is characterized by larger average size and provides membranes with better properties compared to CFA produced at low temperatures. Higher water recovery flux and separation efficiency were obtained with CFA membranes sintered at 1,150°C, given the larger pore size and lower porosity. Further, CFA membranes with an average pore size less than 2 μm are suitable for transport membrane condenser applications (Huang et al. 2022).

9.4.2.3 Membrane Prepared Via Slip Casting

The slip casting method involves the use of a slip which is a suspension of FA to prepare a membrane. To enhance the easy handling and the stability of the slip mixture, it is recommended to add dispersing agents and binders along with a relaxing time (for bubbles removal). Rheothickening behavior was noticed in FA slip in the presence of PVA (Goswami, Pakshirajan, and Pugazhenthi 2022). The effect of slip concentration on membrane properties reveals that at high concentrations, slip results in high viscosity and the formation of active layers with cracks, uneven shrinkage stress, and pores of larger average size. In general, this method is particularly recommended for ceramic composite membranes with complex shapes and structures. In another work, microporous tubular FA support was prepared by extrusion followed by the deposition of two layers of refined FA by the slip-casting process (Fang et al. 2011). The additives used for the support are methylcellulose and water while Lomar-D and DSX 3290 (now known as Rheovis® PU 1190) were used for the active layer. In parallel, the authors investigated the effect of slip casting concentration, time, and withdrawal speed on membranes' properties. The optimized membrane is characterized by a defect-free surface with an average pore size of 0.77 μm and permeability of 1.56 10^4 L/(h m bar). The homogeneous surface was obtained when the slip concentration is in the range of 2–3 wt%. The increase of the slip casting concentration leads to the increase of the membrane thickness (from 13.0 μm to 32.1 μm) which can be attributed to the high particle sticking probability. Conjointly, the increase of viscosity increases the wet layer thickness through capillary action according to the film-coating mechanism. This mechanism acts as the predominant phenomenon up to a specific withdrawal speed (Goswami et al. 2022). Compared to commercial ceramic membranes, the prepared FA membranes demonstrated higher water flux and lower permeability resistance due to the spherical and smooth shape of CFA which decreases the resistance.

9.4.1.4 Application as a Sintering Additive and Membrane Filler

The introduction of additives which can improve the materials' densification, surface interface, and enhance the formation of eutectic composition leading to a reduction in ceramics' sintering temperature are all benefits of using CFA. In fact, CFA was used to prepare valuable silicon carbide (SiC) based porous ceramic membrane at low sintering temperature (1,000°C) with the following characteristics: (i) pure water flux of 5,261 L/(h m bar), (ii) open porosity of 44.7%, and (iii) mean pore size of 3.7 μm and a high oil removal efficiency of 91% (the initial oil concentration was 1,657 mg/L) (Das, Kayal, and Innocentini 2021). The use of FA as a filler affects the mechanical properties of ceramics depending on different parameters such as ceramic raw materials, and the interfacial interaction with different composite matrices. Besides mechanical improvement, CFA in membranes was found to enhance thermal stability, water uptake, and proton conductivity. It also optimizes the pore size distribution, heterogeneity, and the oxygen mass transport coefficient (Manocha, Prasad, and Manocha 2012).

9.4.2 Refractories

In refractory materials' production, the main used components are clay and chamotte while the two most common preparation methods are solid-state reaction and wet chemicals (e.g. sol-gel, hydrolysis, and thermal synthesis). Recent studies have targeted the use of FA for complete or partial replacement of these reactants. It was found that adding CFA to fired clay to prepare refractory bricks improves the thermophysical properties of the product (Sukkae, Suebthawilkul, and Cherdhirunkorn 2018). FA (10–50 wt%) and clay were milled and uniaxially pressed at 150 MPa to form disk samples. The samples were afterward sintered in the temperature range of 1,100°C–1400°C for 60 minutes. The addition of FA decreases the low density of the sintered disks (at 1,250°C) due to the low density of FA (≈2.2) compared to clay (≈2.6). Higher FA content leads to pores with larger size and higher porosity which contribute to high water absorption and lower thermal conductivity (1.6–0.8 w/(m k) as well as a decrease in both bulk density and strength (26–6 MPa).

According to literature, CFA and kaolin are considered as excellent refractory materials (Kamara, Wang, and Ai 2020). For commercial ceramic compositions containing 60 wt% kaolin, 20 wt% chamotte and 20 wt% sawdust, FA at 5 and 10 wt% was added and evaluated. The samples were molded by hand pressing while adding water and PVA at the same weight percentage of 15 wt% each. After modeling, the cylinder samples were dried and sintered at 1,350°C. It was found that the amount of FA is critical to refractory production, a diminution in the mechanical strength is caused by a decrease in mullite formation during sintering. The addition of FA at high temperature demonstrates low reactivity with the refractory ceramic components due to the low interaction with kaolin's crystalline phase. Besides, a slight increase of thermal conductivity was obtained with FA. Based on the authors' claims, the obtained strength (1.45 MPa) is comparable to the strength of the refractory material based uniquely on a commercial composition (1.89 MPa) (Krasnyi et al. 2021). Given the high content of aluminum and silicon oxides in FA, the latter is used in the production of refractories based on cordierite, mullite, and/or forsterite.

Pure cordierite was synthesized using a cost-effective method applying FA as a raw material at low synthesis temperatures. Magnesium oxide, CFA, and dopants (e.g. ZrO_2, CeO_2, TiO_2) at different content compositions ranging from 5 to 20 wt% were explored for cordierite preparation. The obtained mixtures were first wet milled with isopropanol for 3 hours and then dried at 100°C and dry milled for 2 hours to enhance homogeneity. This mixture was blended with boron oxide and PVA and then compacted at 240 MPa into pellets of 10 mm in diameter. To dry out the volatiles (e.g. binder), the pallets were calcinated at 400°C and sintered in the temperature range of 800°C–1,350°C for 3 hours. The use of FA in cordierite synthesis was found to enhance the mechanical properties of the material as a catalytic substrate. The addition of dopants to the cordierite matrix was proved to enhance flexural strength, hardness, and fracture toughness. Cordierite with ZrO_2 dopant recorded the highest hardness and flexural strength (196.72 MPa) (Senthil Kumar et al. 2019).

Geopolymer concrete is a novel and eco-friendly type of construction material to substitute Portland cement concrete. The geopolymer is mainly prepared through the reaction of aluminate and silicate with an alkaline activator such as NaOH which reduces CO_2 emission by 50%–80% and thus reducing the burden of cement production. Given its pozzolanic nature and low price, CFA was used for geopolymers production. It was proven that FA with high Si/Al ratio greatly affects the geopolymers' thermal performance. For instance, geopolymer prepared from FA with Si/Al ≥ 5 in the glass displays compressive strength gains, lower mass loss, and volumetric shrinkage when subjected to high temperatures (1,000°C). Higher Si/Al ratio leads to higher crystallization temperatures (Klima et al. 2022). FA characterized with Si/

Al < 2 demonstrates low stability and high strength losses under the same thermal condition.

Low-strength geopolymers are obtained with FA having a relatively large particle size and low amorphous content. Highly amorphous FA is more effective in accelerating the pozzolanic reaction pace. At low curing temperatures (below 30°C) or when the activation is only done via hydroxide solution, the geopolymers were found to have pores larger than 200 nm. The sensitivity of FA-based systems, based on thermodynamic calculations, has shown the considerable influence of the type and amount of alkali in the system on the melting temperature. Geopolymers activated by potassium showed lower strength and volumetric shrinkage compared to the sodium-based geopolymer due to the formation of more porous gels using potassium. However, with potassium the geopolymers exhibit high-temperature resistance. The FA-based geopolymers demonstrate a high percentage of small pores compared to metakaolin-based materials (Klima et al. 2022). Further, it was reported that the crystalline phases in the phase composition of activated FA for high-temperature applications remain similar to raw FA, which makes designing a working guideline difficult (due to discrepancies in the FA characteristics and its reactivity).

In refractory materials, it was found that FA coming from the fluid bed combustion system should not be used as an additive given its high content of CaO compared to high-temperature FA.

9.4.3 Construction Materials

9.4.3.1 FA-Based Building Bricks

The new technology for manufacturing construction bricks utilizes FA as a raw material (~50–80 wt%). Performance indicators include flexural strength, compressive strength, frost resistance, drying shrinkage, and carbonation coefficient. This innovative manufacturing technology consumes a significant amount of FA and low cement content, enabling the production of a cost-effective, high-performance product that could replace clay-based red bricks (Wang et al. 2016). Mechanical strength is improved due to the FA's void-filling effect, reinforcing the density and internal structure of the matrix, and preventing damage from freeze-thaw cycles by reducing the amount of water absorbed by the matrix.

9.4.3.2 FA-Based Ceramic Tiles

Similarly, the chemical composition and phase constitution between FA and porcelain allow them to substitute traditional raw materials in ceramic tile production. The primary raw materials for ceramic tiles include clay (30–55 wt%), feldspar (20–40 wt%), and quartz (5–30 wt%). Through the process of sintering of the typical ternary mixture, clay is gradually transformed into needle-shaped mullite crystals to ensure the strength of the sintered samples. Clay also acts as a binder for other matrix constituents, providing the necessary plasticity for molding. Feldspar serves as a fluxing agent by reacting with other constituents to form a liquid that promotes matrix densification and shrinkage. Quartz is relatively inert during the sintering process and acts as an inert filler to prevent shrinkage and excessive deformations of ceramic tiles

(Luo et al. 2021). Therefore, FA is used to replace quartz, which is relatively inert and stable during the sintering process. FA can partially replace feldspar, which results in increased liquid production due to its high content of alkali metals (Na and K) and alkaline-earth metals (Ca and Mg). However, the substitution rate is kept below 50 wt% to maintain high performance. When the substitution rate of clay exceeds 25 wt%, noticeable collapse of the mechanical properties of ceramic products was observed due to the decrease in mullite crystals in the ceramic matrix.

9.4.3.3 FA-Based Ceramic Foams Thermal Insulators

FA is used in an eco-friendly and clean process to prepare architectural ceramics such as ceramic foams. The preparation of this type of ceramic involves prior mechanochemical activation of FA using NaOH solution. The solution after activation is used as a raw material to prepare ceramic foams with new thermal insulation properties, such as low apparent density, high porosity, and low thermal conductivity. This latter is achieved by molding the activated solution and then sintering it at 600°C. The resulting material consists of zigzag-shaped cavity walls, confirming its status as an architectural thermal insulation material. Given that conduction, convection, and radiation are components of thermal processes, this foamy structure contains air that blocks thermal conduction, and the zigzag cavity walls prevent thermal convection and radiation (Luo et al. 2019).

9.4.3.4 FA-Based Insulation Board

FA-based thermal insulation materials could be a promising route for consuming FA in large quantities and producing high-added value materials. They can be used as construction materials in the thermal insulation of roofs, pipelines, and composite walls. FA is utilized for its filling property in the structure. It can adjust the desired viscosity and porous structure while enhancing thermal insulation behavior. It reacts with magnesium to form MgO-SiO_2-H_2O-type hydrates and with calcium to produce CaO-SiO_2-H_2O-type hydrates, promoting the development of mechanical strength. Key factors governing the performance of insulation materials are apparent density and particle morphology. When the apparent density is high, the thermal conductivity coefficient is also high. The fine particle size of FA improves compressive strength, while large and irregular particles lead to wide and irregular pores in the structure that negatively impact mechanical strength.

9.4.4 Electronic and Optical Applications

As a support material, FA ceramic (FAC) could be efficiently used as a low-cost material in the synthesis of thin films, suitable for the design of solar light modules, due to its optical, physical, and chemical properties. The synthesized glass-ceramic exhibits remarkable properties in terms of mechanical resistance, lower porosity, and water absorption. In terms of opto-electronic characterization, the photovoltaic performance showed an open-circuit voltage of 168.2 mV, a short-circuit density of 8.11 mA/cm^3, with an efficiency of 0.4% (Fraga et al. 2018). FAC can be applied as a particulate material for thermal energy storage (TES) systems due to its optical, physical, and chemical properties. In conjunction with sustainability and resources efficiency, FAC as a TES material could be regarded as a low-cost technology with interesting properties in terms of low corrosivity, high freezing resistance, and high operating temperature (up to 800°C), compared to the conventional molten salts used as TES material (~585°C) (Agalit, Zari, and Maaroufi 2020). Recent studies emphasized the high performance of FAC in terms of absorptance of the weighted solar with values ranging from 55% to 87%. Such performance is comparable with the carbobead particles. The degree of crystallinity, surface roughness, and content of iron have shown a great impact on the weighted absorptance. A minimum value of 5 wt% of Fe content is required to maintain the absorptance higher than 70% (Glosser et al. 2023). Furthermore, FAC proved a promising performance as a TES material, but further research is required to extend its applications with the study of other properties including thermal conductivity, specific heat capacity, and density.

FA-based ceramic composite can also be used as a high-performance dielectric and magnetic material for the absorption of electromagnetic waves (EMW) (Bora et al. 2018). The FAC composite could be synthesized via carbothermal reduction method using FA and activated carbon. The results demonstrated that the composite could absorb over 100% of the EMW passing through it (see mechanism in Figure 9.2). This outstanding EMW attenuation capacity resulted from the high interfacial polarization and the impedance matching of the heterostructure. The high electrical conductivity in the composite creates a free electron flow within the composite network, which promotes the electron hopping between the micropores (Ayeni et al. 2022).

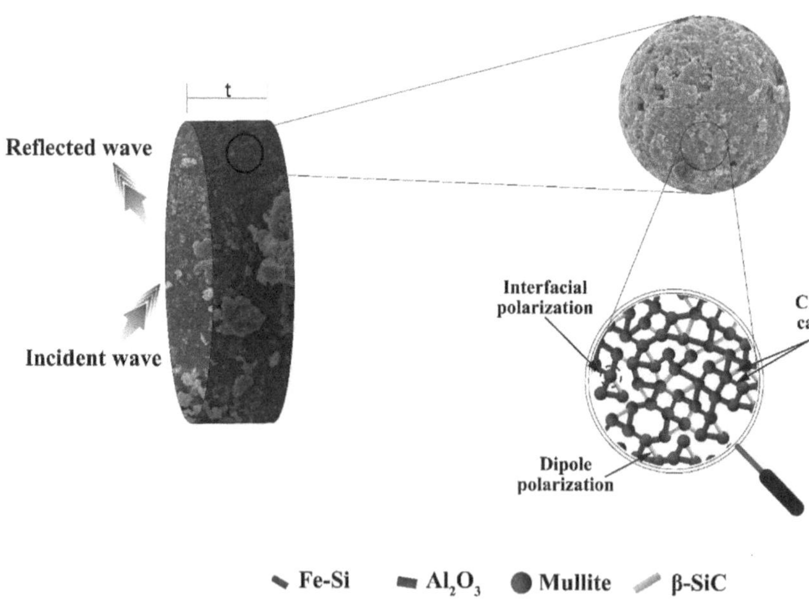

FIGURE 9.2 Mechanism of EMW absorption by FA ceramic composite.

9.4.5 Photocatalysis Applications

FAC can be used in other applications including the environmental remediation of pollutants through photocatalysis technologies. In this regard, CFA could be effectively used as a substrate for the TiO_2 photocatalyst (Favier and Harja 2020). It was reported that the immobilization of TiO_2 nanoparticles onto the CFA substrate enhances the photodegradation of organic dyes (methylene blue and methyl orange). Characterizations revealed a synergistic effect between CFA and TiO_2, accelerating the photocatalytic activity with efficient dyes decolorization (Katsika et al. 2018). Other investigations explored the use of FA as a photocatalyst carrier of nano-TiO_2. The photocatalyst composite was prepared by wet mechanical grinding. The combination mechanism was illustrated by the formation of Si-O-Ti bonds at the interface between FA and TiO_2 nano-spheres (Figure 9.3). It was demonstrated that the degradation of organic pollutants reached over 100% under UV illumination (Tu et al. 2022).

The mixture of red mud and FA was also explored for the preparation of porous ceramic for potential application as a substrate carrier of TiO_2 photocatalyst. The TiO_2 deposition was conducted via ultrasonic dispersion and sol-gel methods. The importance of porosity level was highlighted in the ceramic carrier since it directly affects the deposition rate of TiO_2 and accordingly the photocatalytic performance (Özcan, Birol, and Kaya 2021). Furthermore, the photocatalyst composites containing FAC/FA demonstrate superior performance in pollutants photodegradation compared to conventional photocatalysts. This promotes the use of unconventional resources derived from FA waste in the framework of sustainability and circular economy.

9.4.6 Other Applications

FAC can be incorporated into copper composite nanoparticles to improve the mechanical properties, thermal expansion, and wear resistance of copper as a high electrically conductive material. The synthesized composite (Cu/Gr/FA) revealed considerable

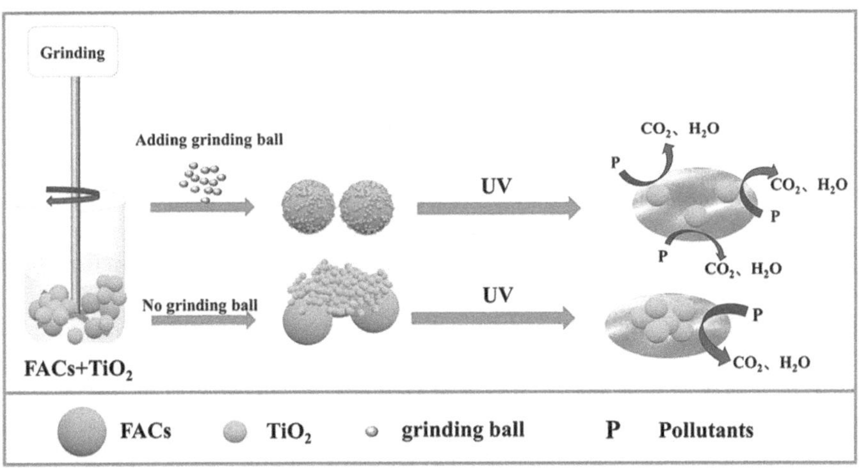

FIGURE 9.3 Procedure of FAC-TiO_2 photocatalyst preparation and its action mechanism.

enhancement in the microhardness, ultimate stress, and Young's modulus with values of 25%, 20%, and 50%, respectively, compared to the copper matrix. Additionally, the wear resistance and the coefficient of thermal expansion were highly reduced (67% and 30%, respectively). Moreover, the developed ceramic composite could be applied in electronic circuits and power generation and transmission (El-Zaidia et al. 2024).

In packaging applications, mullite whiskers derived from FAC can be utilized as packaging toughening material for electronics with high performance compared to conventional alumina ceramics. Mullite whiskers derived-FAC can be synthesized via different methods including melting salt synthesis. The produced mullite whiskers were successfully incorporated in the synthesis of mullite-Al_2O_3-quartz ceramic composite. The results demonstrated efficient mechanical strength and reduced dielectric constant and loss of the ceramic composite with values of 5.16 and 0.006, respectively (Gao et al. 2022).

9.5 CHALLENGES AND OPPORTUNITIES AND FUTURE PERSPECTIVES

Even though FA-derived materials have significant potential in different applications, their use is limited due to several challenges. One such challenge is the high cost of FA transportation due to its low bulk density (0.5–1.2), which renders it less economically viable compared to other commercially available products. This leads to a delay of wide-spread industrial adaption. In addition, the variability of FA materials depends on the coal origin, production processes as well as FA collection period. As a result, these affect the properties and chemical composition of the produced FA. In fact, the reproducibility of processing parameters including sintering temperature of FA-derived ceramic membranes was found to be difficult and depends on FA type. Ca, Fe, and other elements present in FA can induce solid solution phenomenon during the process of sintering, which activates the mullite lattice and reduces the sintering temperature. In parallel, it was proven that FA contains several complex and unfavorable components such as heavy metals (e.g. V, Cr, Ni, Pb, and Mn), radioactive elements (e.g. Rn, Ra, and U) and other impurities making it difficult to guarantee uniform structure and characteristics of the fabricated FA-based materials. This leads to issues with quality and limits the upscaling of production, especially for pharmaceutical, biotechnological, beverage, and food industries that require strict regulations. Indeed, many authors have pointed out that heavy metals in FA dissolve in acidic conditions which limits the use of FA-based membranes for acidic solutions. The leachability or dissolution of toxic elements (especially V, Cr, and Sb) from CFA membranes may occur thus leading to water contamination. Limited studies have investigated the leachability of heavy metals from FA-based membranes in which the content of metals was found to be under the permitted limits. These findings are relative to the origin of FA and can change accordingly, which explains the necessity of studying the FA composition and the membrane effluents to ensure safety for society.

The use of FA as support in membrane fabrication is limited to small scale and requires the addition of alumina to improve mechanical performance. Further studies on high alumina content still need deeper research. In fact, some FA materials require the addition of inorganic oxides to achieve desirable crystalline phases such as mullite, cordierite, and anorthite. Furthermore, there are limited comprehensive studies about the intrinsic mechanism for these membranes.

Apart from the abovementioned points, the valorization of CFA in the preparation of concrete and cement products is limited to 40 wt%, which requires the development of new recycling methodologies. One major drawback is the potential presence of radioactive elements namely ^{226}Ra, ^{232}Th, and ^{40}K. The large consumption of FA in building materials may expose human health to alpha and gamma radiation. Furthermore, it is highly recommended to address potential treatment pathways to reduce its level of radioactivity as a future research direction.

In addition, the presence of moisture incites CFA reaction with free lime, calcium hydroxide, and atmospheric CO_2, hence damaging the concrete structure. The analyzed literature shows substantial efforts to enhance the manufacturing of sustainable cement and the performance of FA-based alkali-activated materials (AAMs). The pore structure and density of FA-based AAMs enable moisture discharge and limit defects during the heating process. However, limited studies have investigated the use of boiler ash in the production of AAMs. In refractory products, the incorporation of FA as a source of mullite is hindered by its high glass content and the associated softening, which are the main factors to high-temperature deformation of various oxide ceramics. Further, the application of FA geopolymers as fire-resistant materials for precast concrete requires high curing temperatures which limits their use in large industrial applications. Other potential applications of FA/FAC in electronics, optical, and photocatalysis have emerged and demonstrated considerable performance regardless of the purity of FA/FAC materials. The existing few studies highlighted promising alternative consuming applications of FA/FAC, which require further investigations to prove the feasibility on a large scale.

In summary, many challenges and research gaps in the valorization of FA in ceramic membranes still need to be addressed such as: (i) pretreatment methods to improve FA purity and reduce particle size, (ii) understanding the different mechanisms governing filtration and how they interact with each other, (iii) evaluating hazard leaching from FA-based membranes during long-term operation, and (iv) comparing properties of membranes prepared with the same materials and under identical working conditions.

REFERENCES

Agalit, H., N. Zari, and M. Maaroufi. 2020. 'Suitability of Industrial Wastes for Application as High Temperature Thermal Energy Storage (TES) Materials in Solar Tower Power Plants – A Comprehensive Review'. *Solar Energy* 208:1151–1165. doi: 10.1016/j.solener.2020.08.055.

Ayeni, J. O., S. Ma, X. Wang, and X. Hou. 2022. 'Fabrication Method, Dielectric Properties, and Electromagnetic Absorption Performance of High Alumina Fly Ash-Based Ceramic Composites'. *Ceramics International* 48(15):21268–21282. doi: 10.1016/j.ceramint.2022.04.063.

Bora, P. J., M. Porwal, K. J. Vinoy, Kishore, P. C. Ramamurthy, and G. Madras. 2018. 'Industrial Waste Fly Ash Cenosphere Composites Based Broad Band Microwave Absorber'. *Composites Part B: Engineering* 134:151–163. doi: 10.1016/j.compositesb.2017.09.062.

Chafiq Elidrissi, Z., D. El Machtani Idrissi, Y. Kouzi, B. Achiou, S. Tahiri, M. Ouammou, and S. Alami Younssi. 2023. 'Effective Conversion of Fly Ash Waste into Na-P1 Zeolite and Its Application on the Adsorption of Cr(VI)'. *Inorganic Chemistry Communications* 156:111192. doi: 10.1016/j.inoche.2023.111192.

Chanthakett, A., M. T. Arif, M. M. K. Khan, and M. Subhani. 2024. 'Chapter 4 – Hydrogen Production from Municipal Solid Waste Using Gasification Method'. Pp. 103–131 in *Hydrogen Energy Conversion and Management*, edited by M. M. K. Khan, A. K. Azad, and A. M. T. Oo. Elsevier.

Das, D., N. Kayal, and M. D. De Mello Innocentini. 2021. 'Permeability Behavior and Wastewater Filtration Performance of Mullite Bonded Porous SiC Ceramic Membrane Prepared Using Coal Fly Ash as Sintering Additive'. *Transactions of the Indian Ceramic Society* 80(3):186–192. doi: 10.1080/0371750X.2021.1934122.

El-Zaidia, M. M., M. Z. Zaki, H. M. Abomostafa, and M. A. Taha. 2024. 'Comprehensive Studies for Evaluating Promising Properties of Cu/Graphene/Fly Ash Nanocomposites'. *Scientific Reports* 14(1):2236. doi: 10.1038/s41598-024-52563-w.

Fang, J., G. Qin, W. Wei, and X. Zhao. 2011. 'Preparation and Characterization of Tubular Supported Ceramic Microfiltration Membranes from Fly Ash'. *Separation and Purification Technology* 80(3):585–591. doi: 10.1016/j.seppur.2011.06.014.

Favier, L., and M. Harja. 2020. 'TiO2/Fly Ash Nanocomposite for Photodegradation of Organic Pollutant'. Pp. 1–24 in *Handbook of Nanomaterials and Nanocomposites for Energy and Environmental Applications*, edited by O. V. Kharissova, L. M. T. Martínez, and B. I. Kharisov. Cham: Springer International Publishing.

Fraga, D., E. Barrachina, I. Calvet, T. Stoyanova, and J. B. Carda. 2018. 'Developing CIGS Solar Cells on Glass-Ceramic Substrates'. *Materials Letters* 221:104–106. doi: 10.1016/j.matlet.2018.03.111.

Gao, J.-m., B. Wang, Z. Du, Y. Guo, and F. Cheng. 2022. 'Molten Salt Synthesis of Mullite Whiskers Entirely Derived from Fly Ash for Electronic Packaging Toughening Ceramic Applications'. *Journal of Materials Research and Technology* 21:3719–3731. doi: 10.1016/j.jmrt.2022.11.010.

Glosser, D., P. Striby, M. Bellusci, R. Bottem, S. Amroun, and P. Suraneni. 2023. 'Characterization of Optical and Physiochemical Properties of Reclaimed Fly Ash for Use as Sensible Thermal Energy Storage Material in Concentrating Solar Power Systems'. *Solar Energy* 263:111944. doi: 10.1016/j.solener.2023.111944.

Goswami, K. P., K. Pakshirajan, and G. Pugazhenthi. 2022. 'Process Intensification through Waste Fly Ash Conversion and Application as Ceramic Membranes: A Review'. *Science of the Total Environment* 808:151968. doi: 10.1016/j.scitotenv.2021.151968.

Huang, J., H. Zhang, Y. Zhang, D. Liang, and H. Chen. 2022. 'Recycle Coal Fly Ash for Preparing Tubular Ceramic Membranes Applied in Transport Membrane Condenser'. *Separation and Purification Technology* 282:119972. doi: 10.1016/j.seppur.2021.119972.

Kamara, S., W. Wang, and C. Ai. 2020. 'Fabrication of Refractory Materials from Coal Fly Ash, Commercially Purified Kaolin, and Alumina Powders'. *Materials* 13(15):3406. doi: 10.3390/ma13153406.

Katsika, E., A. Moutsatsou, V. Karayannis, M. Volioti, and D. Tsoukleris. 2018. 'Synthesis and Characterization of Lignite Fly Ash Ceramic Substrates Coated with TiO2 Slurry, and Evaluation in Environmental Applications'. *Journal of the Australian Ceramic Society* 54(4):711–719. doi: 10.1007/s41779-018-0201-8.

Klima, K. M., K. Schollbach, H. J. H. Brouwers, and Qingliang Yu. 2022. 'Thermal and Fire Resistance of Class F Fly Ash Based Geopolymers – A Review'. *Construction and Building Materials* 323:126529. doi: 10.1016/j.conbuildmat.2022.126529.

Krasnyi, B. L., K. I. Ikonnikov, D. O. Lemeshev, and A. S. Sizova. 2021. 'Fly Ash as Technogenic Raw Material for Producing Refractory and Insulating Ceramic Materials (Review)'. *Glass and Ceramics* 78(1–2):48–56. doi: 10.1007/s10717-021-00347-3.

Luo, Y., Y.h. Wu, S.uh. Ma, S.l. Zheng, and P. K. Chu. 2019. 'An Eco-Friendly and Cleaner Process for Preparing Architectural Ceramics from Coal Fly Ash: Pre-Activation of Coal Fly Ash by a Mechanochemical Method'. *Journal of Cleaner Production* 214:419–428. doi: 10.1016/j.jclepro.2018.12.292.

Luo, Y, Y. Wu, S. Ma, S. Zheng, Y. Zhang, and P. K. Chu. 2021. 'Utilization of Coal Fly Ash in China: A Mini-Review on Challenges and Future Directions'. *Environmental Science and Pollution Research* 28(15):18727–18740. doi: 10.1007/s11356-020-08864-4.

Manocha, L. M., G. Prasad, and S. Manocha. 2012. 'Structural, Mechanical and Frictional Studies on Carbon-Fly Ash-Ceramic Composites'. *Transactions of the Indian Ceramic Society* 71(2):86–94. doi: 10.1080/0371750X.2012.716232.

Mathapati, M., K. Amate, C. D. Prasad, M. L. Jayavardhana, and T. H. Raju. 2022. 'A Review on Fly Ash Utilization'. *Materials Today: Proceedings* 50:1535–1540. doi: 10.1016/j.matpr.2021.09.106.

Özcan, M., B. Birol, and F. Kaya. 2021. 'Investigation of Photocatalytic Properties of TiO2 Nanoparticle Coating on Fly Ash and Red Mud Based Porous Ceramic Substrate'. *Ceramics International* 47(17):24270–24280. doi: 10.1016/j.ceramint.2021.05.138.

Rafieizonooz, M., E. Khankhaje, and S. Rezania. 2022. 'Assessment of Environmental and Chemical Properties of Coal Ashes Including Fly Ash and Bottom Ash, and Coal Ash Concrete'. *Journal of Building Engineering* 49:104040. doi: 10.1016/j.jobe.2022.104040.

Senthil Kumar, M., M. Vanmathi, G. Senguttuvan, R. V. Mangalaraja, and G. Sakthivel. 2019. 'Fly Ash Constituent-Silica and Alumina Role in the Synthesis and Characterization of Cordierite Based Ceramics'. *Silicon* 11(6):2599–2611. doi: 10.1007/s12633-018-0049-0.

Sukkae, R., S. Suebthawilkul, and B. Cherdhirunkorn. 2018. 'Utilization of Coal Fly Ash as a Raw Material for Refractory Production'. *Journal of Metals, Materials and Minerals* 28(1):116–123. doi: 10.14456/JMMM.2018.16.

Tu, Y., J. Zhong, H. Ding, H. Zhang, G. Lv, J. Zhang, and X. Hou. 2022. 'Preparation of Fly Ash Supporting Nano-TiO2 Composite Photocatalyst by a Wet Mechanical Grinding Method'. *Chemical Physics Letters* 805:139978. doi: 10.1016/j.cplett.2022.139978.

Wang, C., G. Xu, X. Gu, Y. Gao, and P. Zhao. 2021. 'High Value-Added Applications of Coal Fly Ash in the Form of Porous Materials: A Review'. *Ceramics International* 47(16):22302–22315. doi: 10.1016/j.ceramint.2021.05.070.

Wang, L., H. Sun, Z. Sun, and E. Ma. 2016. 'New Technology and Application of Brick Making with Coal Fly Ash'. *Journal of Material Cycles and Waste Management* 18(4):763–770. doi: 10.1007/s10163-015-0368-9.

Yang, Y., F. Liu, Q. Chang, Z. Hu, Q. Wang, and Y. Wang. 2019. 'Preparation of Fly Ash-Based Porous Ceramic with Alumina as the Pore-Forming Agent'. *Ceramics* 2(2):286–295. doi: 10.3390/ceramics2020023.

Zimar, Z., D. Robert, A. Zhou, F. Giustozzi, S. Setunge, and J. Kodikara. 2022. 'Application of Coal Fly Ash in Pavement Subgrade Stabilisation: A Review'. *Journal of Environmental Management* 312:114926. doi: 10.1016/j.jenvman.2022.114926.

10 Usage of Fly Ash and Bio-Coal Fly Ash for Bioenergy Production

Namita K. Talapatra, Rahul Gautam, Arnab Ghosh, Simran Kaur Dhillon, Siddhartha Pandey and Amit K Jaiswal

10.1 INTRODUCTION

The last century has witnessed tremendous cravings for energy, while initially it was sunlight, wood fires, biomass firing which were the sources of energy. Energy and climate changes are interconnected with each other and are mostly associated with excessive dependence on finite fossil fuels. The rapid growth of the global economy has accelerated industrialization and urbanization processes, leading to a significant surge in industrial energy consumption and greater reliance on energy sources. This heightened energy demand has resulted in two major consequences: energy scarcity and substantial carbon emissions, leading to the alternation of the atmosphere and environment (Mogili et al., 2022; Stephenson, 2018). Therefore, the linked increased energy demand and climate change have powered the global economy with the altered atmosphere leading to climate change that will cost hugely (Stephenson, 2018).

Carbon dioxide emissions are one of the main greenhouse gases (GHGs); they cause global warming triggered by GHG emissions, which release trap heat in the Earth's atmosphere and do not allow heat to escape the blanket of these GHGs (Talapatra et al., 2021). It is estimated that soon Earth's surface temperature will rise 2°C if GHG emissions are not dealt with. The Paris Agreement's primary goal is to keep the global temperature rise well below 1.5°C, to deal with the threat of climate change (Corbos et al., 2023). These land temperatures rising and industrial era warming were characterized by shifts in climate patterns, increased sea levels, etc. Coastal cities are submerging, which pose a direct threat to human habitat (Talapatra & Ghosh, 2022). One of the major reasons for this is thermal power plants where energy is derived from the fossil fuel coal. They lead to environmental emissions of CO_2, contributing to global warming and climate change. This is high time to act on this situation and to counter the climate changes. It is evident that depending solely on traditional mitigation approaches will not be enough to achieve the goals set in the Paris Agreement. It is evident that human-induced climate change has forced to explore alternative avenues for mitigation (Durá-Alemañ et al., 2023).

DOI: 10.1201/9781003498124-10

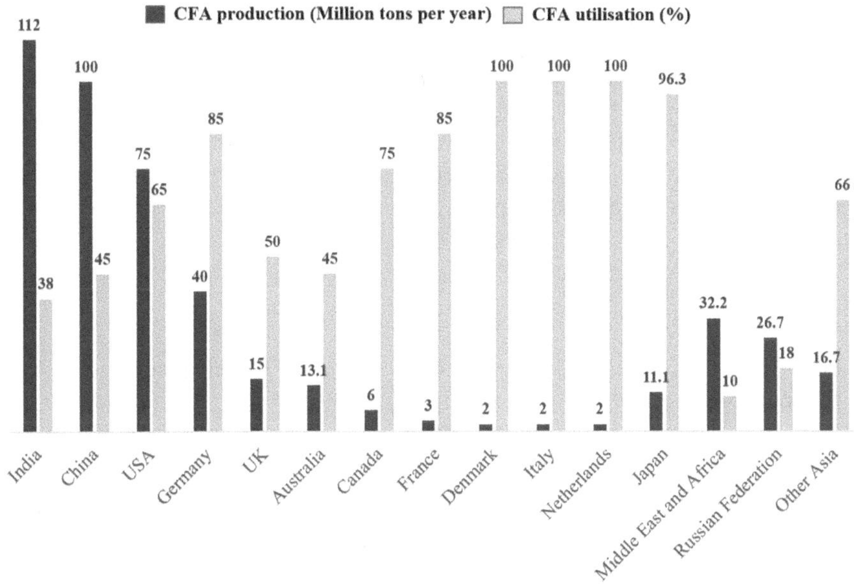

FIGURE 10.1 Production vs utilization of fly ash globally, originated from different countries.

Global warming and diminishing fossil reserves have advocated the renewable energy resources and circular economy for waste minimization (Mittal et al., 2022). Also, growing international conflicts and subsequent geopolitical instabilities have brought energy security back to prominence as a crucial policy issue. Also, bioenergy is essential to meet the Paris Agreement's terms (Amit et al., 2020; Nayak & Ghosh, 2019). The share of bioenergy in the existing energy mix is currently substantial and is likely to increase further over the next few decades. Thermal power plants generate tremendous amount of fly ash (FA) every year. Also, there is an increase in the trend of the usage of bio-coal. The burning of bio-coal also contributes to the fly ash (FA). This FA from thermal power plants and bio-coal can be tapped for bioenergy production; it could be a step forward towards circular economy. This chapter explores, focuses and discusses the use of FA from different sources into bioenergy which is crucial to tackle the disposal and the environmental issues. Also, the reusability of the FA will serve as a step towards waste minimization and mitigation. Figure 10.1 depicts the global production of FA in various regions vs percentage usage.

10.2 FA, ITS SOURCES, QUANTIFICATION, AND CHARACTERIZATION

FA is a residual prominent product from the combustion of solid fuels. It is a fine powdery material with heterogeneous composition of unburned carbon, alongside metal oxides with various other inorganic substances. This fine powdery material is mostly a product of coal-fired power plants. The equipment such as electrostatic precipitators, filter fabric collectors, etc. trap the small particulate matter from the combustion gases coming out of the thermal power plants and allow them to settle. These ashes are mainly composed of oxides of Si, Al, Fe, S, Mg, etc.

Due to its varied composition based on temperature and efficiency of the power plants, characterizing of FA is challenging. According to the American Society for Testing and Materials (ASTM) guidelines, FA is classified into class F and class C based on Al_2O_3, Fe_2O_3, SiO_2, and SO_3 percentages (Akhtar et al., 2019). Though it majorly results from thermal power plants, there are other sources as well, which are discussed below.

10.2.1 FA Sources

FA has been named based on the source of production. The belowmentioned types are the sources of the FA.

10.2.1.1 Coal FA

It is a product of firing coal in coal power plants. The particles in the combustion gases have a tendency to be transported away but by using mechanical or electrostatic equipment, these particulate matter is trapped from the flue gases. At temperatures around 1,200°C–1,500°C, non-organic particles in coal solidify to produce solid and hollow spheres with smooth glassy surfaces that range in size from 0.5 to 200 μm (Yao et al., 2015). The composition of FA varies significantly depending on factors such as the type of coal, operating conditions, combustion conditions, as cooling methods for flue gases, collectors, etc. This wide diversity in FA composition makes it challenging to characterize accurately.

According to the ASTM, based on the size distribution, FA can be characterized into class C and class F. If anthracite or bituminous coal was burnt in power plants, Class F FA with 70% combined SiO_2, $A12O_3$, and Fe_2O_3 content was emitted, while class C FA is resulted from sub-bituminous or lignite coal combustion, with the combined total of Fe_2O_3, Al_2O_3, and SiO_2 surpassing 50%. Therefore, it is the type of coal and metal oxides content used for ASTM classification (Alterary & Marei, 2021). The common techniques for FA characterization include scanning microscopy, fluorescence, X-ray diffraction, and energy-dispersive spectroscopy (Alterary & Marei, 2021).

10.2.1.2 Municipal Solid Waste FA

With population explosion and rapid urbanization, there is increased generation of Municipal Solid Waste (MSW), which is predominately incinerated or used in landfilling. But landfilling takes a very long time to treat MSW and also require a huge land. Therefore, MSW consists of organic materials and inorganic components are incinerated (Jianguo et al., 2004). The composition of MSW is varied based on factors like location, lifestyle, education, awareness, recycling practices, etc. Key constituents like food waste, plastics, and paper products not only make up a significant portion of MSW but also contribute heavily to the presence of heavy metals. These heavy metals released in the ash is a critical factor in incineration; release is dependent on technologies employed, temperature, and the type of incinerator used (Hasselriis & Licata, 1996). MSW FA is a grey fine matrix, has highly soluble salts, and pH typically exceeds 12(Lima et al., 2008).

10.2.1.3 Ash from Co-Combustion of Wood and Oil

Combustion and co-combustion of wood and oil lead to thick grey ash with high pH, mainly due to the oxides of iron. They have lesser Cl content and thus less solubility. It is a good source of cement production and verification processes (Lima et al., 2008).

10.2.1.4 Industrial Boilers

Besides coal-based power plants, industrial boilers from various industries e.g. paper-making industry, chemical industry, steel manufacturing, and metal processing industry contribute to FA. These industries burn coal for energy and heat and this results in the generation of FA.

10.2.1.5 Ashes from Bio-Coal

Bio-coal is carbon-neutral, thus aimed to replace fossil coal. It is a product of thermal conversion of biomass anaerobically. This solid biofuel has characteristics similar to coal. This bio-coal also leads to FA though the portion of ashes from bio-coal is much lesser than FA from other sources (Figure 10.2).

10.3 FA QUANTIFICATION AND CONCERNS OF FA GENERATION

Globally, it is estimated that coal-fired power plants generate around 750 million to 1 billion tonnes of FA annually, though specific figures can vary year by year based on factors like coal consumption patterns, the efficiency of power plants, and the installation of pollution control technologies. India is the largest FA producer followed by China. It is estimated that India produces 200 MMT of FA annually and is expected to cross 600 MMT by 2032 (Dharmesh & Narayan, 2020; Sharma & Yadav, 2021). The utilization of this FA varies greatly by region, with some countries having high utilization rates due to strong regulations and infrastructure that integrates FA into various industrial processes. India is one of the largest producers and consumers of coal in the world, and consequently, it is a major producer of FA. The country's reliance on coal for electricity generation results in substantial quantities of FA, estimated to be over 200 million tonnes per year. With the expansion of the coal power sector and

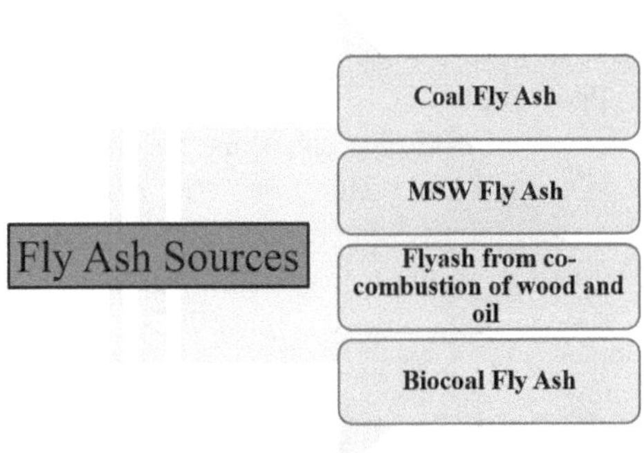

FIGURE 10.2 Major fly ash sources.

the installation of more efficient pollution control technologies, this number has been increasing. India ranks as the third-largest coal producer globally, with coal-based thermal power plants accounting for approximately 70% of the total installed capacity for power generation (Sharma & Yadav, 2021). However, the bituminous and sub-bituminous coals commonly utilized contain ash content exceeding 40%.

FA finds applications in wastewater treatment for removing dyes, inorganic and organic chemicals, and hazardous contaminants (Mushtaq et al., 2019). Moreover, it shows highly promising potential in the construction sector, particularly in concrete and cement production. When incorporated into cement in finely reduced nanosize, FA enhances concrete durability, reduces pore size, and improves resistance to environmental conditions (Ram & Mohanty, 2022).

FA poses significant risks to local ecosystems if disposed improperly or mishandled. The erosion and leachate formation from FA lead to heavy metal pollution (Chen et al., 2024; Dharmesh & Narayan, 2020; Sharma & Yadav, 2021). Because FA is lightweight, it can become airborne in addition to taking up large amounts of land. FA that has been disposed of can release dangerous metals into the soil, plants, groundwater, and surface water. As a result, the two main obstacles faced by coal-based thermal power plants are the need for sufficient land for the disposal of FA and the mitigation of pollution, specifically suspended particle matter (SPM) and the transfer of heavy metals into groundwater and the food chain (Chen et al., 2024; Jambhulkar et al., 2018).

The heavy metal content like Ni, Cd, As, Cr, Pb of the FA can cause extreme environmental pollution and health issues. It can gradually accumulate in the respiratory tract and cause lung poisoning and fibrosis. It accounts for almost 20 million respiratory cases globally.

It was estimated that 1.5 lakh premature deaths are caused due to the FA in India, including underaged children. Even a few studies claimed that FA is 100 times more harmful than nuclear radiations. It also co-related with the destruction of vegetation including mangroves, decrease in crop yield, and groundwater pollution (Mushtaq et al., 2019; Ram & Mohanty, 2022).

To mitigate the environmental impact of FA and reduce land requirements for disposal, the Ministry of Environment and Forest has issued various notifications aiming for 100% utilization of FA (Hussain et al., 2023). Since 1994, the Ministry of Science and Technology (Government of India) has implemented diverse technologies for the safe and productive utilization of FA, leading to a significant increase in FA utilization from 6.64 million tonnes in 1996–1997 to 168.40 million tonnes in 2018–2019 (Rastogi & Paul, 2020).

Additionally, it serves as an effective substitute for reclaiming low-lying areas, aiding in topsoil layer protection and restoration. With the increasing electricity and energy demand in India and globally, its production is expected to rise accordingly. Therefore, it is essential to access the role and applicability of potential in the production of bioenergy for sustainable future (Figure 10.3).

10.4 FA AS A VECTOR OF ENERGY

This chapter is devoted to use of the FA in the field of bioenergy as biomass-derived fuels are environmentally friendly (Tan et al., 2020). One of the sources is algal

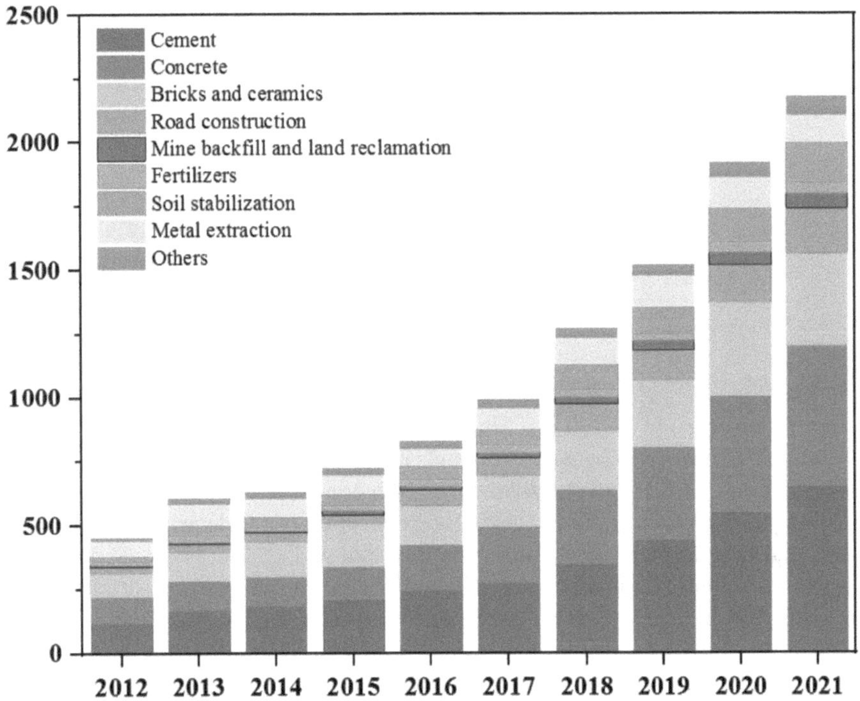

FIGURE 10.3 Major uses of fly ash in different sectors.

biomass, which is one of the most prominent sources that contains large amounts of lipids suited for biodiesel production. These biomass-based fuels are considered as third-generation biofuels (Amit & Ghosh, 2018; Cheng et al., 2016). These alternative fuels must be economically competitive, environmentally acceptable, and readily available (Balat, 2011). The waste-to-energy route curb on GHG emissions and can provide a clean environment by limiting the harmful emissions of CO_2, SOx, NOx emissions, and in the greenhouse effects (Dey et al., 2020; Shumbulo & Ki, 2018). Biofuels from renewable sources are economically, ecologically, and environmentally viable (Manmai et al., 2020). Bio-remediation of the waste materials by microalgae is upcoming, which has proved to be an effective tool in the waste to energy route for pollutant removal as well as an energy source (Burger & Abraham, 2019; Chandra, Amit, & Ghosh, 2019; Chandra, Iqbal, et al., 2019). A variety of biofuels from biomass includes biogas, bioethanol, biodiesel, biohydrogen, bioelectricity, etc., and can be produced from various feedstock edible substrates like corn, sugarcane, and sweet sorghum, and even from non-edible resources like rice straw and wood waste, are first-generation and second-generation biofuels, respectively (Tajmirriahi et al., 2021). Among these fuels, biodiesel has been widely accepted by the public because of its benefits, such as low emissions of pollutants and GHG and renewability.

Bioethanol, another major bioenergy source, is produced from different organic substrates such as sugarcane (first-generation), lingo-cellulosic biomass (second-

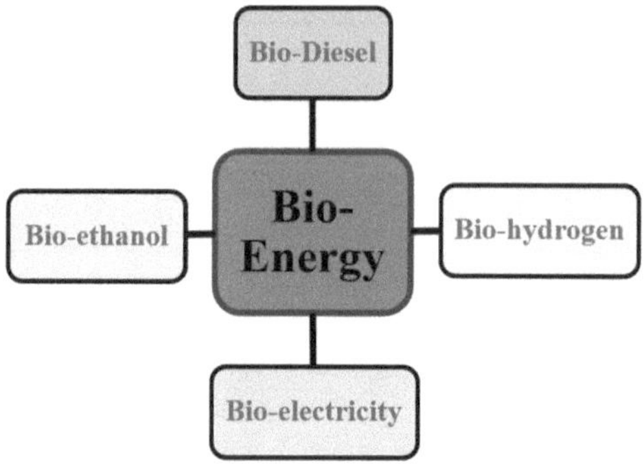

FIGURE 10.4 Usage of fly ash in different bioenergy sectors.

generation) (Koçar & Civaş, 2013) and microalgal biomass (third-generation) (Miranda et al., 2012) with the advantages of biodegradability and lower pollutant emissions as compared to fossil fuels. The production of bioelectricity needs lot of efforts and improvement for optimization (Gautam et al., 2021; Nayak et al., 2018). The optimization of various parameters and design can play an important role in improving the productivity of Microbial Fuel Cells (MFCs) (Figure 10.4).

10.5 FA IN BIODIESEL PRODUCTION

Biodiesel has garnered significant attention as a renewable and eco-friendly alternative to conventional fossil fuels. A pivotal aspect of biodiesel production involves utilizing catalysts to aid the transformation of vegetable oils or animal fats into fatty acid methyl esters (FAME) and glycerol via transesterification reactions. Among the array of catalysts explored, FA has emerged as a promising candidate for catalysing biodiesel production. Despite being commonly perceived as waste, this fine particulate material exhibits noteworthy catalytic properties. Research demonstrates that specific types of FA can effectively expedite the conversion of raw materials into biodiesel and glycerol. It is cost-effective, eco-friendly, and has wide availability. However, the effectiveness of FA catalysts depends on various factors as its source, composition, and pre-treatment methods. The integration of FA into biodiesel production not only offers a sustainable solution for waste management by repurposing a discarded material but also showcases potential enhancements in catalytic activity, thus reducing the environmental impact of the overall production process (Ćwik et al., 2019; Kotwal et al., 2009).

FA effectiveness mainly trunks around its high CaO content (González et al., 2010; Muñoz et al., 2020) and other compounds that have shown efficacy in catalysing FAME production (Devaraj et al., 2019; Marinković et al., 2016; Mittal et al., 2022; Mittal & Kumar Ghosh, 2022). Kotwal et al. demonstrated the utilization of fly ash by impregnating it at high temperatures (500°C for 5 hours) to convert it into a support for the K_2O catalyst, achieving an impressive 87.5% FAME yield from sunflower oil at 170°C and 8 hours of reaction time (Kotwal et al., 2009). Recently, Pavlovic et al. introduced a novel catalyst composed of lignite coal FA and chicken eggshells, which was obtained through alkali activation and hydration-dehydration processes (Pavlović et al., 2020). This catalyst exhibited promising performance, yielding up to 99% FAME. Such innovative approaches highlight the potential of utilizing waste materials for sustainable and efficient biodiesel production. In their investigation, Muñoz et al. explored the utilization of FA as a versatile acid-base catalyst in the biodiesel production process using waste frying oil (WFO) as the feedstock (Muñoz et al., 2020). Employing response surface methodology (RSM), the study aimed to optimize the conditions for achieving maximum FAME yield. Remarkably, the study achieved a 100% FAME yield under specific parameters, including a methanol-to-oil molar ratio of 3.1:1, 11.2% fly ash concentration, a temperature of 59°C, agitation at 245 rpm, and a six-hour reaction time. The research revealed that FA functions as a versatile catalyst, facilitating both transesterification and hydro esterification reactions due to its chemical composition containing constituents with acidic and basic properties. Despite this, the experimental FAME yield under optimized conditions reached 78.1%, attributed to the rapid deactivation of the catalyst. Additionally, experiments conducted in an anhydrous medium resulted in lower FAME yields compared to the control, indicating the potential catalysis of hydrolysis and esterification reactions alongside transesterification by fly ash. These findings suggest that FA serves as a multipurpose catalyst, capable of simultaneously catalysing various reactions, thereby presenting a cost-effective alternative for biodiesel production. Overall, the study provides valuable insights into the practical application of FA as a versatile catalyst in biodiesel production, offering an economically viable alternative to conventional catalysts such as lipases and acids. In their study, Go et al. underscored the advantages of solid catalysts over chemical and biological counterparts, particularly emphasizing their reusability and ease of separation from the reaction mixture (Go & Yeom, 2019). Focusing on optimizing reaction conditions for biodiesel production, the research specifically delves into the utilization of a solid catalyst derived from FA, highlighting its successful recycling and attainment of high biodiesel conversion rates. The researchers devised a method involving the preparation of the solid catalyst by treating raw fly ash (RFA) with acid, combining it with metal hydroxides (NaOH, KOH, or $Ca(OH)_2$), and subsequently calcining it at 700°C for 3 hours. Notably, the study showcased the solid catalyst's ability to be reused for biodiesel production, maintaining consistently high conversion rates across multiple operational cycles. Furthermore, the research sheds light on the promising potential of solid catalysts derived from coal FA for fostering sustainable biodiesel production, thereby contributing to the advancement of environmentally friendly fuel technologies. Despite the promising results, there is a need for further research to address the rapid deactivation of the FA catalyst and to explore its potential for scaling up biodiesel production. The effectiveness of fly ash catalysts can vary depending on factors such as the source, composition, and treatment methods. Optimization of catalyst preparation techniques, including impregnation, activation, and modification processes, is necessary to enhance catalytic activity and selectivity. Additionally, the compatibility of FA catalysts with different feedstocks and reaction conditions requires further investigation to ensure consistent performance.

In conclusion, FA holds great potential as a catalyst for biodiesel production, offering advantages such as availability, cost-effectiveness, and environmental sustai

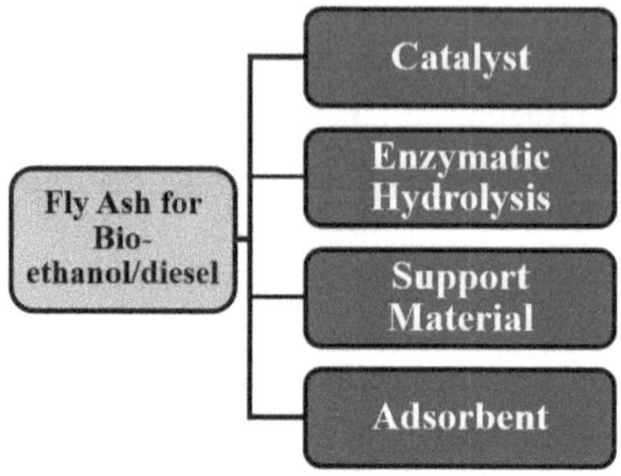

FIGURE 10.5 Role of fly ash for bioethanol and biodiesel production.

nability. However, further research and optimization are required to address challenges related to catalyst variability and performance. With ongoing advancements and innovation in catalyst design and synthesis, fly ash-based catalysts are poised to play a significant role in advancing sustainable biodiesel production methods, contributing to the transition towards a greener and more energy-efficient future (Figure 10.5).

10.6 FA IN BIOETHANOL PRODUCTION

FA has been researched and applied in various innovative ways, including its potential use in the bioethanol production process as well. Preliminary fly ash-amended soil for growing sugarcane results in better yield and productivity (Awoyemi & Adeleke, 2017). In bioethanol production, FA can be utilized as a low-cost adsorbent or catalyst. The high content of silica and alumina in FA makes it a suitable catalyst due to its thermal stability and acidity which can enhance the breakdown of biomass into simpler molecules (Mais et al., 2002). FA has been used as a catalyst in bioethanol production for the thermochemical conversion processes such as gasification and pyrolysis. These processes are involved in converting biomass into syngas (a mixture of carbon monoxide, hydrogen, and carbon dioxide), which can then be fermented to produce ethanol (Freitas et al., 2019). It has also been implemented as an adsorbent in bioethanol industry due to its porous structure; it can be used to remove impurities and contaminants during the fermentation process, which can increase the purity and yield of the ethanol produced. Additionally, FA has been tested in the adsorption of CO_2 during fermentation to potentially enhance ethanol yield, as CO_2 can inhibit the fermentation process if concentrations become too high affecting the pH. Also, FA plays a role in enzymatic hydrolysis, where it may help in breaking down cellulosic biomass into sugars before fermentation to ethanol. The mineral content in FA can potentially enhance the activity of hydrolytic enzymes (Freitas et al., 2019; Qiu et al., 2022).

10.7 FA IN GREEN HYDROGEN PRODUCTION

The use of FA in green hydrogen production is a relatively new and innovative area of research. Green hydrogen is typically produced through the electrolysis

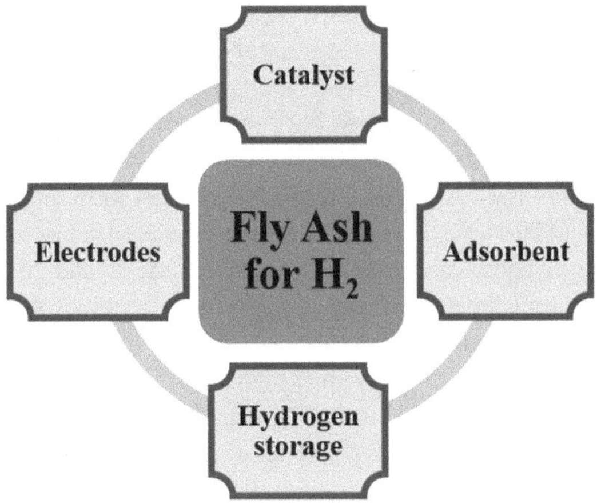

FIGURE 10.6 Role of fly ash in biohydrogen production.

of water using electricity generated from renewable energy sources, but new methods are also attracting the scientific community e.g. dark fermentation, bio-electrochemical methods, photolysis etc. (Alavi-Borazjani et al., 2019; Gautam et al., 2023). One of the promising alternatives for hydrogen production is Microbial Electrolysis Cell (MEC) which is based on BES. FA enhances the efficiency and reduces the cost of the electrolysis process by using FA rich in oxides of Si, Al as electrodes in electrolyzers. Also, it is a suitable catalyst support in water electrolysis by supporting metal catalysts on FA, which can provide stability and increase the surface area for the reactions. There are studies where fly ash/Ni acts as a catalyst for selective hydrogen production (Bepari et al., 2017; Onder & Ozay, 2020). Also, it can be applied to purify hydrogen produced, as its porosity allows it to trap impurities and contaminants (Czarna-Juszkiewicz et al., 2020). Therefore, FA has been explored for its potential in hydrogen storage applications by using zeolites based on FA (Czarna-Juszkiewicz et al., 2020; Paulsen & Moeslund, 2020). These materials and their structures are naturally engineered to adsorb and release hydrogen, playing a role in hydrogen production and its storage (Figure 10.6).

10.8 BENEFITS AND CHALLENGES OF USING FA AND BIO-COAL FA FOR BIOENERGY

FA is a potential waste of the coal-based thermal power plants, with various applications. The emerging field of usage of FA is for the bioenergy production. The major usage of FA is in construction and building industry where it is used as an alternative for cement. The usage of FA reduces CO_2 emissions.

However, the usage of the FA in the domain of bioenergy is new and exciting to explore. It offers dual benefits: (a) it reduces the disposal volume and land requirement and (b) aids in the production of bioenergy. The FA usage for biodiesel production is beneficial and results in high yields.

However, it is in a very naive and early stage of commercialization. For the biodiesel production, FA is used as either a catalyst or catalyst support; issue is the selectivity and low degree of reusability.

The higher demands for green renewable energy, optimistically considering FA as the green material to make the renewable energy extraction greener.

Other potent uses of the FA for bioenergy are in bioethanol production and green hydrogen production. The fly ash in these bioenergy routes act as either as a catalyst, support, or adsorbent. The issues are with low mechanical strength and high permeability. The catalyst reusability and stability are questionable. Also, the use of FA sometimes leads to toxicity due to its composition.

The FA is not always used directly in the process; it also requires pre-treatment, which increases the cost as well as the energy usage affecting the process energy and cost efficiency (Kotwal et al., 2009; Pavlović et al., 2020). These processes also affect the heavy metal removal from the source; however, the BES system is more focused for energy extraction. There are changes of interference of the heavy metals in the course of energy extraction.

REFERENCES

Akhtar, M. N., Akhtar, J., & Tarannum, N. (2019). Physiochemical characterization and dematerialization of coal class F flyash residues from thermal power plant. *Civil Engineering Journal (Iran)*, *5*(5), 1041–1051. https://doi.org/10.28991/cej-2019-03091310

Alavi-Borazjani, S. A., Capela, I., & Tarelho, L. A. C. (2019). Dark fermentative hydrogen production from food waste: Effect of biomass ash supplementation. *International Journal of Hydrogen Energy*, *44*(48), 26213–26225. https://doi.org/10.1016/j.ijhydene.2019.08.091

Alterary, S. S., & Marei, N. H. (2021). Fly ash properties, characterization, and applications: A review. *Journal of King Saud University – Science*, *33*(6), 101536. https://doi.org/10.1016/j.jksus.2021.101536

Amit, Nayak, J. K., & Ghosh, U. K. (2018). An approach for phycoremediation of different wastewaters and biodiesel production using microalgae. *Environmental Science and Pollution Research*, *25*(19), 18673–18681. https://doi.org/10.1007/s11356-018-1967-5

Amit, Nayak, J. K., & Ghosh, U. K. (2020). Microalgal remediation of anaerobic pretreated pharmaceutical wastewater for sustainable biodiesel production and electricity generation. *Journal of Water Process Engineering*, *35*(September 2019), 101192. https://doi.org/10.1016/j.jwpe.2020.101192

Awoyemi, O. M., & Adeleke, E. O. (2017). Bioethanol production from switchgrass grown on coal fly ash-amended soil. *World Journal of Agricultural Research*, *5*(3), 147–155. https://doi.org/10.12691/wjar-5-3-4

Balat, M. (2011). Production of bioethanol from lignocellulosic materials via the biochemical pathway: A review. *Energy Conversion and Management*, *52*(2), 858–875. https://doi.org/10.1016/j.enconman.2010.08.013

Bepari, S., Pradhan, N. C., & Dalai, A. K. (2017). Selective production of hydrogen by steam reforming of glycerol over Ni/Fly ash catalyst. *Catalysis Today*, *291*, 36–46. https://doi.org/10.1016/j.cattod.2017.01.015

Burger, R. E. A., & Abraham, E. (2019). *2nd WaterEnergyNEXUS – International Conference Maximizing Water-Food-Energy Nexus synergies at Basin Scale*. November 2019, 6–8. https://doi.org/10.1007/978-3-030-13068-8

Chandra, R., Amit, & Ghosh, U. K. (2019). Effects of various abiotic factors on biomass growth and lipid yield of Chlorella minutissima for sustainable biodiesel production. *Environmental Science and Pollution Research*, *26*(4), 3848–3861. https://doi.org/10.1007/s11356-018-3696-1

Chandra, R., Iqbal, H. M. N., Vishal, G., Lee, H. S., & Nagra, S. (2019). Algal biorefinery: A sustainable approach to valorize algal-based biomass towards multiple product recovery. *Bioresource Technology*, *278*(November 2018), 346–359. https://doi.org/10.1016/j.biortech.2019.01.104

Chen, Y., Fan, Y., Huang, Y., Liao, X., Xu, W., & Zhang, T. (2024). A comprehensive review of toxicity of coal fly ash and its leachate in the ecosystem. *Ecotoxicology and Environmental Safety*, *269*(August 2023), 115905. https://doi.org/10.1016/j.ecoenv.2023.115905

Cheng, Y.-L., Lee, C.-Y., Huang, Y.-L., Buckner, C. A., Lafrenie, R. M., Dénommée, J. A., Caswell, J. M., Want, D. A., Gan, G. G., Leong, Y. C., Bee, P. C., Chin, E., Teh, A. K. H., Picco, S., Villegas, L., Tonelli, F., Merlo, M., Rigau, J., Diaz, D.,…, & Mathijssen, R. H. J. (2016). We are IntechOpen, the world's leading publisher of Open Access books Built by scientists, for scientists TOP 1%. *Intech*, *11*(tourism), 13. https://www.intechopen.com/books/advanced-biometric-technologies/liveness-detection-in-biometrics

Corbos, R. A., Bunea, O. I., & Jiroveanu, D. C. (2023). The effects of the energy crisis on the energy-saving behavior of young people. *Energy Strategy Reviews*, *49*(2), 101184. https://doi.org/10.1016/j.esr.2023.101184

Ćwik, A., Casanova, I., Rausis, K., & Zarębska, K. (2019). Utilization of high-calcium fly ashes through mineral carbonation: The cases for Greece, Poland and Spain. *Journal of CO2 Utilization*, *32*(April), 155–162. https://doi.org/10.1016/j.jcou.2019.03.020

Czarna-Juszkiewicz, D., Cader, J., & Wdowin, M. (2020). From coal ashes to solid sorbents for hydrogen storage. *Journal of Cleaner Production*, *270*, 122355. https://doi.org/10.1016/j.jclepro.2020.122355

Devaraj, K., Veerasamy, M., Aathika, S., Mani, Y., Thanarasu, A., Dhanasekaran, A., & Subramanian, S. (2019). Study on effectiveness of activated calcium oxide in pilot plant biodiesel production. *Journal of Cleaner Production*, *225*, 18–26. https://doi.org/10.1016/j.jclepro.2019.03.244

Dey, P., Pal, P., Kevin, J. D., & Das, D. B. (2020). Lignocellulosic bioethanol production: Prospects of emerging membrane technologies to improve the process – A critical review. *Reviews in Chemical Engineering*, *36*(3), 333–367. https://doi.org/10.1515/revce-2018-0014

Durá-Alemañ, C. J., Moleón, M., Pérez-García, J. M., Serrano, D., & Sánchez-Zapata, J. A. (2023). Climate change and energy crisis drive an unprecedented EU environmental law regression. *Conservation Letters*, *16*(3), 2–4. https://doi.org/10.1111/conl.12958

Freitas, J. V., Ruotolo, L. A. M., & Farinas, C. S. (2019). Adsorption of inhibitors using a CO2-activated sugarcane bagasse fly ash for improving enzymatic hydrolysis and alcoholic fermentation in biorefineries. *Fuel*, *251*(February), 1–9. https://doi.org/10.1016/j.fuel.2019.04.032

Gautam, R., Nayak, J. K., Ress, N. V, Steinberger-wilckens, R., & Kumar, U. (2023). Bio-hydrogen production through microbial electrolysis cell : Structural components and influencing factors. *Chemical Engineering Journal*, 455(November 2022), 140535. https://doi.org/10.1016/j.cej.2022.140535

Gautam, R., Nayak, J. K., Talapatra, K. N., Amit, & Ghosh, U. K. (2021). Assessment of different organic substrates for Bio-Electricity and Bio-Hydrogen generation in an Integrated Bio-Electrochemical System. *Materials Today: Proceedings*, *xxxx*, 6–10. https://doi.org/10.1016/j.matpr.2021.06.223

Go, Y. W., & Yeom, S. H. (2019). Fabrication of a solid catalyst using coal fly ash and its utilization for producing biodiesel. *Environmental Engineering Research*, *24*(2), 324–330. https://doi.org/10.4491/EER.2018.029

González, A., Moreno, N., Navia, R., & Querol, X. (2010). Study of a Chilean petroleum coke fluidized bed combustion fly ash and its potential application in copper, lead and hexavalent chromium removal. *Fuel*, *89*(10), 3012–3021. https://doi.org/10.1016/j.fuel.2010.04.032

Hasselriis, F., & Licata, A. (1996). Analysis of heavy metal emission data from municipal waste combustion. *Journal of Hazardous Materials*, *47*(1–3), 77–102. https://doi.org/10.1016/0304-3894(95)00107-7

Hussain, A., Ali, J., & Faizan, S. (2023). Exploring the scientific research on coal fly ash and agriculture: Knowledge mapping and future research directions. *Environmental Science and Pollution Research International*, *30*(58), 121292–121305. https://doi.org/10.1007/s11356-023-31075-6

Jambhulkar, H. P., Shaikh, S. M. S., & Kumar, M. S. (2018). Fly ash toxicity, emerging issues and possible implications for its exploitation in agriculture; Indian scenario: A review. *Chemosphere*, *213*, 333–344. https://doi.org/10.1016/j.chemosphere.2018.09.045

Jianguo, J., Jun, W., Xin, X., Wei, W., Zhou, D., & Yan, Z. (2004). Heavy metal stabilization in municipal solid waste incineration flyash using heavy metal chelating agents. *Journal of Hazardous Materials*, *113*(1–3), 141–146. https://doi.org/10.1016/j.jhazmat.2004.05.030

Koçar, G., & Civaş, N. (2013). An overview of biofuels from energy crops: Current status and future prospects. In *Renewable and Sustainable Energy Reviews* (Vol. 28, pp. 900–916). https://doi.org/10.1016/j.rser.2013.08.022

Kotwal, M. S., Niphadkar, P. S., Deshpande, S. S., Bokade, V. V., & Joshi, P. N. (2009). Transesterification of sunflower oil catalyzed by flyash-based solid catalysts. *Fuel*, *88*(9), 1773–1778. https://doi.org/10.1016/j.fuel.2009.04.004

Lima, A. T., Ottosen, L. M., Pedersen, A. J., & Ribeiro, A. B. (2008). Characterization of fly ash from bio and municipal waste. *Biomass and Bioenergy*, *32*(3), 277–282. https://doi.org/10.1016/j.biombioe.2007.09.005

Mais, U., Esteghlalian, A. R., Saddler, J. N., & Mansfield, S. D. (2002). Enhancing the enzymatic hydrolysis of cellulosic materials using simultaneous ball milling. *Applied Biochemistry and Biotechnology – Part A Enzyme Engineering and Biotechnology*, *98–100*, 815–832. https://doi.org/10.1385/ABAB:98-100:1-9:815

Manmai, N., Unpaprom, Y., & Ramaraj, R. (2020). Bioethanol production from sunflower stalk: application of chemical and biological pretreatments by response surface methodology (RSM). *Biomass Conversion and Biorefinery*. https://doi.org/10.1007/s13399-020-00602-7

Marinković, D. M., Stanković, M. V., Veličković, A. V., Avramović, J. M., Miladinović, M. R., Stamenković, O. O., Veljković, V. B., & Jovanović, D. M. (2016). Calcium oxide as a promising heterogeneous catalyst for biodiesel production: Current state and perspectives. *Renewable and Sustainable Energy Reviews*, *56*, 1387–1408. https://doi.org/10.1016/j.rser.2015.12.007

Miranda, J. R., Passarinho, P. C., & Gouveia, L. (2012). Bioethanol production from Scenedesmus obliquus sugars: The influence of photobioreactors and culture conditions on biomass production. *Applied Microbiology and Biotechnology*, *96*(2), 555–564. https://doi.org/10.1007/s00253-012-4338-z

Mittal, V., & Kumar Ghosh, U. (2022). Comparative analysis of two different nanocatalysts for producing biodiesel from microalgae. *Materials Today: Proceedings*, *xxxx*. https://doi.org/10.1016/j.matpr.2022.03.652

Mittal, V., Talapatra, K. N., & Ghosh, U. K. (2022). A comprehensive review on biodiesel production from microalgae through nanocatalytic transesterification process: Lifecycle assessment and methodologies. *International Nano Letters*. https://doi.org/10.1007/s40089-022-00372-2

Mogili, N. V., Murugesan, N., Ayothiraman, S., Gautam, R., Deshavath, N. N., & Reddy Erva, R. (2022). Biohydrogen production from wastewater and organic solid wastes. *Waste-to-Energy Approaches towards Zero Waste*, 165–195. https://doi.org/10.1016/B978-0-323-85387-3.00009-4

Muñoz, R., González, A., Valdebenito, F., Ciudad, G., Navia, R., Pecchi, G., & Azócar, L. (2020). Fly ash as a new versatile acid-base catalyst for biodiesel production. *Renewable Energy*, *162*, 1931–1939. https://doi.org/10.1016/j.renene.2020.09.099

Mushtaq, F., Zahid, M., Bhatti, I. A., Nasir, S., & Hussain, T. (2019). Possible applications of coal fly ash in wastewater treatment. *Journal of Environmental Management*, *240*(February), 27–46. https://doi.org/10.1016/j.jenvman.2019.03.054

Nayak, J. K., Amit, & Ghosh, U. K. (2018). An innovative mixotrophic approach of distillery spent wash with sewage wastewater for biodegradation and bioelectricity generation using microbial fuel cell. *Journal of Water Process Engineering*, *23*. https://doi.org/10.1016/j.jwpe.2018.04.003

Nayak, J. K., & Ghosh, U. K. (2019). Post treatment of microalgae treated pharmaceutical wastewater in photosynthetic microbial fuel cell (PMFC) and biodiesel production. In *Biomass and Bioenergy* (Vol. 131). https://doi.org/10.1016/j.biombioe.2019.105415

Onder, A., & Ozay, O. (2020). Fly ash as catalyst support material in the hydrolysis of ethylenediamine bisborane for hydrogen production: The use of coal-fired power plant waste. *International Journal of Hydrogen Energy*, *45*(20), 11651–11661. https://doi.org/10.1016/j.ijhydene.2020.02.093

Paulsen, R. R., & Moeslund, T. B. (2020). *Image Storage and Compression*, *340*, 27–32. https://doi.org/10.1007/978-3-030-39364-9_3

Pavlović, S. M., Marinković, D. M., Kostić, M. D., Janković-Častvan, I. M., Mojović, L. V., Stanković, M. V., & Veljković, V. B. (2020). A CaO/zeolite-based catalyst obtained from waste chicken eggshell and coal fly ash for biodiesel production. *Fuel*, *267*(January), 117171. https://doi.org/10.1016/j.fuel.2020.117171

Qiu, B., Yang, C., Shao, Q., Liu, Y., & Chu, H. (2022). Recent advances on industrial solid waste catalysts for improving the quality of bio-oil from biomass catalytic cracking: A review. *Fuel*, *315*(August 2021), 123218. https://doi.org/10.1016/j.fuel.2022.123218

Ram, A. K., & Mohanty, S. (2022). State of the art review on physiochemical and engineering characteristics of fly ash and its applications. *International Journal of Coal Science and Technology*, *9*(1). https://doi.org/10.1007/s40789-022-00472-6

Rastogi, A., & Paul, V. K. (2020). A critical review of the potential for fly ash utilisation in construction-specific applications in India. *Environmental Research, Engineering and Management*, *76*(2), 65–75. https://doi.org/10.5755/J01.EREM.76.2.25166

Shah, D., & Narayan, S. (2020). *Fly-Ash-Report-Final_July-23.pdf.*

Sharma, R., & Yadav, R. N. (2021). Fly ash generation-utilization, government initiatives in India and other diverse applications: A review. *Journal of Advanced Scientific Research*, *12*(01), 9–20. https://doi.org/10.55218/jasr.202112102

Shumbulo, E., & Ki, D. (2018). Microalgae to biofuels: 'Promising' alternative and renewable energy, review. *81*(May 2017), 743–755. https://doi.org/10.1016/j.rser.2017.08.042

Stephenson, M. (2018). Energy and climate change: An introduction to geological controls, interventions and mitigations. In *Energy and Climate Change: An Introduction to Geological Controls, Interventions and Mitigations*, 1–186. https://doi.org/10.1016/B978-0-12-812021-7.00009-9

Tajmirriahi, M., Momayez, F., & Karimi, K. (2021). The critical impact of rice straw extractives on biogas and bioethanol production. *Bioresource Technology*, *319*(August 2020), 124167. https://doi.org/10.1016/j.biortech.2020.124167

Talapatra, N., Gautam, R., Mittal, V., & Ghosh, U. K. (2021). A comparative study of the growth of microalgae-bacteria symbiotic consortium with the axenic culture of microalgae in dairy wastewater through extraction and quantification of chlorophyll. *Materials Today: Proceedings*, *xxxx*. https://doi.org/10.1016/j.matpr.2021.06.227

Talapatra, N., & Ghosh, U. K. (2022). New concept of biodiesel production using food waste digestate powder: Co-culturing algae-activated sludge symbiotic system in low N and P paper mill wastewater. *Science of the Total Environment*, *844*(July), 157207. https://doi.org/10.1016/j.scitotenv.2022.157207

Tan, J., Dan, H., & Li, J. (2022). Use of municipal waste incineration fly ashes (MSWI FA) in metakaolin-based geopolymer. *Environmental Science and Pollution Research*, *29*(53), 80727–80738.

Yao, Z. T., Ji, X. S., Sarker, P. K., Tang, J. H., Ge, L. Q., Xia, M. S., & Xi, Y. Q. (2015). A comprehensive review on the applications of coal fly ash. *Earth-Science Reviews*, *141*, 105–121.

11 Challenges and Opportunities in Utilization of Bio-Coal Ashes

Kedar Mehta, Divya Tirva, Maanit Ghodasara, Dharmesh Sur and Abhinay Thakur

11.1 INTRODUCTION

Bio-coal, frequently referred to as bio-coal or biochar, is a notable feature of progressive development in the field of renewable energy and sustainable materials. Produced mostly from biomass by methods such as pyrolysis and torrefaction, bio-coal provides a sustainable substitute for traditional fossil fuels (Lehmann & Joseph, 2009; Kambo & Dutta 2015; Akhator et al., 2023). In response to the rising worldwide energy demand and growing environmental concerns, bio-coal has emerged as a very promising approach for reducing greenhouse gas emissions and mitigating climate change. Biogenic coal is produced by subjecting organic materials to thermal breakdown in an environment with restricted oxygen, leading to the formation of a solid residue that is abundant in carbon. This material demonstrates characteristics akin to those of conventional coal, but with a reduced ecological footprint. The use of bio-coal holds promises as a viable and environmentally friendly energy alternative, therefore facilitating a decrease in dependence on fossil fuels and the consequent carbon emissions (Zhang et al. 2013; Demirbaş 2004).

An inherent trait of bio-coal is its remarkable adaptability. It finds utility in a wide range of applications, including energy production and agricultural amendment. The co-firing of bio-coal with conventional coal in power plants can effectively decrease the total carbon emissions in energy conversion. Moreover, the application of bio-coal to soils can serve as a carbon sink, therefore improving soil fertility and effectively storing carbon for extended durations. One more topic of considerable attention is the sustainability of bio-coal production. The raw resources used in bio-coal production, such as agricultural wastes, forestry waste, and other biomass products, are frequently plentiful and renewable (Lehmann et al. 2006; Sohi et al. 2009). Consequently, the absence of competition between bio-coal production and food production, as well as the prevention of deforestation, renders it a feasible choice for sustainable energy. The burning of bio-coal, which is generated by the process of pyrolysis or torrefaction of

DOI: 10.1201/9781003498124-11

biomass, yields fly ash that is abundant in carbon, minerals, and trace elements (Sear 2009). This composition is determined by the specific biomass type, the combustion process parameters, and the inherent characteristics of the bio-coal. In comparison to conventional coal fly ash, bio-coal fly ash is often distinguished by a greater proportion of unburned carbon, reduced quantities of hazardous metals, and a more diverse mineralogical makeup. The disparities above render bio-coal fly ash a promising asset as well as a formidable obstacle in the realms of waste management and material recovery (Tenenbaum 2000; Downie et al. 2012; Rawat & Kumar 2022).

This chapter explores the characteristics, environmental impact, and various applications of bio-coal fly ash, highlighting its role in sustainable development and circular economy. The growing interest in renewable energy sources has led to the increased use of biomass as a co-firing material with coal in thermal power plants. The combustion of this biomass, alongside coal, produces bio-coal fly ash—a byproduct that combines the characteristics of traditional coal fly ash and biomass ash. The utilization of bio-coal fly ash presents a significant opportunity to enhance sustainability across various sectors, including construction, agriculture, and environmental management (Sarkar 1990; Marganingrum & Estiaty 2020; Wang & Wu 2006a, 2006b).

One of the main issues related to bio-coal fly ash is its proper disposal. Inadequate disposal can result in environmental pollution, namely of soil and water resources, as a consequence of the leaching of trace elements. Nevertheless, with proper management, bio-coal fly ash has the capacity to be reutilized in several ways, therefore transforming a waste product into a useful asset. One application of bio-coal fly ash is as a soil amendment, wherein its abundant carbon content and mineral elements can augment soil fertility and foster the adoption of sustainable agriculture methods (Beesley et al. 2010; Wang & Wu 2006a, 2006b). Moreover, its pozzolanic characteristics render it well-suited for application in the construction sector, namely in the manufacturing of cement and concrete, where it can enhance the longevity and robustness of these raw materials. Additional investigation is required to have a deeper understanding of its characteristics and to make recommendations for its safer and more efficient use (Vassilev & Vassileva 2007). Academic research in recent years has investigated the possible applications of bio-coal fly ash in several sectors such as agriculture, building, and environmental remediation (Demirbaş 2005a, 2005b). These studies emphasize both the potential advantages and the constraints linked to the utilization of bio-coal fly ash. Although bio-coal fly ash has demonstrated potential as a soil additive, there are still apprehensions about the potential long-term impacts of its use on soil health and crop fertility. Likewise, although the use of this material in building supplies provides environmental advantages, concerns regarding the unpredictable nature of its composition and the possible emission of hazardous chemicals must be resolved (Querol et al. 2001).

Ultimately, bio-coal fly ash poses both a complexity and a potential advantage within the realm of sustainable energy generation. In tandem with the growing usage of bio-coal, there will be an increasing demand for efficient approaches to handle its wastes. The potential of bio-coal fly ash to contribute to a more sustainable and circular economy, therefore lowering the environmental effect of energy generation and opening new paths for resource recovery, is contingent upon the implementation of suitable research, legislation, and technical innovation (Sommer et al. 2006).

11.2 BIOMASS FEED STOCKS AND THEIR INFLUENCE ON ASH CHARACTERISTICS

11.2.1 Wood-Derived Bio-Coal Ash

Wood-derived bio-coal ash is a residual material produced and utilized for energy generation during the burning of wood biomass, typically in conjunction with coal. The chemistry, mineral concentration, and form of this ash cause it to possess unique characteristics and environmental consequences (Demirbaş 2004). Wood-derived bio-coal ash consists mainly of inorganic byproducts that remain after the combustion of organic components. The primary constituents of this ash consist of silica (SiO_2), which is frequently the most plentiful and contributes to the ash's glassy appearance, and alumina (Al_3O_3), which, albeit being present in lesser amounts, improves the ash's structural integrity. Another important constituent, especially when wood is co-fired with coal, is calcium oxide (CaO), which gives the ash an alkaline quality. Furthermore, the fluxing behaviour monitored during combustion is influenced by potassium oxide (K_2O) and sodium oxide (Na_2O), which are frequently found in biomass ashes. The ash exhibits different concentrations of magnesium oxide (MgO) and iron oxide (Fe_2O_3), which are contingent upon the particular biomass and coal employed. In addition, bio-coal ash may contain some trace elements including zinc (Zn), copper (Cu), lead (Pb), and cadmium (Cd). The environmental significance of these trace elements is heightened when their concentrations reach a considerable level, as their effects are strongly linked to the specific biomass employed in the combustion process (Vassilev et al. 2010; Querol et al. 1997). The morphological properties of bio-coal ash produced from wood differ based on the combustion conditions and the specific feedstock used. In general, the ash displays:

- A porous and lightweight structure is often observed in the ash produced from wood, which is characterized by its high organic content.
- Amorphous particles play a substantial role in the composition of ash, especially when elevated temperatures lead to the rapid melting and cooling of silica and other minerals.
- Spherical particles may form as a result of the melting and subsequent re-crystallization of mineral components induced by combustion.
- Particles exhibiting irregular shapes and rough surfaces are frequently observed, predominantly resulting from incomplete combustion or the existence of unburned carbon (Lewandowski et al. 2006; Baxter et al. 2014).

Table 11.1 shows the typical composition of bio-coal fly ash, reported by various researchers.

11.2.2 Agricultural Residues

Agricultural residues are a substantial reservoir of biomass for energy generation, characterized by several prevalent types extensively employed in different geographical areas. Rice husk, a residual material from the process of rice milling, is

TABLE 11.1
Composition of bio-coal fly ash

Characteristic	Description	References
Chemical composition	Lower concentration of harmful elements; sulphur content around 0.1% (compared to higher levels in coke).	Wu and Yip (2017), and Vassilev et al. (2015a, 2015b)
Moisture content	Average moisture content of 5.5%, beneficial for combustion efficiency.	Demirbaş (2005a, 2005b) and Basu (2013)
Ash content	Typically, around 2.33%, significantly lower than traditional coal (16%), reducing waste issues.	
Volatile matter	Higher volatile matter content (approximately 15.16%) compared to coke (2.3%), aiding ignition.	
Calorific value	Competitive calorific value up to 7,000 kcal/kg, higher than traditional coke (6,800 kcal/kg).	Gani and Naruse (2007) and Phyllis2 Database (2020)
Reactivity and combustion	Maintains average tapping temperatures around 1,560°C, favourable for metallurgical applications.	
Environmental impact	Considered carbon-neutral or carbon-negative, contributing to a lower net carbon footprint.	Junginger et al. (2019) and Cherubini (2010)
Briquette density and strength	Compressive strength ranging from 7.72 to 10.85 N/mm², indicating good structural integrity.	Kaliyan and Morey (2009) and Sultana et al. (2010)
Burning rate and time	Burning rates between 16.10 and 28.32 g/min, with ignition times around 22.23–45.20 seconds.	
Mineral composition of ash	Contains silica (SiO_2), alumina (Al_2O_3), and iron oxides (Fe_2O_3), affecting physical properties.	Vassilev et al. (2010)

abundant in countries that produce rice and is renowned for its significant silica concentration, which greatly affects the properties of the resulting ash. Straw, comprising remnants from crops such as wheat, barley, and oats, is a widely used biomass feedstock in agriculture, preferred for its accessibility and simplicity of processing. Corn stove, which comprises the foliage, stalks and cobs remaining after the harvest of corn, is widely accessible in areas where corn represents a significant agricultural commodity. The fibrous residue that remains after extracting juice from sugarcane, known as sugarcane bagasse, is a crucial biomass feedstock in countries that produce sugar. Moreover, coconut shell and palm kernel shell, which are secondary products of coconut and palm oil manufacturing, are frequently utilized in areas where these crops are the main agricultural resources (Haque et al. 2021). The composition of ash exhibits significant variation based on the specific type of agricultural residue employed as a raw material (Lima et al. 2008; Cheng et al. 2023). The variations observed are determined by the inherent characteristics of the biomass, which include its mineral composition. The ash obtained from agricultural residues, although useful for several purposes, may also include pollutants that present substantial environmental hazards. A potential issue arises from the existence of heavy metals, including cadmium (Cd), lead (Pb), and arsenic (As), which may be present in minute quantities contingent upon the raw materials and cultivation techniques (Sharma et al. 2022; Wang et al. 2021; Singh et al. 2020; Munir et al. 2020). When not effectively controlled, these toxic heavy metals have the potential to seep into the soil and groundwater, resulting in environmental pollution. The appropriate disposition or use of ash is crucial to avoid environmental pollution, especially in regions where significant amounts of biomass are used for energy generation (Zhang et al. 2020; Ahmaruzzaman 2010).

11.2.3 Energy Crops

The utilization of biomass feedstock, namely energy crops, is progressively expanding in the field of bioenergy generation. These specialized energy crops are deliberately grown for their substantial biomass production, rendering them well-suited for energy conversion. The properties of ash generated from these energy crops can differ greatly based on the specific crop, which subsequently affects their utilization in different applications and their ecological consequences. A comprehensive grasp of these attributes is essential for the maximization of energy generation and the mitigation of environmental hazards (Franus et al. 2015). Energy crops refer to plants that are deliberately grown for the purpose of being used as fuel. These crops are selected based on their fast growth rates, high energy content, and resilience to different climatic and soil conditions. Switchgrass (*Panicum virgatum*) is a perennial grass indigenous to North America that is highly regarded for its substantial biomass production and minimal input needs among the commonly cultivated energy crops. Another large perennial grass, Miscanthus (Miscanthus × giganteus), is renowned for its significant biomass production with little requirement for fertilizers and water. Fast-growing woody crops such as willow (*Salix* spp.) and poplar (*Populus* spp.) are commonly cultivated in short rotation coppice systems, which makes them very suitable for the generation of renewable energy. Sorghum bicolour, a multifunctional crop, serves as both animal feed and a source of energy and can be grown in many climatic conditions, including dry regions. The tall perennial grass known as giant reed (*Arundo donax*) flourishes in temperate and subtropical regions, generating substantial biomass yields, which makes it a crucial crop for bioenergy purposes (Vassilev et al. 2015; Querol et al. 2002).

Distinct features of the ash generated from the combustion of energy crops are determined by the chemical composition of the crop and the growing conditions (Gitari et al. 2006; White & Case 1990).

- **Silica content**: Like conventional biomass sources, energy crops such as switch grass and Miscanthus typically contain a significant amount of silica (SiO_2) in their ash. This high silica concentration can result in the development of a glassy, amorphous structure in the ash, which can be advantageous for specific industrial uses such as cement manufacturing.
- **Alkali metal content**: Energy crops frequently contain elevated levels of alkali metals, including potassium (K) and sodium (Na). These constituents can result in reduced melting points in the ash, thereby heightening the susceptibility to slagging and fouling in boilers.
- **Calcium content**: The ash of certain energy crops, especially woody species such as willow and poplar, contains a greater concentration of CaO. Consequently, the alkaline properties of the ash enhance its suitability as a soil amendment for acidic soils.
- **Phosphorus content**: Sorghum and similar crops often contain substantial amounts of phosphorus (P) in their ash, which not only enhances its potential as a fertilizer but also presents difficulties for combustion processes because of the possibility of fouling (Frost et al. 2002; Minkina et al. 2016).

Table 11.2 provides the classification of bio-coal fly ash by particle size. It categorizes bio-coal fly ash according to particle size, carbon and mineral composition, reactivity, and uses. Fine-grained ash (smaller than 75 μm) possesses average carbon and mineral content, exhibits reactivity, and is employed in construction materials, adsorbents, and soil amendments. Coarse-grained ash (larger than 75 μm) is non-reactive and appropriate for construction materials and landfill cover. Carbon-rich ash contains elevated carbon levels and is reactive, utilized in activated carbon applications. Mineral-rich ash features high mineral content and reactivity, applied in soil amendments and fertilizers. Reactive ash displays variable characteristics but is highly reactive, frequently used in cement production. Non-reactive ash has low reactivity and is utilized in construction materials and landfill cover. Pertinent references substantiate each category, emphasizing the varied applications of bio-coal fly ash.

11.3 COMPARISON WITH TRADITIONAL BIOMASS SOURCES

A comparison can be made between the ash properties of energy crops and those of conventional biomass sources, such as agricultural residues (e.g., rice husk, straw) and forest residues (e.g., wood chips, sawdust) (Vassilev et al. 2013a, 2013b; Wang & Wu 2006a, 2006b).

- **Higher alkali metals**: In comparison to conventional biomass, energy crops frequently exhibit elevated levels of alkali metals. These characteristics increase their susceptibility to generating slagging and fouling in combustion systems, necessitating meticulous control during energy generation.

TABLE 11.2
Category of bio-coal fly ash according to size

Category	Particle size	Carbon content	Mineral content	Reactivity	Applications	References
Fine-grained ash	Smaller than 75 μm	Average	Average	Can be reactive	Construction materials, adsorbents, catalysts, soil amendments	Vassilev et al. (2013a, 2013b), Basu (2015), and Wu et al. (2017)
Coarse-grained ash	Larger than 75 μm	Average	Average	Can be non-reactive	Construction materials, landfill cover, mine backfill	
Carbon-rich ash	Higher than average	High	Average	Can be reactive	Activated carbon, carbon-based materials	
Mineral-rich ash	Average	Average	High	Can be reactive	Soil amendments, fertilizers, glass production	
Reactive ash	Varies	Varies	Varies	High	Cement production, adsorbents	
Non-reactive ash	Varies	Varies	Varies	Low	Construction materials, landfill cover, mine backfill	

- **Lower chlorine content**: One notable characteristic of many energy crops, especially those cultivated on marginal lands, is their relatively lower chlorine (Cl) content compared to agricultural deposits such as straw. Reduced chlorine levels mitigate the potential for corrosive emissions at combustion (Baxter 2005).
- **Variable silica content**: Certain energy crops such as Miscanthus have high silica content comparable to rice husk, but others like willow and poplar have lower silica levels, resembling conventional woody biomass. This variability impacts the possible applications of the ash and the optimal boiler types for these raw materials (Sims et al. 2010).
- **Environmental impact**: The specialized character of energy crops enables more precise regulation of inputs such as fertilizers and pesticides, which may result in reduced concentrations of heavy metals and other pollutants in the ash as compared to conventional biomass sources. Nevertheless, this is heavily contingent upon cultivation methods (Demirbaş 2007; Jenkins et al. 1998).

11.4 COMBUSTION PROCESSES AND THEIR IMPACT ON ASH PROPERTIES

Fuel combustion is an essential process in the production of electricity and industrial applications. Fuel combustion is not only the process of oxidizing fuels to liberate energy but also generates ash as a secondary product. The combustion technology employed alters the characteristics of ash, encompassing its morphology and composition. Novel combustion technologies, such fluidized bed combustion (FBC), exert distinct influences on the characteristics of ash (Bridgwater et al. 1999; Koppejan & Van Loo 2012). Table 11.3 demonstrates the comparison of bio-coal production processes: temperature, oxygen requirements, products, and ash formation. It delineates a comparison of three bio-coal production methodologies: pyrolysis, gasification, and combustion, with respect to temperature, oxygen prerequisites, resultant products, and ash generation. Pyrolysis functions at temperatures between 300°C and 600°C in the absence of oxygen, yielding biochar and gases characterized by low ash content and high carbon concentration. Gasification transpires at temperatures between 700°C and 1,000°C with restricted oxygen, producing syngas and char, characterized by a modest ash production and reduced carbon content compared to pyrolysis. Combustion occurs at temperatures over 1,000°C in the presence of abundant oxygen, yielding CO_2, water, and heat, which leads to a large ash output and diminished carbon content. These procedures differ markedly in their outputs and ash properties (Ohenoja et al. 2016; Ohenoja et al., 2020; Lopez, Gartzen, et al. 2018).

TABLE 11.3
Comparison of bio-coal production processes: temperature, oxygen requirements, products, and ash formation

Process	Temperature	Oxygen	Products	Ash formation	References
Pyrolysis	300–600°C	No oxygen	Biochar, gases (e.g., CO, H_2, CH_4)	Low ash yield, high carbon content	Demirbaş (2004) and Bridgwater (2012)
Gasification	700–1,000°C	Limited oxygen	Syngas (CO, H_2), char	Moderate ash yield, lower carbon content than pyrolysis	Basu (2013) and Kaupp (2013) and Goss (1984)
Combustion	>1,000°C	Excess oxygen	CO_2, H_2O, heat	High ash yield, low carbon content	Vassilev et al. (2013a, 2013b) and Jenkins et al. (1998)

11.4.1 Conventional Combustion

Traditional combustion methods, such as grate firing or pulverized coal combustion, entail the combustion of fuel either in a stationary bed or as a finely ground powder. The aforementioned techniques generally yield ash with a rather uniform composition, although they may generate larger ash particles and a greater percentage of unburnt carbon (Vassilev et al. 2010).

11.4.2 Advanced Combustion Technologies

Cutting-edge combustion technologies are being developed to optimize efficiency, minimize emissions, and better control ash characteristics. FBC and oxy-fuel combustion are two notable technologies in this particular engineering application. FBC is the combustion of fuel in a bed of inert particles that are held in suspension by an upward stream of air or gas, resulting in a fluid-like behaviour. This procedure greatly enhances the uniformity of fuel mixing and the efficiency of combustion. FBC is commonly classified into two primary categories: bubbling fluidized bed (BFB) and circulating fluidized bed (CFB), each providing unique benefits in particular domains (Koornneef et al. 2007; Huang et al. 2021).

11.4.3 Impact on Ash Morphology and Composition

The ash morphology in FBC is typically characterized by smaller and more spherical particles in comparison to those generated by traditional combustion. A fluidized environment enhances combustion efficiency and facilitates ash agglomeration. This may lead to reduced quantities of fine particulate matter and a more uniform distribution of ash dispersion (Munir 2010). The formulation of ash derived from FBC can vary greatly from that generated by traditional techniques. The use of FBC enables enhanced combustion of low-grade fuels and has the potential to mitigate the production of specific undesirable compounds, including sulphur dioxide (SO_2) and nitrogen oxides (NOx) (Basu 2006). The incorporation of limestone or other additives into the bed can also impact the composition, resulting in a reduction of sulphur content in the ash.

11.4.4 Impact of Ash Properties

The decrease in fine particulate matter and the possibility of improved regulation of sulphur content in FBC ash can result in diminished environmental effects, such as decreased acid rain generation and reduced air pollution (Joseph & Mathew 2012). The inherent characteristics of ash impact the management and elimination of ash. As an illustration, the smaller and more spherical particles generated in FBC are less prone to causing obstructions in handling equipment and can be more readily controlled in ash disposal systems (Sahoo & Chakraborty 2015).

11.5 MINERALOGICAL ANALYSIS OF BIO-COAL FLY ASH

Similar to traditional coal fly ash, bio-coal fly ash is a multifaceted substance consisting of several mineral phases. The mineralogical makeup of bio-coal fly ash is

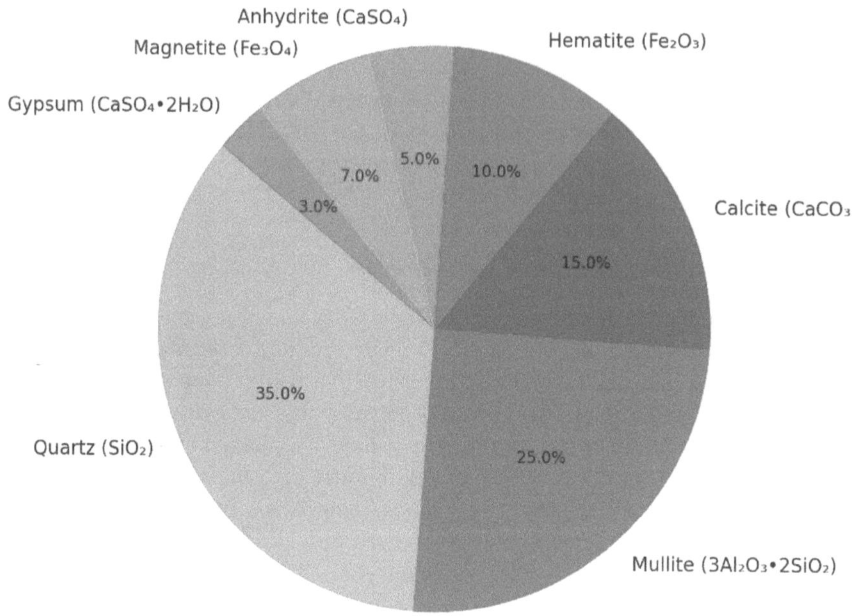

FIGURE 11.1 Composition of bio-coal fly ash based on the percentages of the various mineral phases.

significantly influenced by the biomass type, combustion method, and ash cooling rate. Evaluation of the possible uses of the ash and its environmental impact requires a thorough understanding of the mineral phases present (Minkina et al. 2016; Vassilev et al. 2013a, 2013b; Wang & Wu 2006a, 2006b). Figure 11.1 illustrates the relative percentages of various mineral phases present in bio-coal fly ash. The composition of bio-coal fly ash is essential for comprehending its physical and chemical properties, along with its potential applications across diverse industries.

11.6 OPPORTUNITIES FOR UTILIZATION OF BIO-COAL ASH IN VARIOUS APPLICATIONS

Due to its potential as a raw material for the synthesis of new materials, especially in environmental remediation applications, bio-coal ash, a byproduct of biomass combustion or co-firing in coal power plants, has attracted attention. Bio-coal ash is composed of a blend of organic and inorganic substances, such as different metal oxides, carbonaceous materials, and trace elements. This composition renders it suitable for a wide range of sustainable uses.

11.6.1 Novel Materials Synthesis and Their Application in Environmental Remediation

Significant constituents of bio-coal ash include silica (SiO_3), alumina (Al_2O_3), CaO, iron oxides (Fe_2O_3), and unburned carbon, which present possibilities for innovative material synthesis. These oxides impart the ash with characteristics such as thermal stability, porosity, and surface reactivity, rendering it appropriate for environmental remediation tasks such as adsorption, catalysis, and immobilization of contaminants (Liu et al. 2021; Supraja et al. 2023). There are several methods by which bio-coal ash can be repurposed to aid in environmental remediation.

11.6.1.1 Adsorbent for Pollutant Removal

Due to its porous structure and large surface area, bio-coal ash is highly suitable for adsorption applications. Various contaminants, including heavy metals (e.g., lead, cadmium, mercury) and organic pollutants (e.g., dyes, phenolic compounds), can be effectively eliminated from water and industrial effluents using this technology (Garg et al. 2024). The adsorption of contaminants is facilitated by the presence of metal oxides such as alumina and silica in bio-coal ash, which increase its affinity for these contaminants (Franus et al. 2014).

11.6.1.2 Material for Geopolymer and Concrete Production

Bio-coal ash can be mixed with other substances to create geopolymers, which serve as environmentally friendly substitutes for conventional Portland cement. Geopolymers derived from bio-coal ash demonstrate exceptional mechanical and thermal characteristics, hence contributing to the mitigation of greenhouse gas emissions during their manufacturing. Utilized in construction, these materials can function as a reservoir for heavy metals, thereby inhibiting their release into the environment (Van Deventer et al. 2012; Chaoudhary et al. 2021; Naik et al. 2024).

11.6.1.3 Soil Amendment

Furthermore, bio-coal ash can serve as a soil amendment to enhance soil quality. Due to its alkaline properties, it effectively neutralizes acidic soils and its mineral composition can enhance nutrient accessibility. Moreover, bio-coal ash possesses the capacity to immobilize heavy metals present in polluted soils, thereby decreasing their ability to be absorbed by the body and so ameliorating the hazards linked to soil contamination (Sarfraz et al. 2019; Denis Miroshnichenko et al. 2024).

11.6.1.4 Sustainable Energy Production

The utilization of bio-coal ash in energy-related applications, such as battery electrodes or as a support material for fuel cells, holds promise for the generation of energy from waste materials, thereby making a valuable contribution to the establishment of a circular economy. Table 11.4 classifies bio-coal fly ash according to its source, emphasizing its properties and uses (You et al. 2018).

11.6.2 Opportunities for Utilization of Bio-Coal Ash in Agricultural Applications/Soil Conditioners

Bio-coal ash, a residual material produced by burning biomass or co-firing with coal in energy generation, holds promise for sustainable agricultural uses, namely as a soil improving agent. The mineral and nutrient content of bio-coal ash, including calcium, potassium, magnesium, and phosphorus, can significantly improve soil fertility.

TABLE 11.4
Category of bio-coal fly ash

Category	Origin source	Characteristics	Applications
Agricultural waste-based ash	Agricultural residues (e.g., crop residues, animal waste)	Lower ash content, higher carbon content, nutrient-rich	Soil amendments, fertilizers, biogas production
Forestry waste-based ash	Wood waste (e.g., sawdust, wood chips)	Higher ash content, lower carbon content, mineral-rich	Construction materials, adsorbents, catalysts
Municipal waste-based ash	Municipal solid waste (e.g., food waste, paper, plastics)	Variable ash content, variable carbon content, diverse mineral composition	Construction materials, landfill cover, energy recovery
Industrial waste-based ash	Industrial waste (e.g., wood processing waste, agricultural processing waste)	Varies depending on the specific waste	Construction materials, adsorbents, catalysts

Furthermore, its alkaline characteristics can be employed to remediate acidic soils. Thus, bio-coal ash emerges as a highly promising resource for enhancing agricultural productivity and facilitating waste recycling. The application of bio-coal ash as a soil conditioner serves to enhance the physical, chemical, and biological characteristics of soil (Szatyłowicz & Krasowska 2020). The following characteristics render it efficacious in this position.

pH adjustment: A notable use of bio-coal ash in agriculture is its capacity to function as a liming agent. Alkaline in nature, the ash contains significant amounts of calcium and magnesium oxides, rendering it well-suited for neutralizing acidic soils. The presence of acidic soils poses a significant obstacle to the growth of crops due to their detrimental effects on the accessibility of vital nutrients, such as phosphorus, and the heightened concentration of heavy metals like aluminium. The application of bio-coal ash has been shown to improve soil pH, hence enhancing nutrient accessibility and establishing a more conducive environment for crop development.

Nutrient enrichment: Bio-coal ash is rich in vital nutrients necessary for plant development, including potassium (K), phosphorus (P), calcium (Ca), and traces of iron (Fe) and magnesium (Mg). These nutrients enhance soil fertility, thereby minimizing the reliance on artificial fertilizers. For instance, bio-coal ash contains substantial quantities of potassium, a vital macronutrient for plant development, which thereby enhances plant health and productivity. Scientific studies have demonstrated that the use of bio-coal ash in soil improves the accessibility of essential nutrients, resulting in higher crop yield, particularly in soils lacking sufficient nitrogen. Moreover, the ash's gradual release characteristic can offer a more environmentally beneficial supply of nutrients in comparison to traditional fertilizers (Singh & Bhatnagar 2020).

11.6.2.1 Soil Structure Improvement

The incorporation of bio-coal ash into soil can enhance its physical characteristics, including its structure, water retention, and aeration capability. The permeable characteristics of bio-coal ash enhance soil texture by facilitating increased air circulation and water penetration. This property is especially advantageous in clayey soils, which often exhibit compaction and inadequate drainage. The enhancement of soil porosity by bio-coal ash has the potential to mitigate compaction, facilitate root penetration, and otherwise improve the overall health of the soil structure (Saladi & Ashraf 2020).

11.6.2.2 Heavy Metal Immobilization

Impaired soil quality due to heavy metal contamination can pose a substantial problem in agriculture. The alkaline pH and abundant metal oxides in bio-coal ash enable it to immobilize heavy metals like lead (Pb), cadmium (Cd), and zinc (Zn) by the formation of stable complexes. This phenomenon facilitates the mitigation of the bioavailability and toxicity of heavy metals in the soil, thereby enabling the cultivation of crops in polluted regions with enhanced safety measures. Empirical research has shown that the application of bio-coal ash can effectively reduce the movement of heavy metals in contaminated soils, thereby reducing the potential for plants to absorb these metals. The combination of bio-coal ash and vital nutrients makes it a highly efficient solution for remediating polluted agricultural land (Chen et al. 2020).

11.6.2.3 Carbon Sequestration Potential

Bio-coal ash comprises carbon remnants resulting from the combustion of biomass and so have the potential to facilitate the long-term sequestration of carbon in soil. Incorporating bio-coal ash into agricultural soils can augment the organic carbon content of the soil, thereby strengthening its function as a carbon sink and making a valuable contribution to endeavours aimed at mitigating climate change. Moreover, this facilitates soil health by augmenting microbial activity and nutrient cycling, thereby resulting in heightened plant growth and productivity (Zhang et al. 2018).

11.6.2.4 Use in Organic Farming

Given the increasing awareness of sustainable and organic farming methods, the natural composition of bio-coal ash makes it highly suitable for use in organic agriculture. It can serve as a mineral source or soil amendment, substituting chemical fertilizers and lime products. Its application is in line with the objectives of decreasing synthetic inputs and advocating environmentally friendly farming practices. Nevertheless, it is crucial to closely observe the makeup of bio-coal ash to guarantee its adherence to organic farming regulations, namely regarding its heavy metal content and possible pollutants.

11.6.3 Opportunities for Utilization of Bio-Coal Ash in Construction Industries/Cement and Concrete Applications

11.6.3.1 Supplementary Cementitious Material in Cement Production

Bio-coal ash exhibits pozzolanic characteristics, whereby it undergoes a reaction with calcium hydroxide ($Ca(OH)_2$) in the presence of water to produce calcium silicate hydrates (C-S-H), which are the main chemicals responsible for the strength of cement. In the manufacturing of blended cements, bio-coal ash emerges as a highly advantageous substitute for conventional pozzolanic materials such as fly ash, slag, and silica fume (Sigvardsen et al 2019). By substituting a portion of Portland cement with bio-coal ash, it is possible to:

Reduce the carbon footprint: Cement manufacturing is a significant contributor to global carbon dioxide emissions, mostly because of the calcination process (the transformation of limestone into lime). Augmenting cement with bio-coal ash decreases the need for cement clinker, thereby mitigating CO_2 emissions (Sua-Iam & Makul 2015).

Enhance durability: The pozzolanic reaction enhances the microstructure of concrete, resulting in increased density and reduced permeability. In environments susceptible to sulphate attack or chloride-induced corrosion, this significantly improves the longevity of concrete (Hossain & Lachemi 2019).

Workability and strength: Bio-coal ash enhances the workability of fresh concrete by virtue of its small particle size and spherical form, thereby enabling more efficient packing and reduced water requirements (Sandhu et al 2017). The inclusion of bio-coal ash in hardened concrete promotes improved strength development over time, especially at later stages (28 days or more), because of the ongoing pozzolanic activity. Utilizing bio-coal ash can enhance the production of high-performance concrete by generating superior mechanical characteristics, including higher compressive strength, decreased porosity, and enhanced resistance to chemical attack (de Carvalho Gomes 2020).

11.6.3.2 Bio-Coal Ash as Aggregate Replacement

The inclusion of bio-coal ash as a partial substitute for aggregate in concrete mixtures can decrease the total density of concrete, resulting in the production of lightweight concrete. This particular variety of concrete is advantageous in situations where the reduction of weight is of utmost importance, such as in the construction of tall buildings, bridges, and modular concrete components. The thermal insulation properties of lightweight concrete contribute to the enhancement of energy efficiency in buildings (Rajabipour & Mohammadi 2021). Through the use of bio-coal ash as a substitute for aggregate, the construction sector can decrease its dependence on finite natural resources such as sand and gravel, which are becoming scarcer. Furthermore, the use of bio-coal ash as a recycled material is in accordance with the concepts of sustainable construction and circular economy, in order to decrease the quantity of waste disposed in landfills and mitigate the exhaustion of natural resources (Kou et al. 2007).

11.6.3.3 Geopolymer Concrete

The utilization of bio-coal ash in the production of geopolymer concrete presents a sustainable substitute for conventional Portland cement concrete. Geopolymer concrete is produced by the activation of aluminosilicate aggregates, such as bio-coal ash, using alkaline solutions (e.g., sodium hydroxide and sodium silicate) to create a binding agent that does not require Portland cement. Geopolymer concrete production incurs much reduced CO_2 emissions in comparison to traditional cement production due to the absence of the energy-intensive calcination process. This characteristic renders it a very appealing choice for mitigating the ecological consequences of concrete manufacturing (Wang et al. 2019).

Geopolymer concrete incorporating bio-coal ash demonstrates exceptional mechanical characteristics such as elevated compressive strength, minimal shrinkage, and outstanding resistance to chemical degradation. The aforementioned characteristics render it well-suited for applications in highly demanding settings, such as marine constructions, wastewater treatment plants, and industrial facilities that involve chemical exposure (Maguire et al. 2023).

11.6.4 Bio-Coal Ash in Brick and Block Manufacturing

Bio-coal ash can is incorporated into the manufacturing of bricks, blocks, and pavers as a partial replacement for clay or cement. Its fine particle size, high silica content, and binding properties make it suitable for producing durable and environmentally friendly masonry products (Hossain & Lachemi 2019; Gupta et al. 2020). Bricks made with bio-coal ash have improved thermal insulation properties due to the lightweight nature of the ash. This can lead to more energy-efficient buildings by reducing heating and cooling demands (Zheng et al. 2017). Bio-coal ash can improve the compressive strength and durability of bricks and blocks, making them more resistant to weathering and wear. These properties make bio-coal ash-based bricks suitable for use in various construction applications, from load-bearing walls to decorative facades (Kou & Poon 2013).

11.6.5 Opportunities for Utilization of Bio-Coal Ash in Glass and Rubber Industries

Bio-coal ash, a byproduct of biomass combustion, has diverse applications in manufacturing industries due to its unique chemical composition, which includes silica (SiO_2), alumina (Al_2O_3), and various metal oxides. Its properties make it a valuable resource in the production of glass, rubber, and zeolites, contributing to sustainability and resource efficiency in these industries (Jones & Edwards 2016).

11.6.5.1 Utilization in Glass Manufacturing

Bio-coal ash can be used as a raw material or flux in the production of glass. Its high silica content, combined with other minerals like alumina and CaO, aligns with the requirements of glassmaking. The ash can replace or reduce the use of traditional raw materials such as sand and limestone, thereby lowering production costs and enhancing sustainability (Levis & Themelis 2012; Vlaskin & Kumar 2020; Erol et al. 2017).

- Silica source: The silica in bio-coal ash can be utilized to replace part of the sand used in glass formulation. This is beneficial for producing various types of glass, including container glass, flat glass, and specialty glass (Sharma & Singh 2017).
- Fluxing agent: CaO and alumina in bio-coal ash can act as fluxing agents, reducing the melting temperature of the raw materials. This can lead to

energy savings in the glass melting process and improve the efficiency of production (Kou & Poon 2013).
- Cost reduction: Substituting traditional raw materials with bio-coal ash can reduce the overall cost of glass production, benefiting manufacturers economically (Hossain & Lachemi 2019; Wang et al. 2014).

11.6.5.2 Utilization in Rubber Manufacturing

In rubber manufacturing, bio-coal ash can be used as a filler or reinforcing agent. Its fine particle size and mineral content contribute to improved mechanical properties of rubber products.

- Filler material: The addition of bio-coal ash to rubber compounds can enhance the physical properties of rubber, such as tensile strength, hardness, and abrasion resistance. It can also improve the process ability of rubber during mixing and curing (Haris et al. 2021).
- Reinforcement: The mineral composition of bio-coal ash can provide reinforcement to rubber products, enhancing their performance and durability in applications such as tyres, seals, and gaskets.
- Resource efficiency: Utilizing bio-coal ash as a filler or reinforcement material reduces the reliance on virgin raw materials and helps in the efficient use of industrial byproducts.
- Waste management: Incorporating bio-coal ash into rubber products contributes to waste minimization and supports sustainable manufacturing practices.

Table 11.5 outlines many applications of bio-coal ash across several industries, including chemical, environmental, agricultural, construction, and manufacturing sectors. In the chemical and environmental domains, bio-coal ash is employed for the synthesis of innovative materials, including adsorbents, catalysts, and filters for pollution remediation. In agriculture, it functions as a soil conditioner to improve nutrient retention and water-holding ability. The construction sector gains advantages from its integration with cement and concrete, enhancing strength and durability. Moreover, manufacturing incorporates bio-coal ash in the fabrication of glass, rubber, and zeolites to improve attributes such as strength and abrasion resistance. In the energy and waste management industries, it is included into anaerobic digesters to enhance biogas generation, and it is also utilized as a fill material in landfills and mine backfill to decrease volume and increase stability. In composting applications, bio-coal ash enhances decomposition and increases nutrient content. Every application is underpinned by pertinent research, highlighting the multifaceted potential of bio-coal ash in advancing sustainability across several industries.

11.6.6 Opportunities for Utilization of Bio-Coal Ash in Composting Applications

The combustion of biomass yields bio-coal ash, which presents numerous prospects for augmenting composting procedures and enhancing the quality of compost. The

TABLE 11.5
Applications of bio-coal ash: industry uses, market cap, and processes

Application	Industries	Market cap (estimated)	Process/methods	References
Novel materials synthesis and environmental remediation	Chemical, environmental	High	Synthesis of adsorbents, catalysts, and filters for pollutant removal; soil remediation	Vassilev et al. (2015a, 2015b)
Agricultural applications/ soil conditioners	Agriculture, horticulture	Moderate	Addition to soil to improve nutrient retention, water-holding capacity, and soil structure	Demeyer et al. (2001)
Construction industries/ cement and concrete industries	Construction, materials	High	Partial replacement of cement or aggregate in concrete for improved strength and durability	Gunninget al. (2014)
Glass, rubber, and zeolite manufacturing	Manufacturing	Moderate	Use as a filler or additive to improve properties like strength, abrasion resistance, and ion exchange capacity	Querol et al. (2002)
Anaerobic digestion and biogas technology	Energy, waste management	Moderate	Addition to digesters to enhance biogas production and reduce organic matter	Triolo et al. (2011)
Land fill and mine backfill	Waste management, mining	Low	Use as a fill material to reduce landfill volume and improve stability in mine backfill	Zhang et al. (2019)
Composting applications	Waste management, agriculture	Low	Addition to compost piles to accelerate decomposition and improve nutrient content	Smith and Jasim (2009)

physical and chemical characteristics of the material can enhance the efficiency of composting and the generation of compost of superior quality (Smith et al. 2014). The following are the primary domains in which bio-coal ash can be employed in composting applications:

- **pH adjustment**: It is essential to maintain an ideal pH range (usually between 6.0 and 8.0) to ensure sufficient microbial activity during composting. The inherent alkalinity of bio-coal ash serves to counteract acidic environments and avert pH declines, thereby impeding microbial activity and decelerating the composting process (Zhang et al. 2018).
- **Porosity:** The ash's particulate composition and comparatively large surface area can enhance the aeration capability of compost piles or windrows. Optimized aeration promotes the exchange of oxygen, a crucial factor for

aerobic microbial activity and effective breakdown of organic substances (Singh et al. 2021).

- **Moisture retention**: Ash has the ability to effectively absorb surplus moisture, thereby assisting in the maintenance of ideal moisture levels in composting mixtures. Effective control of moisture is essential for sustaining microbial activity and avoiding anaerobic conditions (Yang & Zhang 2019; Zhao & Gao 2020).
- **Nutrient source**: The mineral composition of bio-coal ash, which includes potassium (K), calcium (Ca), and magnesium (Mg), can enhance compost by providing vital nutrients that promote plant development. This has the potential to enhance the fertility and efficacy of the end compost product (Levis & Themelis 2012).
- **Structure and texture**: The addition of bio-coal ash to compost can enhance its texture and structure, so increasing its suitability for a wide range of agricultural uses. It can facilitate the formation of a well-aerated, crumbly consistency that enhances soil integration and stimulation of plant root growth (Zhang et al. 2018, Singh et al. 2021) Organic ash derived from bio-coal can effectively reduce odours linked to composting:
- **Odour absorption**: The absorbent characteristics of the ash can effectively capture and neutralize odorous substances generated during composting, such as hydrogen sulphide (H_2S) and ammonia (NH_3), thereby enhancing the cleanliness and ease of the composting process (Zhao & Gao 2020; Singh et al. 2021).

The incorporation of bio-coal ash together with various compost feedstocks optimizes the composting process (Zhao & Gao 2020). Integrating bio-coal ash into the process of composting organic waste can yield advantageous outcomes. Bio-coal ash efficiently controls the carbon-to-nitrogen (C/N) ratio of compost mixtures, which is essential for enhancing microbial activity and enabling the process of efficient decomposition. The attainment of this equilibrium has the potential to optimize the entire composting process and improve the quality of the compost (Xu et al. 2024).

Compounding bio-coal ash with agricultural residues enhances composting conditions. The incorporation of bio-coal ash in the composting of agricultural residues can efficiently manage significant amounts of organic waste, improve composting efficiency, and produce very advantageous compost for soil regeneration (Singh et al. 2021).

11.7 CHALLENGES IN THE CURRENT SCENARIO

While bio-coal ash offers several opportunities for use in environmental remediation, some challenges must be addressed to optimize its application.

11.7.1 Heavy Metal Consideration and Its Leaching

The prevalence of heavy metals in bio-coal fly ash is a major issue of concern. The levels of these metals can differ based on the specific biomass type, combustion parameters, and possible impurities present in the raw materials. The heavy metals commonly present in bio-coal fly ash are lead (Pb), cadmium (Cd), arsenic (As), chromium (Cr), and mercury (Hg). These metals can arise from polluted biomass feedstock or indirectly from the combustion process, where elevated temperatures and incomplete combustion can impact their concentration and chemical composition (Jones & Edwards 2016).

Leaching is a notable concern in which soluble compounds, such as heavy metals, is removed from the ash and may potentially pollute the local environment. Key determinants of leaching are pH levels, moisture content, and ash composition. High leaching rates can be attributed to the enhanced solubility of heavy metals under acidic conditions. Similarly, increased moisture levels can accelerate the leaching process by promoting the migration of heavy metals. The mobility of heavy metals can be influenced by the existence of soluble salts and other constituents in the ash (Wu & Wang 2021).

The ecological ramifications of heavy metal leaching are paramount. Pollution of soil and water caused by contaminated leachate can have detrimental effects on plant growth, soil health, and water quality. These risks are managed by regulatory standards, which include guidelines from the Environmental Protection Agency (EPA) and European Union regulations. These standards establish restrictions on the concentrations of heavy metals and the maximum potential for leaching (Jones & Edwards 2016).

11.7.2 Heterogeneity

Management and utilization of bio-coal fly ash are significantly challenged by its heterogeneity. Variability in the physical and chemical characteristics of bio-coal fly ash, a byproduct of biomass combustion, arises from multiple factors. The variability of this phenomenon can have an impact on the efficacy and safety of numerous applications, such as environmental remediation, construction, and agricultural use (Singh et al. 2021). An inherent obstacle is the lack of uniformity in the makeup of bio-coal fly ash. The characteristics of the ash can significantly differ depending on the specific biomass feedstock, the mechanism of combustion, and the conditions under which combustion occurs. Various biomass sources, including wood, agricultural residues, and municipal waste, can individually provide distinct chemical elements and compounds to the ash structure. Thus, the nutrient, heavy metal, and other elemental content of bio-coal fly ash from various sources can vary, so influencing its appropriateness for certain uses. The heterogeneity in ash properties can also impact its physical attributes, including particle dimensions, compactness, and permeability (Zhao & Gao 2020).

11.7.3 Challenges and Considerations in Agricultural Use

Although bio-coal ash presents numerous benefits for agricultural applications, there are certain criteria and possible obstacles to be taken into account:

- **Toxicity of trace elements**: Bio-coal ash can include minute quantities of heavy metals, including arsenic, mercury, and cadmium, at varying levels based on the biomass origin and combustion conditions. Elevated levels of these elements in the food chain can have detrimental effects on plants, animals, and humans. Hence, the meticulous evaluation of the bio-coal ash composition and effective control of application rates are crucial to avoid negative consequences (Liu et al. 2020).

- **Leaching potential**: While bio-coal ash has the ability to immobilize heavy metals, there is a possibility of leaching of toxic chemical compounds into groundwater under specific circumstances, particularly when used excessively. Optimal application techniques and precise dosage management are crucial to minimize this risk (Wu & Wang 2021).
- **Potential contamination**: Bio-coal ash has the potential to include minute quantities of heavy metals or other pollutants, contingent upon the origin of the biomass. Prior to its use in building materials, it is imperative to verify that the ash complies with safety and environmental criteria (Singh et al. 2021).

11.8 OVERVIEW AND PROSPECTS

Bio-coal ash is derived from diverse biomass sources including agricultural waste, plant pulp, and municipal refuse. Bio-coal ash exhibits notable variations depending on its source and combustion conditions, leading to a diverse array of classifications characterized by distinct physical and chemical attributes. Its intrinsic unpredictability underscores the need for careful assessment and regulation when considering its applications. An extensive array of potential applications exists for bio-coal ash. The unique properties of bio-coal ash make it a promising candidate for environmental remediation in the field of new materials technology, providing an effective means to address pollution and other environmental concerns (Ugwu et al 2024). From an agricultural perspective, bio-coal ash has the potential to serve as a highly effective soil conditioner, enhancing both the fertility and structure of the soil. Its application in the production of cement and concrete confers benefits to the construction industry, thereby promoting sustainable building practices. Moreover, bio-coal ash finds application in the manufacturing of glass, rubber, and zeolite, as well as in anaerobic digestion and biogas processes, landfill and mine backfill techniques, and composting. However, the utilization of bio-coal ash is not without limitations. The presence and process of heavy metal contamination and leaching pose significant risks to both the aquatic environment and human well-being (Kumar & Kundu 2018). The heterogeneity in the properties of bio-coal ash introduces intricacy to its application, affecting both consistency and safety. To efficiently harness the benefits of bio-coal ash while reducing its risks, it is crucial to address further challenges such as regulatory adherence and the need for advanced treatment technology. Progress necessitates the implementation of a comprehensive strategy. The adoption of standardized biomass feedstock and combustion processes can significantly reduce inconsistencies and enhance the consistency of bio-coal ash (Seo et al. 2022). Development and implementation of sophisticated treatment technologies and effective management strategies are crucial in addressing the problems of heavy metal contamination and leaching (Singh et al. 2019). It is crucial to engage in continuous research and innovation to optimize the use of bio-coal ash in various applications and ensure its safe and eco-friendly integration into existing systems.

REFERENCES

Adeleke, A. A., Odusote, J. K., Ikubanni, P. P., Orhadahwe, T. A., Lasode, O. A., Ammasi, A., & Kumar, K. (2021). Ash analyses of bio-coal briquettes produced using blended binder. *Scientific Reports, 11*(1), 547. https://doi.org/10.1038/s41598-020-79510-9

Akhator, P. E., & Aiwunu-Akuete, C. O. (2023). A review of biochar production and application in the agroforestry sector. *NIPES-Journal of Science and Technology Research, 5*(4). https://doi.org/10.5281/zenodo.10341715

Ahmaruzzaman, M. (2010). A review on the utilization of fly ash. *Progress in Energy and Combustion Science, 36*(3), 327–363. https://doi.org/10.1016/j.pecs.2009.11.003

Arriagada, J., & Martinez, M. (2018). Synthesis of zeolites from bio-coal ash: Application in wastewater treatment. *Microporous and Mesoporous Materials, 273,* 105–115. https://doi.org/10.1016/j.micromeso.2018.01.028

Basu, P. (2006). *Combustion and Gasification in Fluidized Beds.* CRC Press. https://doi.org/10.1201/9781420005158

Basu, P. (2018). *Biomass Gasification, Pyrolysis, and Torrefaction: Practical Design and Theory.* Academic Press.

Baxter, L. (2005). Biomass-coal co-combustion: opportunity for affordable renewable energy. *Fuel, 84*(10), 1295–1302.

Baxter, X. C., Darvell, L. I., Jones, J. M., Barraclough, T., Yates, N. E., Shield, I., & Williams, A. (2014). Study of miscanthus combustion properties by TGA and XRD. *Fuel, 117,* 306–313. https://doi.org/10.1016/j.fuel.2013.09.031

Beesley, L., Moreno-Jiménez, E., & Gomez-Eyles, J. L. (2010). Effects of biochar and greenwaste compost amendments on mobility, bioavailability, and toxicity of inorganic and organic contaminants in a multi-element polluted soil. *Environmental Pollution, 158*(6), 2282–2287. https://doi.org/10.1016/j.envpol.2010.03.017

Bridgwater, A. V., Meier, D., & Radlein, D. (1999). An overview of fast pyrolysis of biomass. *Organic Geochemistry, 30*(12), 1479–1493. https://doi.org/10.1016/S0146-6380(99)00120-5

Cherubini, F. (2010). GHG balances of bioenergy systems–overview of key steps in the production chain and methodological concerns. *Renewable Energy, 35*(7), 1565–1573. https://doi.org/10.1016/j.renene.2009.11.035.

Choudhary, J., Kumar, B., & Singh, S. (2021). Assessment of engineering and environmental suitability of waste bituminous concrete containing waste biomass ash. *International Journal of Pavement Research and Technology, 14,* 751–763. https://doi.org/10.1007/s42947-020-0242-6

Demirbas, A. (2004). Combustion characteristics of different biomass fuels. *Progress in Energy and Combustion Science, 30*(2), 219–230. https://doi.org/10.1016/j.pecs.2003.10.004

Demirbaş, A. (2005a). Combustion characteristics of different biomass fuels. *Progress in Energy and Combustion Science, 30*(2), 219–230. https://doi.org/10.1016/j.pecs.2003.10.004

Demirbaş, A. (2005b). Potential applications of renewable energy sources, biomass combustion problems in boiler power systems and combustion-related environmental issues. *Progress in Energy and Combustion Science, 31*(2), 171–192. https://doi.org/10.1016/j.pecs.2005.02.002

Demirbaş, A. (2007). Effects of moisture and hydrogen content on the heating value of fuels. *Energy Sources, Part A: Recovery, Utilization, and Environmental Effects, 29*(7), 649–655. https://doi.org/10.1080/009083190957801

de Carvalho Gomes, S. (2020). *Effect of recycled natural water treatment sludge and biochar on the mechanical performance and hydration kinetics of Portland cement composite.* University of Technology Sydney (Australia).

Erol, M., Küçükbayrak, S., & Ersoy-Meriçboyu, A. (2007). Characterization of coal fly ash for possible utilization in glass production. *Fuel, 86*(5-6), 706–714. https://doi.org/10.1016/j.fuel.2006.09.009

Franus, W., Wdowin, M., & Franus, M. (2014). Synthesis and characterization of zeolites prepared from industrial fly ash. *Environmental Monitoring and Assessment, 186*, 5721–5729. https://doi.org/10.1007/s10661-014-3815-5

Franus, W., Wiatros-Motyka, M. M., & Wdowin, M. (2015). Coal fly ash as a resource for rare earth elements. *Environmental Science and Pollution Research, 22*(12), 9464–9474. https://doi.org/10.1007/s11356-015-4111-9

Koppejan, J., & Van Loo, S. (2012). *The Handbook of Biomass Combustion and Co-firing.* Routledge.

Kloprogge, J. T., Komarneni, S., Yanagisawa, K., Frost, R. L., & Fry, R. (1998). Infrared study of some synthetic and natural beidellites. *Journal of Materials Science Letters, 17*, 1853–1855. https://doi.org/10.1023/A:1006694629015

Kloprogge, J. T., Komarneni, S., Yanagisawa, K., Frost, R. L., & Fry, R. (1998). Infrared study of some synthetic and natural beidellites. *Journal of Materials Science Letters, 17*, 1853–1855.

Gani, A., & Naruse, I. (2007). Effect of cellulose and lignin content on pyrolysis and combustion characteristics for several types of biomass. *Renewable Energy, 32*(4), 649–661. https://doi.org/10.1016/j.renene.2006.02.017

Gupta, V., Pathak, D. K., Siddique, S., Kumar, R., & Chaudhary, S. (2020). Study on the mineral phase characteristics of various Indian biomass and coal fly ash for its use in masonry construction products. *Construction and Building Materials, 235*, 117413. https://doi.org/10.1016/j.conbuildmat.2019.117413

Dechnik-Vázquez, Y. A., García-Barrios, L., Ramírez-Marcial, N., van Noordwijk, M., & Alayón-Gamboa, A. (2019). Assessment of browsed plants in a sub-tropical forest frontier by means of fuzzy inference. *Journal of Environmental Management, 236*, 163–181.

Jenkins, B., Baxter, L. L., Miles Jr, T. R., & Miles, T. R. (1998). Combustion properties of biomass. Fuel Processing Technology, 54(1-3), 17–46.

Jones, D. L., & Edwards, A. C. (2016). The use of bio-coal ash for leachate treatment in landfills. *Waste Management, 55*, 100–109. https://doi.org/10.1016/j.wasman.2016.05.010

Joseph, B., & Mathew, G. (2012). Influence of aggregate content on the behavior of fly ash based geopolymer concrete. *Scientia Iranica, 19*(5), 1188–1194. https://doi.org/10.1016/j.scient.2012.07.006

Kaliyan, N., & Morey, R. V. (2009). Densification characteristics of corn stover and switchgrass. *Fuel Processing Technology, 91*(11), 1307–1314. Link: https://elibrary.asabe.org/abstract.asp?aid=27380

Kambo, H. S., & Dutta, A. (2015). A comparative review of biochar and hydrochar in terms of production, physicochemical properties, and applications. *Renewable and Sustainable Energy Reviews, 45*, 359–378. https://doi.org/10.1016/j.rser.2015.01.050

Kaupp, A. (2013). *Small Scale Gas Producer-Engine Systems.* Springer Science & Business Media.

Koornneef, J., Junginger, M., & Faaij, A. (2007). Development of fluidized bed combustion—An overview of trends, performance and cost. *Progress in energy and combustion science, 33*(1), 19–55. https://doi.org/10.1016/j.pecs.2006.07.001

Kou, S. C., Poon, C. S., & Chan, D. (2007). Influence of fly ash as cement replacement on the properties of recycled aggregate concrete. *Journal of materials in civil engineering, 19*(9), 709–717. https://doi.org/10.1061/(ASCE)0899-1561(2007)19:9(709)

Lima, A. T., Ottosen, L. M., Pedersen, A. J., & Ribeiro, A. B. (2008). Characterization of fly ash from bio and municipal waste. *Biomass and Bioenergy, 32*(3), 277–282. https://doi.org/10.1016/j.biombioe.2007.09.005

Lehmann, J., Gaunt, J., & Rondon, M. (2006). Bio-char sequestration in terrestrial ecosystems–A review. *Mitigation and Adaptation Strategies for Global Change, 11*(2), 395–419. https://doi.org/10.1007/s11027-005-9006-5

Lehmann, J., & Joseph, S. (Eds.). (2009). *Biochar for Environmental Management: Science and Technology.* Earthscan.

Levis, J. W., & Themelis, N. J. (2012). The role of bio-coal ash in composting: Benefits and considerations. *Waste Management, 32*(10), 2005–2013. https://doi.org/10.1016/j.wasman.2012.05.022

Lewandowski, I., & Schmidt, U. (2006). Nitrogen, energy, and land use efficiency of miscanthus, reed canary grass and annual crops for bioenergy production. *Biomass and Bioenergy, 31*(6), 426–442. https://doi.org/10.1016/j.biombioe.2006.02.002

Liu, J., et al. (2020). Leaching behavior of heavy metals from biomass fly ash: Implications for environmental protection. *Science of the Total Environment, 739*, 139927. https://doi.org/10.1016/j.scitotenv.2020.139927

Munir, M. A. M., Liu, G., Yousaf, B., Ali, M. U., Abbas, Q., & Ullah, H. (2020). Synergistic effects of biochar and processed fly ash on bioavailability, transformation and accumulation of heavy metals by maize (Zea mays L.) in coal-mining contaminated soil. *Chemosphere, 240*, 124845. https://doi.org/10.1016/j.chemosphere.2019.124845

Munir, S. (2010). A review on biomass-coal co-combustion: current state of knowledge. *Proceedings of the Pakistan Academy of Sciences, 47*(4), 265–287. https://paspk.org/wp-content/uploads/proceedings/2d1c6bb5proc47-4-8.pdf

Cheng, J., Zhang, S., Fang, C., Ma, L., Duan, J., Fang, X., & Li, R. (2023). Removal of heavy metal ions from aqueous solution using biotransformed lignite. *Molecules, 28*(13), 5031. https://doi.org/10.3390/molecules28135031

López, G., Artetxe, M., Amutio, M., Alvarez, J., Bilbao, J., & Olazar, M. (2018). Recent advances in the gasification of waste plastics: A critical overview. *Renewable and Sustainable Energy Reviews, 82*, 576–596. https://doi.org/10.1016/j.rser.2017.09.032

Nzihou, A. (Ed.). (2020). *Handbook on Characterization of Biomass, Biowaste and Related By-products.* Switzerland: Springer.

Garg, S., Bhavya Surendran, V. S., & Avanthi, A. (2024). Recent Trends and Applications of Biochar and Nanoparticles from Plant Biomass. In *Medicinal and Aromatic Plants: Current Research Status, Value-Addition to Their Waste, and Agro-Industrial Potential (Vol II)* (pp. 203–216). Cham: Springer Nature Switzerland. https://doi.org/10.1007/978-3-031-64601-0_14

Marganingrum, D., Hidawati, & Djaja, S. D. S. (2022). A preliminary study of fuel mixtures of industrial sludge, bottom ash, and municipal solid waste for co-firing in coal boilers. *Energy, Ecology and Environment*, 1–13. http://dx.doi.org/10.25105/urbanenvirotech.v3i2.5110

Minkina, T., Fedorenko, G., & Zapevalov, D. (2016). Application of SEM and EDS methods for the analysis of mineralogical and chemical composition of soils. *Journal of Soils and Sediments, 16*, 2306–2314. https://doi.org/10.1007/s11368-016-1442-3

Miroshnichenko, D., Zhylina, M., & Shmeltser, K. (2024). Modern uses of biochar in various technologies and industries. A review. *Chemistry & Chemical* Technology, 18(2). http://science2016.lp.edu.ua/sites/default/files/Full_text_of_%20papers/full_text_1330.pdf

Naik, S. S., Pandey, S., Pawar, S. N., Shinde, B. H., & Prakash, C. (2024). Innovative and interactive methodology for development of geopolymer mortar using fly ash of agricultural waste briquettes. *International Journal on Interactive Design and Manufacturing (IJIDeM), 18*(8), 5643–5651. https://doi.org/10.1007/s12008-023-01280-1

Ohenoja, K., Tanskanen, P., Peltosaari, O., Wigren, V., Österbacka, J., & Illikainen, M. (2016). Effect of particle size distribution on the self-hardening property of biomass-peat fly ash from a bubbling fluidized bed combustion. *Fuel Processing Technology, 148*, 60–66. https://doi.org/10.1016/j.fuproc.2016.02.023

Ohenoja, K., Pesonen, J., Yliniemi, J., & Illikainen, M. (2020). Utilization of fly ashes from fluidized bed combustion: A review. *Sustainability, 12*(7), 2988. https://doi.org/10.3390/su12072988

Querol, X., Alastuey, A., López-Soler, A., Plana, F., Andrés, J. M., Juan, R., ... & Ruiz, C. R. (1997). A fast method for recycling fly ash: microwave-assisted zeolite synthesis. *Environmental Science & Technology, 31*(9), 2527–2533. https://pubs.acs.org/doi/abs/10.1021/es960937t

Querol, X., Moreno, N., Umana, J. C., Alastuey, A., Hernandez, E., Lopez-Soler, A., & Plana, F. (2001). Synthesis of zeolites from coal fly ash: An overview. *International Journal of Coal Geology, 50*(1–4), 413–423. https://doi.org/10.1016/S0166-5162(02)00124-6

Querol, X., Moreno, N., Umana, J. C., Alastuey, A., Hernández, E., Lopez-Soler, A., & Plana, F. (2002). Synthesis of zeolites from coal fly ash: An overview. *International Journal of Coal Geology, 50*(1–4), 413–423. https://doi.org/10.1016/S0166-5162(02)00124-6

Rawat, S., & Kumar, S. (2022). Critical review on processing technologies and economic aspect of bio-coal briquette production. *Preparative Biochemistry & Biotechnology, 52*(8), 855–871. https://doi.org/10.1080/10826068.2021.2001754

Sarfraz, R., Hussain, A., Sabir, A., Ben Fekih, I., Ditta, A., & Xing, S. (2019). Role of biochar and plant growth promoting rhizobacteria to enhance soil carbon sequestration—A review. *Environmental Monitoring and Assessment, 191*, 1–13. https://doi.org/10.1007/s10661-019-7400-9

Sarkar, A. K. (1990). A preliminary study of the properties of bio-coal fly ash. *Cement and Concrete Research, 20*(4), 751–759. https://doi.org/10.1016/0008-8846(90)90006-Q

Saladi, N., & Ashraf, W. (2020). Ground and sieved bio ash versus coal fly ash: Comparative analysis of pozzolanic reactivity. *Journal of Materials in Civil Engineering, 32*(12), 04020377. https://doi.org/10.1061/(ASCE)MT.1943-5533.0003443

Sear, L. K. A. (2009). *Properties and Use of Coal Fly Ash: A Valuable Industrial by-Product.* Thomas Telford. https://www.icevirtuallibrary.com/doi/abs/10.1680/pauocfa.30152.fm

Sandhu, R. K., & Siddique, R. (2017). Influence of rice husk ash (RHA) on the properties of self-compacting concrete: A review. *Construction and Building Materials, 153*, 751–764. https://doi.org/10.1016/j.conbuildmat.2017.07.165

Singh, A. K., Masto, R. E., Hazra, B., Esterle, J., & Singh, P. K. (2020). *Ash from Coal and Biomass combustion.* Cham, Switzerland: Springer. https://link.springer.com/book/10.1007/978-3-030-56981-5

Singh, B., & Ghosh, M. (2019). Heavy metals in bio-coal ash: Implications for environmental safety and waste management. *Journal of Hazardous Materials, 379*, 120849. https://doi.org/10.1016/j.jhazmat.2019.120849

Singh, K., & Bhatnagar, A. (2020). Soil amendment potential of bio-coal ash: A source of nutrients and remediation of heavy metals. *Journal of Environmental Management, 275*, 111245.

Sigvardsen, N. M., Kirkelund, G. M., Jensen, P. E., Geiker, M. R., & Ottosen, L. M. (2019). Impact of production parameters on physiochemical characteristics of wood ash for possible utilisation in cement-based materials. *Resources, Conservation and Recycling, 145*, 230–240.

Smith, A. M., Ekpo, U., & Ross, A. B. (2020). The influence of pH on the combustion properties of bio-coal following hydrothermal treatment of swine manure. *Energies, 13*(2), 331. https://doi.org/10.3390/en13020331

Smith, A. M., & Ross, A. B. (2016). Production of bio-coal, bio-methane and fertilizer from seaweed via hydrothermal carbonisation. *Algal Research, 16*, 1–11. https://doi.org/10.1016/j.algal.2016.02.026

Smith, S. R., & Jasim, S. (2009). Small-scale home composting of biodegradable household waste: overview of key results from a 3-year research programme in West London. *Waste Management & Research, 27*(10), 941–950.

Sohi, S. P., Lopez-Capel, E., Krull, E., & Bol, R. (2009). Biochar's roles in soil and climate change: A review of research needs. *CSIRO Land and Water Science Report, 05/09*, 64.

Sommer, S. G., Clough, T. J., & Petersen, S. O. (2006). Greenhouse gas emissions from agricultural biomass combustion. *Environmental Science & Technology, 40*(3), 870–876. https://doi.org/10.1021/es051771j

Supraja, K. V., Kachroo, H., Viswanathan, G., Verma, V. K., Behera, B., Doddapaneni, T. R. K. C., ... & Jain, R. (2023). Biochar production and its environmental applications: recent developments and machine learning insights. *Bioresource Technology, 387*, 129634. https://doi.org/10.1016/j.biortech.2023.129634

Sua-Iam, G., & Makul, N. (2015). Utilization of coal-and biomass-fired ash in the production of self-consolidating concrete: a literature review. *Journal of Cleaner Production, 100*, 59–76. https://doi.org/10.1016/j.jclepro.2015.03.038

Szatyłowicz, E., & Krasowska, M. (2020). Assessment of heavy metals leaching from fly ashes as an indicator of their agricultural use. *Desalination and Water Treatment, 199*, 288–296. https://doi.org/10.5004/dwt.2020.25743

Sultana, A., Kumar, A., & Harfield, D. (2010). Development of agri-pellet production cost and optimum size. *Bioresource Technology, 101*(6), 2103–2110. https://doi.org/10.1016/j.biortech.2010.02.011

Tenenbaum, D. J. (2000). Will the circle be unbroken: A history of the Circular Economy concept. *Environmental Health Perspectives, 108*(3), A107–A112. https://doi.org/10.1289/ehp.108-a107

Triolo, J. M., Sommer, S. G., Møller, H. B., Weisbjerg, M. R., & Jiang, X. Y. (2011). A new algorithm to characterize biodegradability of biomass during anaerobic digestion: influence of lignin concentration on methane production potential. *Bioresource Technology, 102*(20), 9395–9402.

Junginger, H. M., Mai-Moulin, T., Daioglou, V., Fritsche, U., Guisson, R., Hennig, C., ... & Wild, M. (2019). The future of biomass and bioenergy deployment and trade: a synthesis of 15 years IEA Bioenergy Task 40 on sustainable bioenergy trade. *Biofuels, Bioproducts and Biorefining, 13*(2), 247–266. https://doi.org/10.1002/bbb.1993

Ugwu, K. E., Ezema, C. G., Ibeto, C. N., & Okafor, I. F. (2024). Properties of biocoal briquettes from mesoporous coals and rice husk and their effects on environmental pollution. *Discover Environment, 2*(1), 106. https://doi.org/10.1007/s44274-024-00141-2

Van Deventer, J. S., Provis, J. L., & Duxson, P. (2012). Technical and commercial progress in the adoption of geopolymer cement. *Minerals Engineering, 29*, 89–104. https://doi.org/10.1016/j.mineng.2011.09.009

Vassilev, S. V., Baxter, D., Andersen, L. K., & Vassileva, C. G. (2010). An overview of the chemical composition of biomass. *Fuel, 89*(5), 913–933. https://doi.org/10.1016/j.fuel.2009.10.022

Vassilev, S. V., Baxter, D., Andersen, L. K., & Vassileva, C. G. (2013a). An overview of the organic and inorganic phase composition of biomass. *Fuel, 94*, 1–33. https://doi.org/10.1016/j.fuel.2011.09.030

Vassilev, S. V., & Vassileva, C. G. (2007). A new approach for the classification of coal fly ashes based on their origin, composition, properties, and behavior. *Fuel, 86*(10–11), 1490–1512. https://doi.org/10.1016/j.fuel.2006.11.020

Vassilev, S. V., Vassileva, C. G., Baxter, D., & Andersen, L. K. (2013b). Mineral and chemical composition of biomass ash. *Fuel, 105*, 19–39. https://doi.org/10.1016/j.fuel.2012.09.041

Vassilev, S. V., Vassileva, C. G., & Vassilev, V. S. (2015a). Composition, properties, and challenges of biomass-based ash from energy conversion. *Fuel, 158*, 370–450. https://doi.org/10.1016/j.fuel.2015.05.034

Vassilev, S. V., Vassileva, C. G., & Vassilev, V. S. (2015b). Advantages and disadvantages of composition and properties of biomass in comparison with coal: An overview. *Fuel, 158*, 330–350. https://doi.org/10.1016/j.fuel.2015.05.050

Vlaskin, M. S., & Kumar, V. (2020). Assessing the effectiveness of the hydrothermal carbonization method to produce bio-coal from wet organic wastes. *Thermal Engineering, 67*(7), 441–450. https://doi.org/10.1134/S0040601520070071

Wang, L., Lurina, M., Hyytiäinen, J., & Mikkonen, E. (2014). Bio-coal market study: Macro and micro-environment of the bio-coal business in Finland. *Biomass and Bioenergy, 63*, 198–209. https://doi.org/10.1016/j.biombioe.2014.01.044

Wang, F., Lin, Y., & Zhang, X. (2021). Evaluation of leaching potential of bio-coal ash in water treatment processes. *Water Research, 188*, 116–129.

Wang, L., Zhang, H., & Zhang, L. (2019). Utilization of bio-coal ash for improving anaerobic digestion: A review. *Renewable and Sustainable Energy Reviews, 110,* 175–183. https://doi.org/10.1016/j.rser.2019.04.059

Wang, S., & Wu, H. (2006). Environmental-benign utilization of fly ash as low-cost adsorbents. *Journal of Hazardous Materials, 136*(3), 482–501. https://doi.org/10.1016/j.jhazmat.2006.01.067

White, S. C., & Case, E. D. (1990). Characterization of fly ash from coal-fired power plants. *Journal of Materials Science, 25*, 5215–5219. https://doi.org/10.1016/j.renene.2006.02.017

Wu, H., & Yip, K. (2017). Chemical and mineral transformations of biomass ash. *Progress in Energy and Combustion Science, 62*, 31–57. https://doi.org/10.1016/j.pecs.2017.05.002

Wu, S., & Wang, X. (2021). Heavy metal leaching from bio-coal ash: Mechanisms and management strategies. *Environmental Science & Technology, 55*(12), 8072–8081. https://doi.org/10.1021/acs.est.1c01583

Haris, M., Hamid, Y., Usman, M., Wang, L., Saleem, A., Su, F., ... & Li, Y. (2021). Crop-residues derived biochar: synthesis, properties, characterization and application for the removal of trace elements in soils. *Journal of Hazardous Materials, 416*, 126212.

Xu, Q., Zhang, Y., Yang, R., Li, J., Chen, J., Wang, J., ... & Shi, K. (2024). The utilization of Lysinibacillus bacterial powder to induce Fe plaque formation mitigates cadmium and chromium levels in rice. *Journal of Hazardous Materials, 463*, 132825.

Xu, Q., Zhang, T., Niu, Y., Mukherjee, S., Abou-Elwafa, S. F., Nguyen, N. S. H., ... & Shaheen, S. M. (2024). A comprehensive review on agricultural waste utilization through sustainable conversion techniques, with a focus on the additives effect on the fate of phosphorus and toxic elements during composting process. *Science of The Total Environment*, 173567. https://doi.org/10.1016/j.scitotenv.2024.173567

You, S., Lim, Y. J., Dai, Y., & Wang, C. H. (2018). On the temporal modelling of solar photovoltaic soiling: Energy and economic impacts in seven cities. Applied Energy, 228, 1136–1146.

Zhang, A., Bian, R., Pan, G., Cui, L., Hussain, Q., Li, L.,. . ., & Zheng, J. (2013). Effects of biochar amendment on soil quality, crop yield, and greenhouse gas emission in a Chinese rice paddy: A field study of 2 consecutive rice-growing cycles. *Field Crops Research, 127*, 153–160. https://doi.org/10.1016/j.fcr.2011.09.010

Zhang, H., et al. (2020). The role of silica in bio coal fly ash and its implications for utilization. *Journal of Hazardous Materials, 384*, 121326. https://doi.org/10.1016/j.jhazmat.2019.121326

Zhang, J., Li, M., Taheri, A., Zhang, W., Wu, Z., & Song, W. (2019). Properties and application of backfill materials in coal mines in China. *Minerals, 9*(1), 53.

Zhang, S., Zhang, X., & Li, X. (2018). Utilization of bio-coal ash for composting organic waste: A review. *Journal of Cleaner Production, 174*, 1104–1112. https://doi.org/10.1016/j.jclepro.2017.10.159

Seo, J. Y., Tokmurzin, D., Lee, D., Lee, S. H., Seo, M. W., & Park, Y. K. (2022). Production of biochar from crop residues and its application for biofuel production processes–an overview. *Bioresource Technology, 361*, 127740. https://doi.org/10.1016/j.biortech.2022.127740

Maguire, R., Maguire, V., & Maguire, K. (2023). Biochar sequestration in mine backfill. *The International Journal of Climate Change: Impacts and Responses, 15*(2), 1. https://doi.org/10.18848/1835-7156/CGP/v15i02/1-14

Zhou, H., Zhang, Y., Cheng, X., et al. (2020). Catalytic applications of bio-coal ash in advanced oxidation processes: A review. *Environmental Science & Technology, 54*(10), 6350–6365.

12 Economy of Bio-Coal in the World

Abhinay Thakur and Dharmesh Sur

12.1 INTRODUCTION

Bio-coal, also referred to as torrefied or hydrothermal carbon, stands as a crucial innovation in the realm of renewable energy (Ghobadi & Sepasgozar, 2023; Kircher et al., 2023; Narayana Sarma & Vinu, 2023). It is different from conventional coal, which is mostly made of biological stuff that has been fossilized and greatly increases emissions of greenhouse gases and environmental damage. Thermal conversion is the method used to produce bio-coal, in which biomass feedstocks including forestry trash, energy crops, and agricultural wastes are heated without the presence of oxygen (S. F. Ahmed et al., 2023; Licandeo, Flores, & Feijoo, 2023). This process, known as torrefaction, results in the transformation of biomass into a coal-like substance with enhanced energy density, improved handling characteristics, and reduced moisture content. Unlike raw biomass, bio-coal is more energy-dense and can be stored, transported, and utilized more efficiently, making it a viable alternative to traditional coal. The significance of bio-coal stems from its ability to address multiple sustainability challenges simultaneously. Firstly, by utilizing renewable biomass feedstocks, bio-coal production offers a pathway to mitigate greenhouse gas emissions. In contrast to fossil fuels, which, when burned, release carbon dioxide that has been stored for thousands of years into the atmosphere, bio-coal is thought to be carbon-neutral. A closed carbon cycle is created when the carbon dioxide that is released during burning is balanced by the carbon dioxide that is taken up by the biomass during growth. Additionally, bio-coal helps to lessen dependency on fossil fuels, improving the availability of energy and lowering susceptibility to changes in price and supply interruptions. In light of global efforts to diversify energy inventories and lessen dependency on fossil fuels, bio-coal is emerging as a viable substitute that can either supplement or entirely substitute coal-fired power generation. In the European Union (EU), for example, greater investments in bioenergy, including the manufacture of bio-coal, have resulted from efforts to switch to cleaner energy sources (V. Kumar, Younis, Szulejko, & Kim, 2023; Smol & Marcinek, 2023). The EU's statistical office, Eurostat, reports that in 2020, the EU's share of net final energy use was 19.7% from renewable energy sources, with biomass making up a sizeable percentage of this mix. With its environmentally benign alternative to coal, bio-coal helps the EU meet its renewable energy goals and lower greenhouse gas emissions. It is produced from sustainable biomass feedstocks. Figure 12.1 shows the preparation of bio-coal briquette using eco-friendly apparatus.

DOI: 10.1201/9781003498124-12

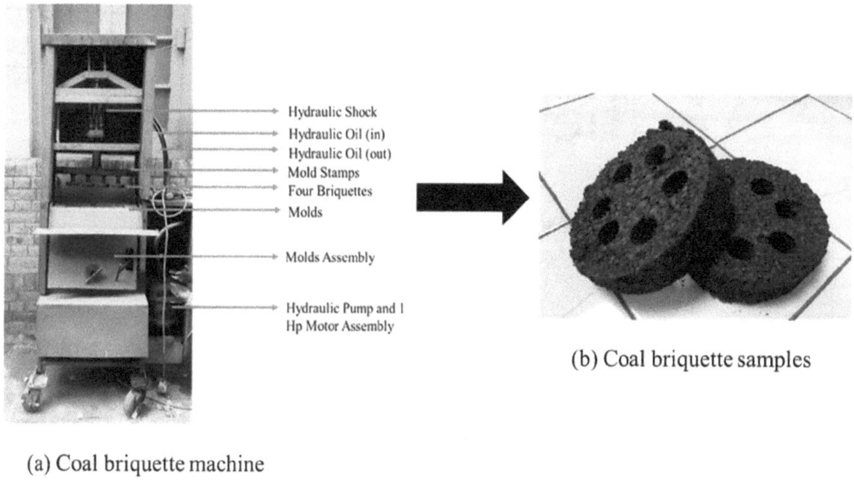

FIGURE 12.1 Depiction of the preparation of bio-coal briquette samples using the coal briquette machine.

Similarly, in the United States, bio-coal is gaining traction as a viable option for power generation and industrial applications. A number of jurisdictions have enacted renewable energy mandates and incentives to encourage the use of bioenergy, including bio-coal, in response to growing worries about climate change and air pollution. The U.S. Energy Information Administration (EIA) estimates that in 2020, electricity generated in the United States came from renewable energy sources at a rate of 20%, with biomass accounting for around 1.4% of this total (Mo et al., 2023). In addition to lowering carbon emissions from coal-fired power stations, bio-coal manufacturing facilities, such as pellet mills and torrefaction plants, are spreading across the nation and generating jobs and economic prospects in rural regions. Furthermore, bio-coal offers a sustainable answer to issues with energy poverty and rural development in developing nations like Brazil and India. The manufacturing of bio-coal from sugarcane bagasse and other agricultural leftovers provides a renewable substitute for fossil fuels in the transportation and power-generating sectors in Brazil, the world's largest producer of sugarcane. As per the Ministry of Mines and Energy of Brazil, bioelectricity produced from biomass contributed to the nation's energy security and rural development in 2020, making up around 9.5% of Brazil's overall energy consumption (Elgarahy et al., 2023; Gunathunga, Gagen, Evans, Erskine, & Southam, 2023). Similar to this, bio-coal production in India, where biomass is plentiful, has the potential to reduce poverty, open up job opportunities, and enhance rural residents' access to energy. The Ministry of New and Renewable Energy (MNRE) of India reports that as of March 2021, the nation's biomass-based power generation capacity was 10.76 GW, including notable contributions from energy crops, forestry leftovers, and agricultural waste (Tauseef Hassan, Wang, Khan, & Zhu, 2023). Bio-coal production initiatives, supported by government subsidies and incentives, are helping to modernize the agricultural sector, increase rural incomes, and reduce dependence on traditional biomass fuels for cooking and heating. By providing farmers with opportunities to cultivate energy crops or utilize agricultural residues for bio-coal production, rural economies can benefit from additional revenue streams and job opportunities, thus contributing to poverty alleviation and sustainable livelihoods. Furthermore, bio-coal's compatibility with currently operating coal-fired power plants is one of its main benefits. Because bio-coal requires little in the way of retrofitting and may be used as a standalone fuel or in conjunction with standard coal, it is an affordable way to switch to cleaner energy sources without having to make large infrastructure investments. This compatibility enables a smoother transition toward renewable energy and facilitates the gradual phasing out of coal-fired power plants, thereby reducing environmental impacts and improving air quality.

Understanding the production costs of bio-coal is paramount for evaluating its economic feasibility and competitiveness in comparison to conventional fuels. Production costs encompass a multitude of factors, ranging from feedstock acquisition and processing expenses to energy consumption and transportation logistics. Every one of these components is essential to figuring out how cost-effective bio-coal manufacturing is overall. One of the key factors influencing production costs is the selection of biomass feedstock. The energy content, cost-effectiveness, and availability of various feedstocks differ. For example, while dedicated energy plants including the switchgrass or miscanthus leaves may require significant effort in growing and harvesting, agricultural wastes like maize stover or rice husks may be plentiful and reasonably priced (Qian, Zhao, Qin, & Song, 2023; Zhao, Wang, et al., 2023). Additionally, the energy content of the feedstock influences the overall efficiency of the bio-coal production process, with higher energy feedstocks yielding more cost-effective outcomes. In addition to feedstock considerations, processing expenses represent a significant component of production costs. The torrefaction process, which converts biomass into bio-coal through thermal treatment, incurs operational and maintenance costs associated with equipment, labor, and energy inputs. Optimizing the torrefaction process to maximize efficiency and minimize energy consumption is essential for reducing production costs and enhancing economic competitiveness.

Moreover, transportation logistics play a pivotal role in determining production costs, particularly for biomass feedstock sourcing and bio-coal distribution. The proximity of biomass sources to production facilities and end-users influences transportation expenses, with longer distances necessitating higher logistical costs. Efficient supply chain management and infrastructure development are crucial for minimizing transportation-related expenditures and ensuring cost-effective bio-coal production and distribution. Many variables, such as legislative frameworks, technology developments, energy regulations, and consumer demand for renewable energy, affect the bio-coal business. While bio-coal offers significant environmental benefits and potential cost savings compared to traditional coal, its market penetration is still in its nascent stage (Hu, Chen, Yang, Li, & Wu, 2023; Tilsted, Bauer, Deere Birkbeck, Skovgaard, & Rootzén, 2023). Challenges related to scalability, cost competitiveness, and market penetration pose barriers to widespread adoption. Nevertheless, favorable policy initiatives, such as carbon pricing mechanisms and renewable energy targets, are driving market growth and incentivizing investment in bio-coal production. Regulatory support and financial incentives can mitigate investment risks and enhance the economic viability of bio-coal projects, thereby accelerating market development and expansion. The global adoption of bio-coal hinges on its ability to compete with traditional coal and other renewable energy sources on cost, performance, and sustainability metrics. Overcoming barriers related to technology scalability, feedstock availability, infrastructure development, and regulatory uncertainty is vital for knowing the entire potential of bio-coal as a sustainable alternative. As awareness of climate change impacts grows and countries commit to carbon reduction targets, the demand for bio-coal is expected to increase, creating opportunities for market expansion and innovation (Kurniawan et al., 2023; Mumtaz, Sobek, Werle, Sajdak, & Muzyka, 2023). To overcome obstacles and realize the global economic and environmental benefits of bio-coal, cooperation between governments, business interests, and research institutions is crucial (Thakur & Kumar, 2023a, 2023b, 2023c, 2024; Thakur, Sharma, Ganjoo, Assad, & Kumar, 2022).

12.2 BIOMASS FEEDSTOCK FOR BIO-COAL PRODUCTION

A process's ability to produce bio-coal depends on its biomass feedstock, which also determines the process's scalability, sustainability in the environment, and economic feasibility. The features, availability, processing techniques, and potential for the generation of bio-coal of several biomass feedstock categories—such as forestry residues, municipal solid waste (MSW), dedicated energy crops, and agricultural residues—are highlighted in this chapter's thorough analysis.

12.2.1 Agricultural Residues

Agricultural residues, including crop residues (such as rice husks, corn straw, and wheat straw) and processing residues (e.g., sugarcane bagasse and cotton gin trash), represent abundant and readily available biomass sources for bio-coal production. These residues are generated from agricultural activities and agro-industrial processes, exploring a cost-effective and sustainable feedstock option. Agricultural residues are typically collected post-harvest or post-processing, thereby reducing waste and providing additional revenue streams for farmers. The Food and Agriculture Organization of the United Nations (FAO) estimates that a significant amount of the global biomass produced each year is made up of agricultural wastes (Davis, Okunlola, Di Lullo, Giwa, & Kumar, 2023). For example, in 2020, global agricultural production generated an estimated 5.9 billion metric tons of crop residues, including cereal straw, rice husks, sugarcane bagasse, and cotton gin trash (Tabibian & Sharifzadeh, 2023). These residues, which are typically left in fields after harvest or post-processing, have immense potential as feedstock for various value-added applications, including bioenergy production, soil amendment, and animal bedding. In the EU, agricultural residues are recognized as valuable biomass resources for renewable energy production and waste management. According to Eurostat, the statistical office of the EU, the usage of agricultural residues for energy purposes has been increasing steadily in recent years. In 2020, the EU produced approximately 302 million metric tons of crop residues, of which a significant portion was used for bioenergy production, including biogas, biofuels, and biomass-based electricity generation. Farmers may increase revenue, meet renewable energy goals, and lower greenhouse gas emissions from agricultural operations by turning agricultural residues into bioenergy.

Similarly, in the United States, agricultural residues are utilized for various bioenergy and bioproduct applications, contributing to rural economic development and waste reduction efforts. Agricultural residues, such as crop residues, forest residues, and animal manure, are important feedstocks for the generation of biofuels like ethanol and biodiesel, according to the U.S. Department of Agriculture (USDA) (Ağbulut et al., 2023). In 2020, the United States produced over 370 million metric tons of crop residues alone, representing a significant biomass resource for bioenergy production and waste management. By participating in biomass supply chains and bioenergy markets, farmers can diversify their income streams and reduce the environmental footprint of agricultural operations. Moreover, agricultural waste is used for a variety of reasons in emerging nations like China and India, such as enhancing soil fertility, producing energy, and feeding animals. In nations like China and India that have sizable agricultural industries, agricultural wastes constitute a significant source of renewable energy, based on the International Energy Agency (IEA) (Nandhini, Sivaprakash, Rajamohan, & Vo, 2023; Okoh & Okpanachi, 2023). In India, for example, crop residues like rice straw and wheat straw are utilized for biomass power generation, biogas production, and cooking fuel substitution, contributing to rural electrification and clean energy access. Similarly, in China, agricultural residues are utilized for bioenergy production, organic fertilizer production, and livestock feed supplementation, helping to improve resource efficiency and reduce agricultural waste pollution (Kaur, Thakur, Pathania, & Sharma, 2022; Nidhi, Kaur, Thakur, & Kumar, 2023; Pathania, Sharma, Vermani, Kaur, & Thakur, 2023; Thakur, Sharma, Sharma, & Kaur, 2023; Thakur, Kaur, & Kaur, 2020).

Agricultural residues, which include materials like straw, stover, and husks, exhibit diverse characteristics influenced by the type of crop, harvesting methods, and regional factors. Typically, these residues are rich in cellulose, hemicellulose, and lignin—key components that make them suitable for thermochemical conversion processes such as torrefaction. For instance, wheat straw contains approximately 35% cellulose, 25% hemicellulose, and 15% lignin. However, the availability of these residues varies seasonally and regionally. In regions like Asia, rice straw is abundant after harvest, while corn stover is prevalent in the U.S. Midwest during autumn (Charai et al., 2023; Shi et al., 2022). Excess residues are frequently burned in the field or left unused, which exacerbates environmental problems including soil erosion and air pollution. To improve handling, storage, and conversion efficiency, agricultural residues undergo preprocessing steps like size reduction, drying, and densification. Size reduction involves grinding residues to a smaller particle size, typically 2–10 mm for corn stover, while drying reduces moisture content to less than 15%. Densification converts residues into pellets or briquettes, which typically have a density of 600–800 kg/m^3. These residues are improved by the thermochemical process of torrefaction, which involves heating them to temperatures between 200°C and 300°C without the presence of oxygen. This phenomenon increases their energy density, grind ability, and hydrophobicity (Li et al., 2019; Naskar & Sharma, 2024; Prismantoko et al., 2024). For example, torrefied wheat straw can achieve an energy density of 20–25 MJ/kg compared to 15–18 MJ/kg for raw straw. Torrefaction also makes residues 30%–40% more grindable and significantly reduces their moisture absorption, thereby improving their storage and handling properties (Muhammad et al., 2024). Globally, torrefaction is employed in Europe, where countries like the Netherlands and Sweden use it to convert agricultural residues into bio-coal for co-firing in power plants. In North America, research explores torrefying corn stover and wheat straw for industrial energy applications, while pilot projects in Asia focus on rice straw to address residue management challenges. The environmental impact of residue management is significant, with burning contributing to air pollution and soil degradation. For example, burning rice straws in India has resulted in serious problems with air quality and health hazards. Torrefaction presents a sustainable alternative by converting residues into high-energy bio-coal, thus mitigating waste and reducing environmental harm while offering a viable renewable energy source.

12.2.2 Dedicated Energy Crops

A sustainable substitute for conventional crops, dedicated energy crops including switch grass, miscanthus leaves, willow, and poplar are cultivated especially for the production of bioenergy. These perennial grasses and woody plants are characterized by rapid growth rates, high biomass yields, and low input requirements, making them suitable for marginal lands and agroforestry systems. In order to produce bio-coal, dedicated energy crops offer a carbon-neutral and renewable feedstock source that lessens reliance on fossil fuels and reduces greenhouse gas emissions. Based on the information provided by the IEA (Charai et al., 2023; Gollakota & Shu, 2023), dedicated energy crops, such as switchgrass, miscanthus, and willow, have significant potential to contribute to bioenergy production and sustainable land use practices. These crops may be produced on marginal soils inappropriate for food crops since they are exclusively planted for energy, avoiding conflict with crops while maximizing biomass yield and resource efficiency. In the EU, dedicated energy crops are recognized as key feedstocks for bioenergy production and renewable energy development. According to Eurostat, the statistical office of the EU, the cultivation of dedicated energy crops has been steadily increasing in recent years, driven by supportive policies and incentives for renewable energy deployment. In 2020, the EU harvested over 2.6 million hectares of land for dedicated energy crop cultivation, producing approximately 34 million metric tons of biomass feedstock (Farooq & Alhalabi, 2023; Seleem, Sameh, Esily, & Ibrahiem, 2023). This biomass was utilized for various bioenergy applications, including biofuels, biogas, and biomass-based electricity generation, contributing to the EU's renewable energy targets and climate change mitigation efforts. Similarly, as an environmentally friendly fuel source for bioenergy production and carbon sequestration, dedicated energy crops are becoming more and more popular in the United States. Dedicated energy crops have the potential to supply a sizeable amount of the country's renewable energy demands while lowering greenhouse gas emissions from the use of fossil fuels, according to the U.S. Department of Energy (DOE) (Andooz, Eqbalpour, Kowsari, Ramakrishna, & Ansari Cheshmeh, 2023). In 2020, the United States cultivated over 3 million hectares of land for dedicated energy crop production, generating approximately 60 million metric tons of biomass feedstock. The country's energy portfolio was diversified and carbon emissions from the power and transportation sectors were decreased by using this biomass for a variety of bioenergy applications, including the production of cellulosic ethanol, biomass power generation, and bioproducts (Kaur et al., 2023; Nidhi & Kaur, 2022; Rana et al., 2023; Thakur, Sharma, Kaur, & Kaur, 2019).

Moreover, in developing countries such as Brazil and India, dedicated energy crops are seen as promising solutions to energy security challenges and rural development opportunities. In Brazil, for example, dedicated energy crops such as sugarcane and eucalyptus are widely cultivated for bioenergy production, particularly ethanol and bioelectricity generation. The Brazilian Ministry of Agriculture, Livestock, and Supply reports that in 2020, the country harvested more than 10 million hectares of farmland specifically for the development of energy crops, resulting in the production of almost 720 million metric tons of biomass feedstock (Dai, Su, Liu, & Xu, 2023; Hong Nham, Mai Hoa, & Ha, 2023). By producing bioenergy from this biomass, Brazil was able to meet its renewable energy targets and lessen its reliance on imported fossil fuels. Similar to this, India cultivates special energy crops such as sweet sorghum and jatropha for the goal of producing biofuel and promoting rural development. The MNRE of India states that in an effort to improve job opportunities, increase energy security, and support sustainable land use practices, the nation has started extensive plantation initiatives for certain energy crops (Liu, Guo, Fan, Huang, & Liu, 2023). In 2020, India cultivated over 2.5 million hectares of land for dedicated energy crop cultivation, producing approximately 30 million metric tons of biomass feedstock (Hosein, Whale, Simsek, & Urmee, 2023; Li, Umar, Mirza, & Yue, 2023). This biomass was utilized for biofuel production, rural electrification, and decentralized energy generation, contributing to India's energy transition and climate change mitigation goals.

The special qualities of energy crops, which are grown with the intention of producing energy, make them more suitable for the generation of bio-coal. These crops have a low percentage of moisture, a high amount of ash, and high cellulose and lignin contents. For instance, switchgrass, commonly used in the United States, contains about 30%–35% cellulose and 10%–20% lignin, while miscanthus, widely grown in Europe, features around 40% cellulose and 15% lignin. The moisture content of these crops is generally low, with switchgrass having a moisture content ranging from 8% to 12% when harvested (Guo, Teng, Yang, Sun, & Liu, 2021; Wibowo, Ketwong, Cholwatthanatanakorn, Ding, & Areeprasert, 2024). Additionally, the high ash content of these crops, such as miscanthus with 6%–8% ash, impacts the quality of the bio-coal produced. Dedicated energy crops are typically grown on marginal lands that are not suitable for food production. This practice helps avoid competition with food crops and preserves arable land for agricultural use. For example, switchgrass is cultivated on degraded or low-yield lands in the United States, while miscanthus is planted on marginal lands across Europe (Qin, Zhou, Zhou, & Guo, 2022; Vafaei & Ghahremaninezhad, 2024). The availability of these crops is influenced by factors such as land use, climate conditions, and agricultural practices, with significant potential for expansion through sustainable cultivation methods and genetic improvements. To convert dedicated energy crops into bio-coal, preprocessing and torrefaction processes are employed. Due to their woody nature, these crops often require additional preprocessing steps. Size reduction involves grinding the biomass to a finer particle size, typically 2–10 mm for crops like miscanthus. Drying is also crucial, with technologies such as rotary dryers used to achieve a moisture content of less than 15% before torrefaction. The torrefaction process enhances the energy density, grindability, and hydrophobicity of these crops. For example, torrefied switchgrass can reach an energy density of 20–25 MJ/kg compared to 15–18 MJ/kg for raw switchgrass. Advancements in biomass pretreatment technologies, such as steam explosion and hydrothermal processing, further improve the suitability of dedicated energy crops for bio-coal production (Li et al., 2024). Steam explosion, which uses high-pressure steam to disrupt the lignocellulosic structure, can increase cellulose accessibility by 30%–40%. Hydrothermal processing, which involves treating biomass with water at high temperatures and pressures, enhances the reactivity of crops like miscanthus, making them more efficient for bio-coal production. Globally, dedicated energy crops are being utilized in various regions. In the United States, switchgrass is a prominent energy crop grown on marginal lands in states like Kentucky and Nebraska, supported by research initiatives to optimize yield and processing techniques. In Europe, miscanthus is extensively cultivated in the UK and Germany, with ongoing research aimed at improving its conversion processes. In Asia, particularly in China and India, the giant reed is being explored for its potential as a dedicated energy crop, with investments in research focused on enhancing cultivation and processing methods.

12.2.3 Forestry Residues

Forestry residues, including logging residues (e.g., branches, tops, and stumps), forest thinnings, and sawmill residues, are generated from forest management and timber processing activities. These residues are plentiful and widely dispersed in wooded areas, providing a sustainable and environmentally friendly supply of fuel for the production of bio-coal. The FAO of the United Nations reports that logging residues alone are anticipated to produce billions of cubic meters of waste each year. Forest management and timber processing activities around the world produce forestry residues (Kabir, Akter, Huang, Tijing, & Shon, 2023; Mohamed, Hamid, Montoya-Bautista, & Li, 2023). These residues, including branches, tops, stumps, and bark, are abundant and widely distributed in forested regions, making them an attractive feedstock source for bio-coal production. Forest leftovers are acknowledged as an important biomass resource in North America for the production of renewable energy and sustainable forest management. According to the U.S. Forest Service (USFS), logging residues, forest thinnings, and sawmill residues are generated from timber harvesting and processing operations across the continent. In the United States alone, forestry residues are estimated to exceed 200 million dry tons annually, representing a significant potential feedstock source for bioenergy production (Guo et al., 2023; Pan et al., 2023). Forest-dependent communities can diversify their sources of income, lower their risk of wildfires, and advance sustainable forest management techniques by using forestry wastes for the production of bio-coal. Likewise, forestry residues are essential to the generation of bioenergy and the fight against climate change in Europe. According to Eurostat, the statistical office of the EU, forestry residues are abundant in forested regions across Europe, including branches, tops, stumps, and bark left after timber harvesting and processing activities. In 2020, the EU produced over 260 million cubic meters of logging residues alone, representing a substantial biomass resource for bio-coal production (Afshan & Yaqoob, 2023; Sadiq et al., 2023). European nations may lessen their reliance on fossil fuels, cut down on greenhouse gas emissions, and increase the ability of forest ecosystems to withstand climate change by using forestry wastes for the production of bioenergy.

Moreover, in countries with extensive forest resources such as Canada and Russia, forestry residues are a critical feedstock source for bio-coal production and sustainable forest management. According to Natural Resources Canada, forestry residues are generated from timber harvesting operations, forest restoration activities, and wood processing industries across Canada. In 2020, Canada produced over 60 million cubic meters of logging residues, representing a significant biomass resource for bioenergy production and carbon sequestration (Thanapal, Annamalai, Lawrence, & Sweeten, 2023; Zheng, Zhu, Dong, Chen, & Zhu, 2023). Similarly, in Russia, forestry residues are abundant in boreal forests, offering a renewable and carbon-neutral feedstock source for bio-coal production and climate change mitigation. In developing countries such as Brazil and Indonesia, forestry residues are utilized for bioenergy production, rural development, and environmental conservation. According to the Brazilian Forest Service, logging residues and forest thinnings are generated from timber harvesting activities in the Amazon rainforest and other forested regions of Brazil. In 2020, Brazil produced over 200 million cubic meters of logging residues, representing a significant biomass resource for bioenergy production and sustainable land use practices (de Jong, Andrade Torres, Beisl Vieira de Melo, Mendes-Santana, & Valverde Pontes, 2023; Xue et al., 2023). Similarly, in Indonesia, forestry residues are generated from timber harvesting operations and forest restoration activities, providing a renewable and carbon-neutral feedstock source for bio-coal production and forest conservation efforts.

Forestry residues, often left to decompose in forests, represent a largely underutilized resource that can contribute to wildfire risks, habitat degradation, and increased carbon emissions. These residues, which include branches, bark, and wood chips, vary widely in composition and quality based on tree species, forest type, and harvesting methods. Typically, forestry residues are high in lignin and low in moisture, characteristics that make them well-suited for thermochemical conversion processes such as torrefaction (Stojanović et al., 2023; Tian, Tang, Xiu, Zhou, & Xue, 2022). For example, residues from coniferous trees, like pine, often have high lignin content and a moisture content as low as 10%–15%, making them ideal for conversion into bio-coal. The availability of forestry residues is influenced by several factors including forest management practices, timber harvesting intensity, and market demand for wood products. In well-managed forests, such as those in Sweden and Finland, residues are often systematically collected and utilized to reduce waste and improve forest health. In contrast, in regions with less efficient forest management practices or lower demand for wood products, residues are more likely to be left unused, contributing to environmental issues. Preprocessing and torrefaction of forestry residues involve steps similar to those used for agricultural residues and dedicated energy crops. The preprocessing stage often includes chipping and grinding to reduce the bulkiness and improve handling. For example, wood chips from forestry residues are typically reduced to a particle size of 2–5 cm. Drying is also crucial, with the goal of achieving a moisture content below 15% before torrefaction. Torrefaction then enhances the energy density and grindability of these residues, similar to its effects on agricultural residues. For instance, torrefied pine residues can reach an energy density of 22–27 MJ/kg compared to 18–22 MJ/kg for raw pine residues (Mohsen et al., 2023; Ogata et al., 2024; Wang, Wang, Ye, Ren, & Xu, 2021). Due to their bulkiness and heterogeneity, forestry residues often require additional logistics considerations. Effective handling and transportation are essential to optimize their utilization. Integrated forest biomass supply chains and decentralized torrefaction facilities can significantly enhance the efficiency of bio-coal production. For example, in the Pacific Northwest of the United States, integrated biomass supply chains have been developed to manage and utilize forestry residues from large-scale timber operations. These supply chains often include local chipping facilities, drying units, and torrefaction plants that work together to process residues into high-quality bio-coal. In Europe, countries like Germany and Austria have implemented decentralized torrefaction facilities that are strategically located near forestry operations. These facilities help reduce transportation costs and emissions by processing residues close to their source. This approach not only improves the sustainability of bio-coal production but also supports local economies and forest management practices.

12.2.4 Municipal Solid Waste

Residential, commercial, and institutional activities generate a heterogeneous mixture of organic and non-organic elements that make up MSW, sometimes referred to as household waste or rubbish (Zhan, Zhang, Hanjin, Li, & Wu, 2023). MSW is a substantial biomass resource for the manufacturing of bio-coal, providing a long-term answer to waste management and the production of renewable energy. Based on data from the World Bank, the rates of MSW generation have been rising globally, primarily due to urbanization, economic development, and population expansion. In 2020, global MSW generation surpassed 2 billion metric tons annually, with projections indicating further growth in the coming years (Acordi et al., 2023; Moktadir & Ren, 2023). Because there is an excess of garbage, there is a chance to use MSW as a feedstock for the manufacturing of bio-coal, which will lessen the need for fossil fuels and the environmental effects of disposing of waste. MSW is acknowledged as a useful resource for the generation of bioenergy and circular economy efforts in Europe. The EU statistical office, Eurostat, reports that in 2020, the EU produced over 220 million metric tons of MSW, with a large amount of it being landfilled or burned for waste management (Anu, Singh, Raza, Nakonieczny, & Shahzad, 2023; Zhang et al., 2023). Nonetheless, the possibility of MSW as a feedstock for the generation of bioenergy, especially bio-coal, is receiving more and more attention. European nations may handle waste management issues while lowering greenhouse gas emissions, decreasing landfilling, and advancing renewable energy ambitions by turning MSW into bio-coal. Similarly, MSW is starting to be seen as an invaluable asset for waste-to-energy and bioenergy production in the United States. The United States produced around 292 million metric tons of MSW in 2018, of which about 32% was recycled or composted and the rest was disposed of in landfills or waste-to-energy facilities, according to the U.S. Environmental Protection Agency (EPA) (A. A. Ahmed, Nazzal, Darras, & Deiab, 2023; Wu, 2023). By converting MSW into bio-coal, municipalities and waste management companies can create additional revenue streams, reduce landfilling, and contribute to renewable energy goals while minimizing environmental impacts.

Furthermore, because of poor waste management infrastructure and rising urbanization, MSW presents a serious concern in developing nations like China and India. China and India are two of the world's biggest producers of MSW, producing approximately 620 million metric tons and 275 million metric tons yearly, respectively, according to the World Bank (Saeid & Rahmanian, 2023; Subbarao et al., 2023). However, limited landfill space, pollution concerns, and resource scarcity have prompted efforts to explore alternative waste management strategies, including bioenergy production from MSW. By converting MSW into bio-coal, these countries can address waste management challenges, improve air quality, and enhance energy security while promoting sustainable development. Moreover, MSW is used for the generation of bio-coal as part of integrated waste management systems in nations like Sweden and Japan that have sophisticated waste-to-energy technologies. According to the Japan Waste Management Association, Japan generated over 45 million metric tons of MSW in 2020, with approximately 62% being incinerated for energy recovery (Zhao, Huang, Wang, & Wang, 2023). Japan may further improve its waste management procedures, lower greenhouse gas emissions, diversify its energy sources, and advance a circular economy by turning MSW into bio-coal. Similar to this, MSW is burned in Sweden in order to recover energy, and the leftover ash is used to produce bio-coal. Sweden can minimize trash creation, lessen its reliance on fossil fuels, and advance sustainable waste management techniques by optimizing resource recovery from MSW.

However, MSW presents unique challenges related to contamination, heterogeneity, and regulatory compliance, requiring specialized processing and conversion technologies. MSW is characterized by its diverse composition, including organic fraction (e.g., food waste, yard waste) and non-organic fraction (e.g., plastics, paper, glass). The availability of MSW varies depending on population density, waste generation rates, and waste management practices. In urban areas, MSW is collected and transported to landfill sites or waste-to-energy facilities, presenting an opportunity to divert organic waste streams for bio-coal production and reduce landfilling and greenhouse gas emissions (Ahmad et al., 2023; Cai et al., 2023). MSW undergoes mechanical and biological preprocessing steps, such as sorting, shredding, and composting, to separate organic waste fractions from non-organic materials. The organic fraction is then subjected to thermal conversion processes like torrefaction to produce bio-coal, while non-organic materials are recycled or disposed of through appropriate waste management practices. Gasification and pyrolysis are examples of advanced MSW conversion technologies that can improve the energy recovery and resource usage efficiency of MSW-derived bio-coal.

12.3 TECHNOLOGICAL ADVANCEMENTS IN BIO-COAL PRODUCTION

Innovations in technology are essential for improving the productivity, scalability, and sustainability of methods used to produce bio-coal. This chapter explores the latest innovations in torrefaction methods, reactor designs, and integration with existing infrastructure, highlighting their contributions to the advancement of bio-coal technology.

12.3.1 Torrefaction Methods

By heating biomass without oxygen, a process known as torrefaction transforms it into a material akin to biochar. This section looks at several torrefaction techniques and how they affect the quality of bio-coal, how well processes work, and how well the environment performs. Conventional torrefaction entails heating biomass in an inert environment, like carbon dioxide or nitrogen, to temperatures between 200°C and 300°C (Hasan, Abedin, Amin, Nekmahmud, & Oláh, 2023; Sardi, Uno, Pasila, Altway, & Mahfud, 2023). This process removes moisture and volatile components from the biomass, resulting in a product with enhanced energy density, grindability, and storage stability. However, traditional torrefaction may require prolonged residence times and high-energy inputs, limiting its scalability and economic viability. Novel techniques for torrefaction, like microwave-assisted torrefaction, hydrothermal torrefaction, and flash torrefaction, have been made possible by developments in torrefaction technology (Sardi Uno, Pasila, Altway, & Mahfud, 2023). These technologies offer improvements in process efficiency, energy consumption, and product quality compared to traditional torrefaction methods. Microwave-assisted torrefaction utilizes electromagnetic radiation to heat biomass rapidly and uniformly, reducing processing times and energy consumption. This technology enhances the homogeneity and reactivity of the torrefied biomass, resulting in a higher-quality bio-coal product.

Hydrothermal torrefaction and flash torrefaction are advanced thermochemical processes used to convert biomass into high-quality bio-coal, each offering unique advantages and applications. Hydrothermal torrefaction involves treating biomass with hot water or steam under pressure, typically at temperatures between 180°C and 250°C and pressures ranging from 1 to 10 MPa (Singh et al., 2022). This method enhances process intensification by utilizing water to accelerate chemical reactions, which improves the efficacy of biomass conversion. The use of water or steam reduces the energy required for heating compared to traditional dry torrefaction methods, making the process more energy-efficient. Additionally, hydrothermal torrefaction is versatile and capable of processing a broad array of feedstocks, including those having excessive moisture content like algae and aquatic plants. For instance, research in Finland has demonstrated the effective conversion of wood residues into high-energy bio-coal using this technology, highlighting its potential for improving the energy density and grindability of biomass. Flash torrefaction, by contrast, utilizes high temperatures and very short residence times, typically less than ten minutes, to rapidly convert biomass through pyrolysis and torrefaction. By attaining rapid pyrolysis at temperatures around 250°C and 350°C, this technique improves the efficiency of converting biomass (Gupta, Pathak, Siddique, Kumar, & Chaudhary, 2020; Sigvardsen & Ottosen, 2019). The brief residence times reduce the total energy input required and minimize energy losses, making flash torrefaction an energy-efficient method. This rapid approach also results in a cost-effective solution for large-scale bio-coal production. For example, in the Netherlands, flash torrefaction has been applied to wood pellets, producing bio-coal with an energy density of up to 30 MJ/kg (Sen, Das, Bandyopadhyay, Dash, & Raut, 2016). Similarly, in the United States, flash torrefaction of agricultural residues like corn stover has led to significant reductions in moisture content and increased energy density, making the bio-coal suitable for power plants and other energy applications.

12.3.2 Reactor Designs and Innovations

Torrefaction reactor design has a major impact on capital costs, product quality, and process efficiency. Rotary kiln reactors are commonly used for continuous torrefaction of biomass feedstocks, offering advantages in terms of process control, heat transfer, and scalability. These reactors utilize a rotating cylindrical drum to convey biomass through a heated chamber, facilitating uniform heating and residence time distribution. Rotary kiln reactors can be configured with direct or indirect heating methods, depending on the desired process conditions and product specifications (Bittencourt & Young, 2023; Paredes, Paredes, & García, 2023). Fluidized bed reactors are characterized by a fluidized particle suspension, allowing for efficient heat and mass transfer between biomass particles and the surrounding gas phase. These reactors offer advantages in terms of temperature control, reaction kinetics, and feedstock flexibility. Fluidized bed reactors can operate at relatively low temperatures and short residence times, minimizing thermal degradation and improving product quality. Additionally, fluidized bed reactors can be integrated with biomass gasification or combustion processes to achieve co-production of bio-coal and syngas or heat. Fixed bed reactors consist of a stationary bed of biomass particles or pellets, allowing for controlled heating and torrefaction under batch or semi-batch conditions (Chhandama, Rai, & Lalawmpuii, 2023; Jin et al., 2023; Shang, Zhao, Zhang, & Wang, 2023; Varbanov et al., 2023). These reactors offer advantages in terms of process flexibility, product quality, and energy efficiency. Fixed bed reactors can be operated at high temperatures and longer residence times, facilitating thorough biomass conversion and torrefaction. However, fixed bed reactors may require periodic reloading and unloading of biomass feedstock, leading to batch-wise operation and lower throughput compared to continuous reactors.

12.3.3 Integration with Existing Infrastructure

The amalgamation of bio-coal manufacturing procedures with extant facilities including industrial establishments and coal-fired power plants offers prospects to augment energy efficacy, mitigate ecological ramifications, and foster sustainable growth. In existing coal-fired power plants, co-firing bio-coal with conventional coal offers an affordable option to lower greenhouse gas emissions and boost the usage of renewable energy sources. Bio-coal can be pulverized and mixed with coal prior to combustion, providing additional fuel flexibility and emission reductions (Foong et al., 2023; Liguo, Ahmad, Khan, Haq, & Khattak, 2023). Co-firing bio-coal with coal can improve combustion efficiency, reduce ash deposition and slagging, and extend the operational lifespan of existing boilers and turbines. Moreover, co-firing bio-coal with coal can help power plants meet regulatory requirements and renewable energy mandates without significant capital investments. Retrofitting existing biomass co-generation facilities with torrefaction units or upgrading conventional biomass boilers to accommodate torrefied biomass can enhance energy efficiency, feedstock flexibility, and product quality. By torrefying biomass feedstocks prior to combustion, co-generation facilities can achieve higher energy densities, lower moisture content, and reduced emissions compared to raw biomass (Aslam et al., 2023). Retrofitting existing biomass co-generation facilities with torrefaction units enables the utilization of low-value biomass feedstocks and increases the competitiveness of bio-coal as a renewable energy source.

Integrated biorefineries represent a comprehensive approach to biomass utilization, combining the production of bio-coal with other value-added processes such as biofuels, biochemicals, and various bioproducts. By co-locating bio-coal production facilities with biofuel refineries or biochemical plants, integrated biorefineries optimize resource use, minimize waste, and enhance economic viability (Gollakota, Volli, & Shu, 2019; Zhang, Wen, & Li, 2021; Zhao et al., 2024). The integration of bio-coal production within biorefineries allows for a streamlined process where the same biomass feedstock can be used to produce multiple products, improving overall efficiency. For instance, a single biomass input can be processed to produce bio-coal, biofuels, and biochemicals, maximizing the value derived from the raw material. This approach not only enhances the economic viability of the operation but also reduces the costs associated with biomass collection, transportation, and processing. A notable example is the Biomass and Renewable Energy Centre (BREC) in the United Kingdom, which integrates bio-coal production with biofuel and biochemical processes. The facility uses wood chips to produce bio-coal through torrefaction, while also generating biochar and syngas as co-products (Lankapati, Lathiya, Choudhary, Dalai, & Maheria, 2023; Ukwattage, Ranjith, & Bouazza, 2013). The biochar is used for soil amendment, enhancing soil fertility and sequestering carbon, while the syngas is utilized for electricity generation, contributing to renewable energy supply. This integrated approach significantly reduces operational costs and increases the profitability of the biorefinery. Co-products from bio-coal production, such as biochar and syngas, offer additional economic and environmental benefits. Biochar, a stable form of carbon, can be used for soil amendment, improving soil health and fertility. It also acts as a carbon sequestration agent, capturing carbon dioxide and mitigating greenhouse gas emissions. Syngas, a mixture of carbon monoxide, hydrogen, and other gases produced during biomass conversion, can be utilized for renewable energy generation, further enhancing the sustainability of the biorefinery operation (Rathnayake, Julnipitawong, Tangtermsirikul, & Toochinda, 2018).

A Biorefinery in the Netherlands showcases the economic and environmental benefits of integrated biorefineries. The facility processes agricultural residues to produce bio-coal and biochar. The biochar is sold to agricultural sectors as a soil enhancer, while the bio-coal is used in power plants. The syngas produced during the process are used to generate electricity, contributing to the facility's energy needs and reducing dependence on fossil fuels. This integration not only provides a revenue stream from the sale of biochar and bio-coal but also reduces energy costs and carbon emissions (Kijo-Kleczkowska, Szumera, Gnatowski, & Sadkowski, 2022). Integrated biorefineries enhance sustainability by reducing waste generation and maximizing resource utilization. These facilities minimize the negative environmental effects of biomass processing and promote the concept of circular economy by merging several industrial processes. In addition to increasing economic viability, the holistic approach lowers operating expenses and generates different revenue streams. The Biomass Research and Development Initiative (BRDI) in the United States provides funding for projects that combine the production of bio-coal with biofuel and biochemical processes, exhibiting notable benefits for the environment and economy. The work carried out at the Richland, Washington-based Pacific Northwest National Laboratory (PNNL) is a notable illustration of this. This facility focuses on converting forest residues—such as wood chips and sawdust—into high-value products using advanced technologies. The PNNL employs torrefaction to produce bio-coal from forest residues, achieving an energy density of 22–26 MJ/kg, compared to 18–22 MJ/kg for raw biomass (Arora, Gosu, Subbaramaiah, & Hameed, 2021; Yadav, Pandey, Kumar, & Singh, 2022). Approximately 60%–70% of the processed biomass is converted into bio-coal. This method, combined with the production of biofuels and biochemicals, reduces the operational costs of bio-coal production by about 20% compared to standalone facilities. The bio-coal is not only more energy-dense but also cost-effective due to its integration with other processes. The facility also produces biofuels, converting 20%–30% of the biomass into fuels such as ethanol and biodiesel, which have market prices ranging from $1.50 to $3.00 per liter. Additionally, around 10%–15% of the biomass is converted into high-value biochemicals like acids and solvents, valued between $5,000 and $20,000 per ton (Thomas et al., 2024; Thulasikanth & Padmanabhan, 2023). This integration of bio-coal, biofuels, and biochemicals results in multiple revenue streams, which enhances the economic viability of the facility. By integrating these processes, PNNL effectively reduces the overall operational costs by 15%–20%, thanks to shared infrastructure and increased efficiencies. The environmental benefits are also significant, with bio-coal production contributing to carbon sequestration—up to 1.5 tons of CO_2 per ton of bio-coal—and reducing waste typically associated with biomass by-products (Sarkar, Maiti, Malik, & Xu, 2022; Su, Chen, & Lin, 2020; Yuan et al., 2022; Zheng, Liao, Liu, Li, & Zhang, 2022).

12.4 CHALLENGES AND FUTURE OPPORTUNITIES

In the landscape of bio-coal production, several intricate challenges must be navigated to ensure its widespread adoption and sustainable growth. One of the foremost challenges lies in securing a consistent and high-quality supply of biomass feedstocks. This challenge is multifaceted, involving issues such as the geographical distribution of feedstock sources, variations in feedstock quality and composition, and logistical constraints related to procurement and transportation (Harnisch & Lehneis, 2023; Perumal, 2023). Addressing these challenges requires a comprehensive approach, including strategic sourcing partnerships, investments in supply chain infrastructure, and the development of robust quality control measures to ensure feedstock consistency and reliability. Another significant challenge in bio-coal production stems from the technical complexities involved in scaling up production processes from laboratory or pilot-scale demonstrations to large-scale commercial operations. These complexities include optimizing reactor designs, refining torrefaction methods, and managing process variability to ensure consistent product quality and performance. Moreover, transitioning from batch-wise or semi-continuous production to continuous, industrial-scale operations requires careful planning, rigorous testing, and iterative process optimization to achieve optimal efficiency and productivity.

Another obstacle facing the bio-coal sector is market penetration. In a dynamic and frequently unstable energy market, it must contend with established fossil fuel companies as well as other renewable energy sources (Marshall, O'Dochartaigh, Prothero, Reynolds, & Secchi, 2023; Olabi et al., 2023). Overcoming market barriers requires not only cost competitiveness but also differentiation based on performance, sustainability, and market acceptance. Moreover, regulatory uncertainties and policy fluctuations can impact market dynamics and investor confidence, necessitating proactive engagement with policymakers and stakeholders to advocate for supportive policies and incentives that promote bio-coal adoption and investment. Environmental concerns represent a critical challenge for bio-coal production, as the sustainability of feedstock sourcing and production practices directly impact its environmental footprint and long-term viability (Al Momani et al., 2023; Harichandan, Kar, & Rai, 2023; He, Wei, Ding, Zheng, & Niu, 2023). In order to ensure sustainable feedstock sourcing, economic and social goals must be balanced with environmental factors like the utilization of land, preservation of biodiversity, and carbon sequestration. In a similar vein, minimizing environmental effects and guaranteeing the sustainability of bio-coal production processes depend on the adoption of ecologically friendly production techniques, such as effective energy use, waste reduction, and emissions reduction.

However, amidst these challenges lie numerous opportunities for advancement and innovation in bio-coal production. Diversifying feedstock sources, including agricultural residues, dedicated energy crops, forestry residues, and MSW, can enhance resource availability, resilience, and sustainability (Cheng, Luo, Jenkins, & Larson, 2023; R. Kumar, Bhaumik, Mahesh Galrani, Xavier Francis, & Dhawan, 2023; Mukherjee et al., 2023). Moreover, technological innovations in torrefaction methods, reactor design, and process optimization hold promise for improving efficiency, reducing costs, and enhancing product quality. Investment, innovation, and market uptake can be encouraged by policy support and market incentives involving carbon pricing processes, mandates for renewable energy, and funding for research (Ibrahim et al., 2023; Muhammad Mustafa Rizvi et al., 2023). Furthermore, sustainability certification and market differentiation strategies can enhance bio-coal's credibility, transparency, and market acceptance, ensuring its role as a sustainable and competitive energy solution for the future.

12.5 CONCLUSION

While the industry faces significant hurdles in terms of feedstock availability, technological complexity, market penetration, and environmental concerns, there is also immense potential for growth and innovation in bio-coal production. The challenge of securing a consistent and high-quality supply of biomass feedstocks requires proactive measures to diversify feedstock sources, optimize supply chain logistics, and implement robust quality control measures. Strategic partnerships with biomass suppliers, investments in biomass collection and processing infrastructure, and the development of biomass certification schemes can help mitigate supply chain risks and ensure feedstock reliability. Additionally, advancements in biomass conversion technologies, such as torrefaction, offer opportunities to improve feedstock utilization efficiency, reduce processing costs, and enhance product quality. There are logistical and technical obstacles to overcome when bringing bio-coal manufacturing techniques from the lab to the commercial level. Careful planning, testing, and optimization are necessary. Innovative reactor designs, process automation technologies, and integrated process control systems can streamline production operations, improve process efficiency, and minimize variability. To further speed up the commercialization and uptake of bio-coal technologies, cooperation between industrial players, academic institutions, and governmental organizations can promote information sharing, technology transfer, and capacity building. The bio-coal sector faces a great deal of market penetration because it must contend with well-established fossil fuel industries as well as alternative renewable energy sources in a competitive and dynamic energy market. Differentiation strategies based on cost competitiveness, performance advantages, and environmental benefits are essential for gaining market share and attracting investment. Moreover, market incentives and regulatory support—such as carbon pricing schemes, renewable energy mandates, and tax rebates can boost the demand for bio-coal and foster an environment that is conducive to investment for industry participants.

Environmental sustainability is a fundamental consideration in bio-coal production, as it directly impacts the industry's long-term viability and societal acceptance. Sustainable feedstock sourcing practices, including land use management, biodiversity conservation, and carbon sequestration, are essential for minimizing environmental impacts and ensuring the integrity of ecosystems. Additionally, implementing environmentally responsible production practices, such as energy efficiency improvements, waste minimization, and emissions reduction, can enhance the sustainability credentials of bio-coal and foster public trust and acceptance. Looking ahead, the future of bio-coal production holds immense promise, provided that the industry can overcome its current challenges and capitalize on emerging opportunities. Bio-coal has the potential to be a key player in the shift to low-carbon, sustainable energy by embracing technological innovation, market-driven methods, and sustainability principles. Reaching this goal and reaching the full potential of bio-coal as a sustainable energy option that benefits people and the environment will require cooperative efforts by industrial players, policymakers, academic institutions, and civil society.

REFERENCES

Acordi, J., Simão, L., Faraco, M. N. S., Borgert, C. H., Olivo, E., Montedo, O. R. K., & Raupp-Pereira, F. (2023). Waste valorization of coal mining waste from a circular economy perspective: A Brazilian case study based on environmental and physicochemical features. *Resources Policy*, *80*, 103243. https://doi.org/10.1016/j.resourpol.2022.103243

Afshan, S., & Yaqoob, T. (2023). Unravelling the efficacy of green innovation and taxation in promoting environmental quality: A dual-model assessment of testing the LCC theory in emerging economies. *Journal of Cleaner Production*, *416*, 137850. https://doi.org/10.1016/j.jclepro.2023.137850

Ağbulut, Ü., Sirohi, R., Lichtfouse, E., Chen, W.-H., Len, C., Show, P. L.,…, & Hoang, A. T. (2023). Microalgae bio-oil production by pyrolysis and hydrothermal liquefaction: Mechanism and characteristics. *Bioresource Technology*, *376*, 128860. https://doi.org/10.1016/j.biortech.2023.128860

Ahmad, M., Yousaf, M., Han, J.-C., Ur Rahman, S., Muhammad Adeel Sharif, H., Wang, L.,…, & Huang, Y. (2023). State-of-the-art analysis of the fuel desulphurization processes: Perspective of CO2 utilization in coal biodesulphurization. *Chemical Engineering Journal*, *478*, 147517. https://doi.org/10.1016/j.cej.2023.147517

Ahmed, A. A., Nazzal, M. A., Darras, B. M., & Deiab, I. M. (2023). Global warming potential, water footprint, and energy demand of shared autonomous electric vehicles incorporating circular economy practices. *Sustainable Production and Consumption*, *36*, 449–462. https://doi.org/10.1016/j.spc.2023.02.001

Ahmed, S. F., Rafa, S. J., Mehjabin, A., Tasannum, N., Ahmed, S., Mofijur, M.,… , & Kamangar, S. (2023). Bio-oil from microalgae: Materials, production, technique, and future. *Energy Reports*, *10*, 3297–3314. https://doi.org/10.1016/j.egyr.2023.09.068

Al Momani, D. E., Al Ansari, Z., Ouda, M., Abujayyab, M., Kareem, M., Agbaje, T., & Sizirici, B. (2023). Occurrence, treatment, and potential recovery of rare earth elements from wastewater in the context of a circular economy. *Journal of Water Process Engineering*, *55*, 104223. https://doi.org/10.1016/j.jwpe.2023.104223

Andooz, A., Eqbalpour, M., Kowsari, E., Ramakrishna, S., & Ansari Cheshmeh, Z. (2023). A comprehensive review on pyrolysis from the circular economy point of view and its environmental and social effects. *Journal of Cleaner Production*, *388*, 136021. https://doi.org/10.1016/j.jclepro.2023.136021

Anu, Singh, A. K., Raza, S. A., Nakonieczny, J., & Shahzad, U. (2023). Role of financial inclusion, green innovation, and energy efficiency for environmental performance? Evidence from developed and emerging economies in the lens of sustainable development. *Structural Change and Economic Dynamics*, *64*, 213–224. https://doi.org/10.1016/j.strueco.2022.12.008

Arora, S., Gosu, V., Subbaramaiah, V., & Hameed, B. H. (2021). Lithium loaded coal fly ash as sustainable and effective catalyst for the synthesis of glycerol carbonate from glycerol. *Journal of Environmental Chemical Engineering*, *9*(5), 105999. https://doi.org/10.1016/j.jece.2021.105999

Aslam, S., Rani, S., Lal, K., Fatima, M., Hardwick, T., Shirinfar, B., & Ahmed, N. (2023). Electrochemical hydrogen production: Sustainable hydrogen economy. *Green Chemistry*, *25*(23), 9543–9573. https://doi.org/10.1039/d3gc02849f

Bittencourt, F. P., & Young, A. F. (2023). Simulation and economic evaluation of a multipurpose pyrolysis plant for bio-oil production from different residual biomasses in Brazil. *Bioresource Technology Reports*, *22*, 101475. https://doi.org/10.1016/j.biteb.2023.101475

Cai, Y., Woollacott, J., Beach, R. H., Rafelski, L. E., Ramig, C., & Shelby, M. (2023). Insights from adding transportation sector detail into an economy-wide model: The case of the ADAGE CGE model. *Energy Economics*, *123*, 106710. https://doi.org/10.1016/j.eneco.2023.106710

Charai, M., Mghazli, M. O., Channouf, S., El hammouti, A., Jagadesh, P., Moga, L., & Mezrhab, A. (2023). Lightweight waste-based gypsum composites for building temperature and moisture control using coal fly ash and plant fibers. *Construction and Building Materials*, *393*, 132092. https://doi.org/10.1016/j.conbuildmat.2023.132092

Cheng, F., Luo, H., Jenkins, J. D., & Larson, E. D. (2023). The value of low- and negative-carbon fuels in the transition to net-zero emission economies: Lifecycle greenhouse gas emissions and cost assessments across multiple fuel types. *Applied Energy*, *331*, 120388. https://doi.org/10.1016/j.apenergy.2022.120388

Chhandama, M. V. L., Rai, P. K., & Lalawmpuii. (2023). Coupling bioremediation and biorefinery prospects of microalgae for circular economy. *Bioresource Technology Reports*, *22*, 101479. https://doi.org/10.1016/j.biteb.2023.101479

Dai, S., Su, M., Liu, Y., & Xu, Z. (2023). Digital economy, resource richness, external conflicts, and ecological footprint: Evidence from emerging countries. *Resources Policy*, *85*, 103976. https://doi.org/10.1016/j.resourpol.2023.103976

Davis, M., Okunlola, A., Di Lullo, G., Giwa, T., & Kumar, A. (2023). Greenhouse gas reduction potential and cost-effectiveness of economy-wide hydrogen-natural gas blending for energy end uses. *Renewable and Sustainable Energy Reviews*, *171*, 112962. https://doi.org/10.1016/j.rser.2022.112962

de Jong, P., Andrade Torres, E., Beisl Vieira de Melo, S. A., Mendes-Santana, D., & Valverde Pontes, K. (2023). Socio-economic and environmental aspects of bio-LPG and bio-dimethyl ether (Bio-DME) production and usage in developing countries: The case of Brazil. *Cleaner and Circular Bioeconomy*, *6*, 100055. https://doi.org/10.1016/j.clcb.2023.100055

Elgarahy, A. M., Priya, A. K., Mostafa, H. Y., Zaki, E. G., Elsaeed, S. M., Muruganandam, M., & Elwakeel, K. Z. (2023). Toward a circular economy: Investigating the effectiveness of different plastic waste management strategies: A comprehensive review. *Journal of Environmental Chemical Engineering*, *11*(5), 110993. https://doi.org/10.1016/j.jece.2023.110993

Farooq, A., & Alhalabi, W. (2023). Evaluation of hydrogen fuel cell-based systematic vehicular application to promote the green economy using LabVIEW. *Results in Engineering*, *20*, 101607. https://doi.org/10.1016/j.rineng.2023.101607

Foong, S. Y., Chan, Y. H., Lock, S. S. M., Chin, B. L. F., Yiin, C. L., Cheah, K. W.,… , & Lam, S. S. (2023). Microwave processing of oil palm wastes for bioenergy production and circular economy: Recent advancements, challenges, and future prospects. *Bioresource Technology*, *369*, 128478. https://doi.org/10.1016/j.biortech.2022.128478

Ghobadi, M., & Sepasgozar, S. M. E. (2023). Circular economy strategies in modern timber construction as a potential response to climate change. *Journal of Building Engineering*, *77*(July), 107229. https://doi.org/10.1016/j.jobe.2023.107229

Gollakota, A. R. K., & Shu, C.-M. (2023). Chapter 4 – Comparisons between fossil fuels and bio-fuels. In K. P. Shadangi, P. K. Sarangi, K. Mohanty, I. Deniz, & A. R. Kiran Gollakota (Eds.), *Bioenergy Engineering* (pp. 67–85). Woodhead Publishing. https://doi.org/10.1016/B978-0-323-98363-1.00021-1

Gollakota, A. R. K., Volli, V., & Shu, C.-M. (2019). Progressive utilisation prospects of coal fly ash: A review. *Science of The Total Environment*, *672*, 951–989. https://doi.org/10.1016/j.scitotenv.2019.03.337

Gunathunga, S. U., Gagen, E. J., Evans, P. N., Erskine, P. D., & Southam, G. (2023). Anthropedogenesis in coal mine overburden; the need for a comprehensive, fundamental biogeochemical approach. *Science of The Total Environment*, *892*, 164515. https://doi.org/10.1016/j.scitotenv.2023.164515

Guo, F., Guo, Y., Chen, L., Jia, W., Zhu, Y., Li, Y.,…, & Wu, J. (2023). Multitudinous components recovery, heavy metals evolution and environmental impact of coal gasification slag: A review. *Chemosphere*, *338*, 139473. https://doi.org/10.1016/j.chemosphere.2023.139473

Guo, Y., Teng, Q., Yang, Z., Sun, B., & Liu, S. (2021). Investigation on bio-desilication process of fly ash based on a self-screened strain of Bacillus amyloliquefaciens and its metabolites. *Journal of Biotechnology*, *341*, 146–154. https://doi.org/10.1016/j.jbiotec.2021.09.016

Gupta, V., Pathak, D. K., Siddique, S., Kumar, R., & Chaudhary, S. (2020). Study on the mineral phase characteristics of various Indian biomass and coal fly ash for its use in masonry construction products. *Construction and Building Materials*, *235*, 117413. https://doi.org/10.1016/j.conbuildmat.2019.117413

Harichandan, S., Kar, S. K., & Rai, P. K. (2023). A systematic and critical review of green hydrogen economy in India. *International Journal of Hydrogen Energy*, *48*(81), 31425–31442. https://doi.org/10.1016/j.ijhydene.2023.04.316

Harnisch, F., & Lehneis, R. (2023). The power grids need to be made ready for a circular and bio-based economy. *Next Sustainability*, *2*, 100010. https://doi.org/10.1016/j.nxsust.2023.100010

Hasan, M., Abedin, M. Z., Amin, M. Bin, Nekmahmud, M., & Oláh, J. (2023). Sustainable biofuel economy: A mapping through bibliometric research. *Journal of Environmental Management*, *336*, 117644. https://doi.org/10.1016/j.jenvman.2023.117644

He, S., Wei, W., Ding, S., Zheng, S., & Niu, T. (2023). Coevolution between science and policy: A systematic analysis on circular economy in China and worldwide. *Environmental Science & Policy*, *139*, 104–117. https://doi.org/10.1016/j.envsci.2022.10.020

Hong Nham, N. T., Mai Hoa, T. T., & Ha, L. T. (2023). Influences of digitalization on sustaining marine minerals: A path toward sustainable blue economy. *Ocean & Coastal Management*, *239*, 106589. https://doi.org/10.1016/j.ocecoaman.2023.106589

Hosein, A. S., Whale, J., Simsek, Y., & Urmee, T. (2023). Exploring energy policy scenarios to transition to a low carbon economy by 2050: A case study on the Northern Territory of Australia. *Energy Policy*, *180*, 113663. https://doi.org/10.1016/j.enpol.2023.113663

Hu, J., Chen, X., Yang, X., Li, R., & Wu, L. (2023). Co-processing of gas oil and bio-oil derived from algae and straw: Techno-economic analysis. *Fuel*, *340*, 127583. https://doi.org/10.1016/j.fuel.2023.127583

Ibrahim, H. A., Ayomoh, M. K., Bansal, R. C., Gitau, M. N., Yadavalli, V. S. S., & Naidoo, R. (2023). Sustainability of power generation for developing economies: A systematic review of power sources mix. *Energy Strategy Reviews*, *47*, 101085. https://doi.org/10.1016/j.esr.2023.101085

Jin, Y., Lenzen, M., Montoya, A., Laycock, B., Yuan, Z., Lant, P.,…, & Malik, A. (2023). Greenhouse gas emissions, land use and employment in a future global bioplastics economy. *Resources, Conservation and Recycling*, *193*, 106950. https://doi.org/10.1016/j.resconrec.2023.106950

Kabir, M. M., Akter, M. M., Huang, Z., Tijing, L., & Shon, H. K. (2023). Hydrogen production from water industries for a circular economy. *Desalination*, *554*, 116448. https://doi.org/10.1016/j.desal.2023.116448

Kaur, H., Thakur, R. C., Pathania, V. S., & Sharma, S. (2022). Effect of Choline-based ionic liquid (Cholinium ethanoate) on volumetric and acoustic properties of aliphatic amino acids (glycine and L-alanine). *Journal of Molecular Liquids*, *346*, 118247. https://doi.org/10.1016/j.molliq.2021.118247

Kaur, H., Thakur, R. C., Singh, N., Kumar, A., Gill, R., Thakur, A., & Sharma, S. (2023). Influence of 1-butyl-3-propylimidazolium bromide on viscometric and conductometric behavior of proteinogenic amino acids (glycine and L-alanine). *AIP Conference Proceedings*, *2800*(1), 20092. https://doi.org/10.1063/5.0163118

Kijo-Kleczkowska, A., Szumera, M., Gnatowski, A., & Sadkowski, D. (2022). Comparative thermal analysis of coal fuels, biomass, fly ash and polyamide. *Energy*, *258*, 124840. https://doi.org/10.1016/j.energy.2022.124840

Kircher, M., Aranda, E., Athanasios, P., Radojcic-Rednovnikov, I., Romantschuk, M., Ryberg, M.,..., & Vorgias, C. E. (2023). Treatment and valorization of bio-waste in the EU. *EFB Bioeconomy Journal, 3*, 100051. https://doi.org/10.1016/j.bioeco.2023.100051

Kumar, R., Bhaumik, Mahesh Galrani, R., Xavier Francis, A., & Dhawan, H. (2023). Optimization of hydrogen fuel cell economy for next generation power supply. *Materials Today: Proceedings*. https://doi.org/10.1016/j.matpr.2023.03.257

Kumar, V., Younis, S. A., Szulejko, J. E., & Kim, K.-H. (2023). Performance of BiO1.5-xIx and composite quasi-photocatalysts for the removal of gaseous elemental Hg0 from coal combustion flue gases: A review. *Renewable and Sustainable Energy Reviews, 185*, 113659. https://doi.org/10.1016/j.rser.2023.113659

Kurniawan, T. A., Othman, M. H. D., Liang, X., Goh, H. H., Gikas, P., Chong, K.-K., & Chew, K. W. (2023). Challenges and opportunities for biochar to promote circular economy and carbon neutrality. *Journal of Environmental Management, 332*, 117429. https://doi.org/10.1016/j.jenvman.2023.117429

Lankapati, H. M., Lathiya, D. R., Choudhary, L., Dalai, A. K., & Maheria, K. C. (2023). Modification and characterization of Mordenite zeolite derived from waste coal fly ash and its application as a heterogeneous catalyst for the n-butyl levulinate synthesis. *Catalysis Communications, 183*, 106772. https://doi.org/10.1016/j.catcom.2023.106772

Li, C., Zhang, Q., Zhang, X., Li, K., Yuan, F., & Sun, Z. (2024). The stabilization ability of NaA zeolite derived from fly ash for lead and cadmium in soil: Mechanisms and evaluation of effectiveness. *Science of the Total Environment, 942*, 173834. https://doi.org/10.1016/j.scitotenv.2024.173834

Li, C.-F., Wu, H.-Q., Chu, Z.-Y., Wang, X.-C., Li, Y.-L., & Guo, J.-H. (2019). Precise determination of radiogenic Sr and Nd isotopic ratios and Rb, Sr, Sm, Nd elemental concentrations in four coal ash and coal fly ash reference materials using isotope dilution thermal ionization mass spectrometry. *Microchemical Journal, 146*, 906–913. https://doi.org/10.1016/j.microc.2019.02.034

Li, T., Umar, M., Mirza, N., & Yue, X.-G. (2023). Green financing and resources utilization: A story of N-11 economies in the climate change era. *Economic Analysis and Policy, 78*, 1174–1184. https://doi.org/10.1016/j.eap.2023.05.007

Licandeo, F., Flores, F., & Feijoo, F. (2023). Assessing the impacts of economy-wide emissions policies in the water, energy, and land systems considering water scarcity scenarios. *Applied Energy, 342*, 121115. https://doi.org/10.1016/j.apenergy.2023.121115

Liguo, X., Ahmad, M., Khan, S., Haq, Z. U., & Khattak, S. I. (2023). Evaluating the role of innovation in hybrid electric vehicle-related technologies to promote environmental sustainability in knowledge-based economies. *Technology in Society, 74*, 102283. https://doi.org/10.1016/j.techsoc.2023.102283

Liu, H., Guo, W., Fan, Z., Huang, F., & Liu, S. (2023). Comparative life cycle energy, water consumption and carbon emissions analysis of deep in-situ gasification based coal-to-hydrogen with carbon capture and alternative routes. *Journal of Cleaner Production, 426*, 139129. https://doi.org/10.1016/j.jclepro.2023.139129

Marshall, D., O'Dochartaigh, A., Prothero, A., Reynolds, O., & Secchi, E. (2023). Are you ready for the sustainable, biocircular economy? *Business Horizons, 66*(6), 805–816. https://doi.org/10.1016/j.bushor.2023.05.002

Mo, W., Du, K., Sun, Y., Guo, M., Zhou, C., You, M.,..., & Xiang, J. (2023). Technical-economic-environmental analysis of biomass direct and indirect co-firing in pulverized coal boiler in China. *Journal of Cleaner Production, 426*, 139119. https://doi.org/10.1016/j.jclepro.2023.139119

Mohamed, B. A., Hamid, H., Montoya-Bautista, C. V, & Li, L. Y. (2023). Circular economy in wastewater treatment plants: Treatment of contaminants of emerging concerns (CECs) in effluent using sludge-based activated carbon. *Journal of Cleaner Production, 389*, 136095. https://doi.org/10.1016/j.jclepro.2023.136095

Mohsen, A., Kohail, M., Alharbi, Y. R., Abadel, A. A., Soliman, A. M., & Ramadan, M. (2023). Bio-mechanical efficacy for slag/fly ash-based geopolymer mingled with mesoporous NiO. *Case Studies in Construction Materials*, *19*, e02283. https://doi.org/10.1016/j.cscm.2023.e02283

Moktadir, M. A., & Ren, J. (2023). Modeling challenges for Industry 4.0 implementation in new energy systems towards carbon neutrality: Implications for impact assessment policy and practice in emerging economies. *Resources, Conservation and Recycling*, *199*, 107246. https://doi.org/10.1016/j.resconrec.2023.107246

Muhammad, T., Jiang, C., Li, Y., Manan, I., Ma, C., Geng, H.,…, & Adnan, M. (2024). Impacts and mechanism of coal fly ash on kitchen waste composting performance: The perspective of microbial community. *Chemosphere*, *350*, 141068. https://doi.org/10.1016/j.chemosphere.2023.141068

Muhammad Mustafa Rizvi, S., kazmi, B., Ali Ammar Taqvi, S., Mobeen Khan, M., Zabiri, H., Qadir, D., & Metwally, A. S. M. (2023). Techno-economic sustainability assessment for bio-hydrogen production based on hybrid blend of biomass: A simulation study. *Fuel*, *347*, 128458. https://doi.org/10.1016/j.fuel.2023.128458

Mukherjee, P. K., Das, B., Bhardwaj, P. K., Tampha, S., Singh, H. K., Chanu, L. D.,…, & Devi, S. I. (2023). Socio-economic sustainability with circular economy—An alternative approach. *Science of the Total Environment*, *904*, 166630. https://doi.org/10.1016/j.scitotenv.2023.166630

Mumtaz, H., Sobek, S., Werle, S., Sajdak, M., & Muzyka, R. (2023). Hydrothermal treatment of plastic waste within a circular economy perspective. *Sustainable Chemistry and Pharmacy*, *32*, 100991. https://doi.org/10.1016/j.scp.2023.100991

Nandhini, R., Sivaprakash, B., Rajamohan, N., & Vo, D.-V. N. (2023). Carbon-free hydrogen and bioenergy production through integrated carbon capture and storage technology for achieving sustainable and circular economy – A review. *Fuel*, *342*, 126984. https://doi.org/10.1016/j.fuel.2022.126984

Narayana Sarma, R., & Vinu, R. (2023). An assessment of sustainability metrics for waste-to-liquid fuel pathways for a low carbon circular economy. *Energy Nexus*, *12*, 100254. https://doi.org/10.1016/j.nexus.2023.100254

Naskar, J., & Sharma, A. K. (2024). Assessment of enhanced strength and stiffness properties of bio-engineered coal fly ash. *Construction and Building Materials*, *413*, 134793. https://doi.org/10.1016/j.conbuildmat.2023.134793

Nidhi, S., & Kaur, H. (2022). Advancement in field of energy storage applications by using ionic liquids. *Journal of Physics: Conference Series*, *2267*(1). https://doi.org/10.1088/1742-6596/2267/1/012043

Nidhi, S., Kaur, H., Thakur, R. C., & Kumar, A. (2023). Effect of temperature on conductance studies of 1,4-dimethyl-4H-1,2,4-triazolium iodide in binary aqueous solutions of benzylamine and benzamide. *AIP Conference Proceedings*, *2800*(1), 20229. https://doi.org/10.1063/5.0163221

Ogata, F., Nagai, N., Uematsu, Y., Kobayashi, Y., Kitamura, N., Saenjum, C., & Kawasaki, N. (2024). Synthesis of fine Na-type zeolite grains from coal fly ash and the assessment of the adsorption capability of lead ions from aqueous solutions. *RSC Sustainability*, *2*(2), 491–498. https://doi.org/10.1039/d3su00421j

Okoh, A. S., & Okpanachi, E. (2023). Transcending energy transition complexities in building a carbon-neutral economy: The case of Nigeria. *Cleaner Energy Systems*, *6*, 100069. https://doi.org/10.1016/j.cles.2023.100069

Olabi, A. G., Shehata, N., Sayed, E. T., Rodriguez, C., Anyanwu, R. C., Russell, C., & Abdelkareem, M. A. (2023). Role of microalgae in achieving sustainable development goals and circular economy. *Science of The Total Environment*, *854*, 158689. https://doi.org/10.1016/j.scitotenv.2022.158689

Pan, X., Ma, X., Zhang, Y., Shao, T., Peng, T., Li, X.,…, & Chen, W. (2023). Implications of carbon neutrality for power sector investments and stranded coal assets in China. *Energy Economics*, *121*, 106682. https://doi.org/10.1016/j.eneco.2023.106682

Paredes, B. M., Paredes, J. P., & García, R. (2023). Integration of biocoal in distributed energy systems: A potential case study in the Spanish coal-mining regions. *Energy*, *263*, 125833. https://doi.org/10.1016/j.energy.2022.125833

Pathania, V., Sharma, S., Vermani, B. K., Kaur, H., & Thakur, R. C. (2023). Investigating the solvation behavior of some perchlorates in non-aqueous medium and predicting their ion-solvation thermodynamics using Born formula. *Journal of Molecular Liquids*, *371*, 121067. https://doi.org/10.1016/j.molliq.2022.121067

Perumal, G. (2023). Characterisation analysis of bio oil generation using Hibiscus cannabinus seeds. *Natural Product Research*. https://doi.org/10.1080/14786419.2023.2261142

Prismantoko, A., Karuana, F., Ghazidin, H., Ruhiyat, A. S., Adelia, N., Prayoga, M. Z. E.,…, & Hariana, H. (2024). Ash deposition behavior during co-combustion of solid recovered fuel with different coals. *Thermal Science and Engineering Progress*, *48*, 102404. https://doi.org/10.1016/j.tsep.2024.102404

Qian, H., Zhao, Y., Qin, F., & Song, G. (2023). Life-cycle environmental burdens of ethylene production in the context of China's chemical feedstock transition from naphtha to coal and shale gas by-product of ethane. *Environmental Impact Assessment Review*, *101*, 107152. https://doi.org/10.1016/j.eiar.2023.107152

Qin, Q., Zhou, J., Zhou, L., & Guo, Q. (2022). Investigation on effect of coal fly ash on properties of corn straw pyrolysis products. *Journal of the Energy Institute*, *100*, 213–224. https://doi.org/10.1016/j.joei.2021.11.011

Rana, S., Thakur, R. C., Kaur, H., Sharma, A., Sharma, R., Thakur, V.,…, & Fatma, I. (2023). Investigating the solvation behavior of some lithium salts in binary aqueous mixtures of 1-ethyl-3-methylimidazolium tetrafluoroborate ([EMIM][BF4]) at equidistant temperatures (T = 298.15, 303.15, 308.15, 313.15, 318.15) K. *Journal of Chemical & Engineering Data*, *68*(6), 1291–1304. https://doi.org/10.1021/acs.jced.3c00037

Rathnayake, M., Julnipitawong, P., Tangtermsirikul, S., & Toochinda, P. (2018). Utilization of coal fly ash and bottom ash as solid sorbents for sulfur dioxide reduction from coal fired power plant: Life cycle assessment and applications. *Journal of Cleaner Production*, *202*, 934–945. https://doi.org/10.1016/j.jclepro.2018.08.204

Sadiq, M., Chavali, K., Kumar, V. V. A., Wang, K.-T., Nguyen, P. T., & Ngo, T. Q. (2023). Unveiling the relationship between environmental quality, non-renewable energy usage and natural resource rent: Fresh insights from ten asian economies. *Resources Policy*, *85*, 103992. https://doi.org/10.1016/j.resourpol.2023.103992

Saeid, P., & Rahmanian, N. (2023). Bio-energy programs in Europe. In *Reference Module in Earth Systems and Environmental Sciences*. Elsevier. https://doi.org/10.1016/B978-0-323-93940-9.00149-3

Sardi, B., Uno, I., Pasila, F., Altway, A., & Mahfud, M. (2023). Low rank coal for fuel production via microwave-assisted pyrolysis: A review. *FirePhysChem*, *3*(2), 106–120. https://doi.org/10.1016/j.fpc.2023.02.002

Sarkar, M., Maiti, M., Malik, M. A., & Xu, S. (2022). Evaluation of the crack-healing performance and durability of bacteria integrated alkali-activated fly ash composites. *Journal of Building Engineering*, *54*, 104642. https://doi.org/10.1016/j.jobe.2022.104642

Seleem, M. S., Sameh, R., Esily, R. R., & Ibrahiem, D. M. (2023). A closer look at bio-hydrogen strategy in post-carbon age and its prospect in Egypt. *Journal of Environmental Management*, *345*, 118773. https://doi.org/10.1016/j.jenvman.2023.118773

Sen, S. K., Das, M. M., Bandyopadhyay, P., Dash, R. R., & Raut, S. (2016). Green process using hot spring bacterium to concentrate alumina in coal fly ash. *Ecological Engineering*, *88*, 10–19. https://doi.org/10.1016/j.ecoleng.2015.12.012

Shang, H., Zhao, W., Zhang, X., & Wang, Z. (2023). The potential of biomass-derived bio-liquid to prevent the spread of SARS-CoV-2 from waste and its production-based life cycle assessment. *Science of The Total Environment*, *869*, 161833. https://doi.org/10.1016/j.scitotenv.2023.161833

Shi, P., Wang, B., Yin, R., Ren, W., Wang, J., Chang, L.,…, & Han, L. (2022). Migration behavior of heavy metals in coal fly ash-derived Cu-SSZ-13 during synthesis and NH_3-SCR application. *Fuel*, *318*. https://doi.org/10.1016/j.fuel.2022.123633

Sigvardsen, N. M., & Ottosen, L. M. (2019). Characterization of coal bio ash from wood pellets and low-alkali coal fly ash and use as partial cement replacement in mortar. *Cement and Concrete Composites*, *95*, 25–32. https://doi.org/10.1016/j.cemconcomp.2018.10.005

Singh, A., Singh, J., Dhami, H. S., Sinha, M. K., Viswanathan, K., & Kumar, R. (2022). Quantifying the influences of coal fly ash on rheological and compaction behaviour of iron powder and mild steel machining chips. *Advanced Powder Technology*, *33*(11), 103819. https://doi.org/10.1016/j.apt.2022.103819

Smol, M., & Marcinek, P. (2023). Chapter 5 – Circular economy—the new innovation wave. In M. N. Vara Prasad, & M. Smol (Eds.), *Sustainable and Circular Management of Resources and Waste Towards a Green Deal* (pp. 65–77). Elsevier. https://doi.org/10.1016/B978-0-323-95278-1.00024-3

Stojanović, J., Savić-Zdravković, D., Jovanović, B., Vitorović, J., Bašić, J., Stojanović, I.,…, & Milošević, Đ. (2023). Histopathology of chironomids exposed to fly ash and microplastics as a new biomarker of ecotoxicological assessment. *Science of the Total Environment*, *903*, 166042. https://doi.org/10.1016/j.scitotenv.2023.166042

Su, H., Chen, H., & Lin, J. (2020). A sequential integration approach using Aspergillus Niger to intensify coal fly ash as a rare metal pool. *Fuel*, *270*, 117460. https://doi.org/10.1016/j.fuel.2020.117460

Subbarao, P. M. V, D' Silva, T. C., Adlak, K., Kumar, S., Chandra, R., & Vijay, V. K. (2023). Anaerobic digestion as a sustainable technology for efficiently utilizing biomass in the context of carbon neutrality and circular economy. *Environmental Research*, *234*, 116286. https://doi.org/10.1016/j.envres.2023.116286

Tabibian, S. S., & Sharifzadeh, M. (2023). Statistical and analytical investigation of methanol applications, production technologies, value-chain and economy with a special focus on renewable methanol. *Renewable and Sustainable Energy Reviews*, *179*, 113281. https://doi.org/10.1016/j.rser.2023.113281

Tauseef Hassan, S., Wang, P., Khan, I., & Zhu, B. (2023). The impact of economic complexity, technology advancements, and nuclear energy consumption on the ecological footprint of the USA: Towards circular economy initiatives. *Gondwana Research*, *113*, 237–246. https://doi.org/10.1016/j.gr.2022.11.001

Thakur, A., & Kumar, A. (2023a). Ecotoxicity analysis and risk assessment of nanomaterials for the environmental remediation. *Macromolecular Symposia*, *410*(1), 1–23. https://doi.org/10.1002/masy.202100438

Thakur, A., & Kumar, A. (2023b). Exploring the potential of ionic liquid-based electrochemical biosensors for real-time biomolecule monitoring in pharmaceutical applications: From lab to life. *Results in Engineering*, *20*(July), 101533. https://doi.org/10.1016/j.rineng.2023.101533

Thakur, A., & Kumar, A. (2023c). Recent trends in nanostructured carbon-based electrochemical sensors for the detection and remediation of persistent toxic substances in real-time analysis. *Materials Research Express*, *10*(3). https://doi.org/10.1088/2053-1591/acbd1a

Thakur, A., & Kumar, A. (2024). Unraveling the multifaceted mechanisms and untapped potential of activated carbon in remediation of emerging pollutants: A comprehensive review and critical appraisal of advanced techniques. *Chemosphere*, 346(November 2023), 140608. https://doi.org/10.1016/j.chemosphere.2023.140608

Thakur, A., Sharma, S., Ganjoo, R., Assad, H., & Kumar, A. (2022). Anti-corrosive potential of the sustainable corrosion inhibitors based on biomass waste: A review on preceding and perspective research. *Journal of Physics: Conference Series, 2267*(1). https://doi.org/10.1088/1742-6596/2267/1/012079

Thakur, R. C., Kaur, R., & Kaur, H. (2020). Transport properties of thiaminehydrochloridein binary aqueous mixtures of galactose. *European Journal of Molecular & Clinical Medicine*, *7*(07), 2020.

Thakur, R. C., Sharma, R., Kaur, V., & Kaur, H. (2019). Molecular interactions of Niacin in binary aqueous solutions of KCl and FeCl 3 at different temperatures: Volumetric Approach. *Think India Journal*, *22*(37), 1604–1613.

Thakur, R. C., Sharma, A., Sharma, R., & Kaur, H. (2023). A comparative analysis of volumetric, viscometric and conductometric properties of Triethylmethylammonium Tetrafluoroborate (TEMABF4) and Tetraethylammonium Tetrafluoroborate (TEABF4) in pure propylene carbonate (PC) and binary aqueous propylene carbonate. *Journal of Molecular Liquids*, *374*, 121244. https://doi.org/10.1016/j.molliq.2023.121244

Thanapal, S. S., Annamalai, K., Lawrence, B., & Sweeten, J. M. (2023). Biomass co-combustion with coal. In *Reference Module in Earth Systems and Environmental Sciences*. Elsevier. https://doi.org/10.1016/B978-0-323-90386-8.00050-4

Thomas, B. S., Dimitriadis, P., Kundu, C., Vuppaladadiyam, S. S. V., Raman, R. K. S., & Bhattacharya, S. (2024). Extraction and separation of rare earth elements from coal and coal fly ash: A review on fundamental understanding and on-going engineering advancements. *Journal of Environmental Chemical Engineering*, *12*(3), 112769. https://doi.org/10.1016/j.jece.2024.112769

Thulasikanth, V., & Padmanabhan, R. (2023). Fabrication of sustainable closed-cell aluminium foams using recycled fly ash and eggshell powder. *Materials Today Communications*, *37*, 107302. https://doi.org/10.1016/j.mtcomm.2023.107302

Tian, Z., Tang, X., Xiu, Z., Zhou, H., & Xue, Z. (2022). The mechanical properties improvement of environmentally friendly fly ash-based geopolymer mortar using bio-mineralization. *Journal of Cleaner Production*, *332*, 130020. https://doi.org/10.1016/j.jclepro.2021.130020

Tilsted, J. P., Bauer, F., Deere Birkbeck, C., Skovgaard, J., & Rootzén, J. (2023). Ending fossil-based growth: Confronting the political economy of petrochemical plastics. *One Earth*, *6*(6), 607–619. https://doi.org/10.1016/j.oneear.2023.05.018

Ukwattage, N. L., Ranjith, P. G., & Bouazza, M. (2013). The use of coal combustion fly ash as a soil amendment in agricultural lands (with comments on its potential to improve food security and sequester carbon). *Fuel*, *109*, 400–408. https://doi.org/10.1016/j.fuel.2013.02.016

Vafaei, B., & Ghahremaninezhad, A. (2024). Self-healing effect of hydrogels in cement slag and fly ash pastes. *Construction and Building Materials*, *438*, 137036. https://doi.org/10.1016/j.conbuildmat.2024.137036

Varbanov, P. S., Wang, B., Ocłoń, P., Radziszewska-Zielina, E., Ma, T., Klemeš, J. J., & Jia, X. (2023). Efficiency measures for energy supply and use aiming for a clean circular economy. *Energy*, *283*, 129035. https://doi.org/10.1016/j.energy.2023.129035

Wang, T., Wang, K., Ye, F., Ren, Y., & Xu, C. (2021). Characterization and thermal properties of a shape-stable Na2CO3-K2CO3/coal fly ash/expanded graphite composite phase change materials for high-temperature thermal energy storage. *Journal of Energy Storage*, *33*, 102123. https://doi.org/10.1016/j.est.2020.102123

Wibowo, H., Ketwong, T., Cholwatthanatanakorn, N., Ding, L., & Areeprasert, C. (2024). Production of bagasse fly ash-derived CO2 adsorbent by physical activation and by nitrogen-functionalization using hydrothermal treatment. *Waste Management*, *177*, 66–75. https://doi.org/10.1016/j.wasman.2024.01.029

Wu, H. (2023). Evaluating the role of renewable energy investment resources and green finance on the economic performance: Evidence from OECD economies. *Resources Policy*, *80*, 103149. https://doi.org/10.1016/j.resourpol.2022.103149

Xue, Z., Feng, Y., Li, H., Zhu, Z., Xu, C., Ju, J., & Yang, Y. (2023). A comprehensive review on progresses of coal and minerals bioflotation in presence of microorganisms. *Journal of Environmental Chemical Engineering*, *11*(6), 111182. https://doi.org/10.1016/j.jece.2023.111182

Yadav, S., Pandey, V. C., Kumar, M., & Singh, L. (2022). Plant diversity and ecological potential of naturally colonizing vegetation for ecorestoration of fly ash disposal area. *Ecological Engineering*, *176*, 106533. https://doi.org/10.1016/j.ecoleng.2021.106533

Yuan, B., Hu, H., Huang, Y., Fu, B., Liu, H., Luo, G.,…, & Yao, H. (2022). Condensation and adsorption characteristics of gaseous selenium on coal-fired fly ash at low temperatures. *Chemosphere*, *287*, 132127. https://doi.org/10.1016/j.chemosphere.2021.132127

Zhan, L., Zhang, X., Hanjin, L., Li, R., & Wu, Y. (2023). Study on a new reed-coal coupled power generation system in Wuliangsuhai Lake wetland—Life cycle assessment and economic and technical analysis. *Fuel*, *353*, 129095. https://doi.org/10.1016/j.fuel.2023.129095

Zhang, J., Wen, K., & Li, L. (2021). Bio-modification of coal fly ash using urease-producing bacteria. *Fuel*, *286*, 119386. https://doi.org/10.1016/j.fuel.2020.119386

Zhang, L., Shen, Q., Pang, C. H., Chao, W., Tong, S., Kow, K. W.,…, & Wei, W. (2023). Life cycle assessment of bio-fermentation ethanol production and its influence in China's steeling industry. *Journal of Cleaner Production*, *397*, 136492. https://doi.org/10.1016/j.jclepro.2023.136492

Zhao, L., Zaborowski, E., Bordoloi, S., Rajagopalan, N., Sharma, B. K., Baroi, C.,…, & Zhang, L. (2024). Characterization of novel polysulfide polymer coated fly ash and its application in mitigating diffusion of contaminants. *Environmental Pollution*, *347*, 123706. https://doi.org/10.1016/j.envpol.2024.123706

Zhao, P., Huang, Z., Wang, P., & Wang, A. (2023). Comparative study on high-efficiency Pb(II) removal from aqueous solutions using coal and rice husk based humic acids. *Journal of Molecular Liquids*, *369*, 120875. https://doi.org/10.1016/j.molliq.2022.120875

Zhao, P., Wang, A., Wang, P., Huang, Z., Fu, Z., & Huang, Z. (2023). Two recyclable and complementary adsorbents of coal-based and bio-based humic acids: High efficient adsorption and immobilization remediation for Pb(II) contaminated water and soil. *Chemosphere*, *318*, 137963. https://doi.org/10.1016/j.chemosphere.2023.137963

Zheng, J., Liao, L., Liu, R., Li, C., & Zhang, Y. (2022). A rational and feasible approach to the co-management of condensates from biomass torrefaction and carbon-rich fly ash from fluidized-bed coal gasification. *Waste Management*, *154*, 312–319. https://doi.org/10.1016/j.wasman.2022.10.016

Zheng, J.-L., Zhu, Y.-H., Dong, Y.-Y., Chen, Y., & Zhu, M.-Q. (2023). Techno-economic analysis and life cycle assessment of industrial production of ammonia via bio-oil conversion. *Energy*, *280*, 128223. https://doi.org/10.1016/j.energy.2023.128223

13 Prospects of Ecosystem Services for Fly Ash

Priyanka Ray, Surajit Das,
Dikshita Goswami and Sanjana Das

13.1 INTRODUCTION

With the escalating concerns over environmental degradation and the urgent need for sustainable development, exploring alternative materials and practices that mitigate ecological harm while fostering socio-economic progress has become imperative. Fly ash (FA), a residue of coal combustion in power plants (Yu et al., 2019) has emerged as a potential resource for addressing both environmental challenges and economic needs. Similar to FA, biomass ash, which is the byproduct from the combustion of biomass waste, in recent times, has also gained some attention (Olatoyan et al., 2023). As society transitions toward a greener future, understanding the prospects of FA utilization on ecosystem services stands as a crucial area of inquiry.

FA, historically considered a waste product, has gained attention due to its versatile properties and potential applications across various sectors. FA possesses unique characteristics that make it suitable for diverse uses, as it contains essential plant nutrients such as calcium, magnesium, and potassium which can improve soil fertility and enhance plant growth. It is alkaline and can help neutralize acidic soils, thereby improving soil pH. The carbon content in FA contributes to its potential for carbon sequestration. When used in agricultural soils (Iyer & Scott, 2001; Kikuchi, 1999) or as a component in construction materials, it can contribute to long-term carbon storage. It is also used as a precursor for zeolite (Querol et al., 2002) It can also absorb pollutants and improve water quality (Ahmaruzzaman, 2009). It can be used in the reclamation of mine sites and degraded lands. This helps in stabilizing soil and promoting vegetation growth. Incorporating FA into concrete and other construction materials improves their strength, durability, and workability (Nath & Sarker, 2011). FA can be applied to slopes and embankments to stabilize the soil, thereby reducing the risk of erosion. The use of FA in land reclamation projects can create new habitats, supporting biodiversity by providing areas for plant and animal colonization.

One of the primary avenues through which FA can positively impact ecosystems is its utilization in infrastructure development (Kolias et al., 2005; Takada et al., 1995; Wang & Wu, 2006). Incorporating FA into concrete and other construction materials enhances their durability and strength and reduces the demand for virgin materials such as cement and aggregates. Consequently, this diminishes the environmental footprint associated with conventional construction practices, including resource

DOI: 10.1201/9781003498124-13

extraction, energy consumption, and emissions of greenhouse gas. Furthermore, the utilization of FA in construction contributes to waste reduction by diverting significant quantities of this byproduct from landfills, thereby mitigating potential environmental hazards and promoting landfill space conservation. Additionally, the incorporation of FA into reclamation projects can aid in the restoration of degraded landscapes, promoting biodiversity conservation and ecosystem resilience.

13.2 PROPERTIES OF FA

13.2.1 Physical Properties

FA is a complex substance comprising fine powder particles, frequently found in spherical form, with its structure being either solid or hollow. Depending on the level of residual carbon present, the color of FA can range from tan to gray to black (Abid et al., 2018). FA falls within the range of sandy silt to silty sand, classified based on grain size distribution. Generally, its particle size distribution resembles silt, typically ranging from 0.5 to 300 μm (Koukouzas et al., 2006; Roy et al., 1981; Yao et al., 2015). Its specific gravity falls in the range of 2.1 to 3.0 (Awoyera & Adesina, 2019). The bulk density of FA is low, while its specific surface area is high (Sair et al., n.d.).

13.2.2 Chemical Properties

FA is composed of three primary constituents: Organic materials, inorganic materials containing both amorphous and crystalline phases, and fluid materials present in both organic and inorganic components. Predominantly, FA consists of aluminosilicate compounds (Iribarne et al., 2001) alongside metallic and calcium oxides. Studies suggest a consistent ratio of $Al_2O_3/(Al_2O_3 + SiO_2)$ in its spheres, signifying their origin from Kaolinite present in coal. In addition to these constituents, it contains trace elements, including Ba, V, Cr, Ni, Pb, Sr, and Zn (Yao et al., 2015), which are often enriched in magnetospheres (Yang et al., 2014). Elements such as K, Al, Cl, P, Na, Fe, Mg, Si, and Ti are moderately abundant in FA, while Cs, Mn, Li, Sn, Ni, and Sr are present in parts per million (ppm) levels. Furthermore, elements like As, Ge, Cu, Sb, Mo, Se, Rb, and Zr are found in parts per billion (ppb) levels.

13.3 FA VARIABILITY

The properties of FA can change significantly based on factors such as temperature and combustion conditions, type of coal burned, the cooling process, and primarily on combustion control. Because of this broad variability, characterizing the composition of FA is very challenging. As a result, the American Society for Testing and Materials (ASTM) classified the FA into two classes: Class C FA and Class F FA. Class F is pozzolanic and is derived from the burning of either bituminous or anthracite coal (Alterary & Marei, 2021), requiring a minimum total content of 70% of Al_2O_3, SiO_2, and Fe_2O_3. However, the Class C ashes are pozzolanic and cementitious, derived from the burning of either lignite or sub-bituminous coal. The characterization of Class C and Class F type FA typically relies on the metal oxide content's analysis, with Class F exhibiting higher loss of ignition (LOI) compared to Class C (Maroto-Valer et al., 2001). The LOI test helps identify the proportion of unburned carbon and metallic oxides present in FA. The main distinction between the two classes of FA lies in their calcium content and their levels of silica, alumina, and iron. The cooling rate of FA particles influences the formation of inorganic matter, with faster cooling resulting in small glassy particles and slower cooling leading to the formation of larger crystalline particles. Based on formation time, FA can also be classified into primary, secondary, and tertiary phases. Primary phases remain unchanged, while secondary phases, such as oxides and silicates, are formed during coal combustion (Alterary & Marei, 2021).

13.4 TESTING AND CHARACTERIZATION TECHNIQUES

The techniques used for FA characterization involve XRD (X-ray diffraction), SEM (scanning electron microscopy), and XRF (X-ray fluorescence) (Kruse et al., 2013). SEM helps in identifying the amorphous phase composition of FA, while XRD is used to qualitatively determine its crystalline phases. Laser diffraction is commonly employed to measure the particle size of FA. XRF plays a crucial part in assessing bulk oxide compositions of various types of FA, particularly in determining the content of the amorphous phase. Additionally, other spectroscopic techniques such as Magic Angle Spinning Nuclear Magnetic Resonance (MAS-NMR), and Fourier transform infra-red (FTIR) spectroscopy are also utilized for FA characterization.

13.5 BENEFITS OF USING FA IN ECOSYSTEM DEVELOPMENT

FA has been used in various sectors. Using FA in developing other sectors implements waste management and boosts ecosystem development. It provides provisional services in food security (Ukwattage et al., 2013) and raw material development like geopolymer production (Andini et al., 2008; Swanepoel & Strydom, 2002). The coal FA, CFA-based material (Tekin et al., 2015), Zeolite X has been used for medicinal purposes (Ramanathan et al., 2020) (Figure 13.1).

13.5.1 Soil Quality Improvement

FA can enter the soil through various channels, including industrial emissions (Klose & Makeschin, 2004) and direct deposition onto the soil surface in its raw form (Pandey et al., 2016). Additionally, it can be incorporated into the topmost soil layer alongside organic minerals like biochar (Weber et al., 2015), sewage sludge, compost (Ram & Masto, 2010), or waste from paper production (Martins et al., 2024). The application methods may also involve introducing ash between existing upper soil layers, which enhances water retention in the upper layer of the soil (0–40 cm) while preserving upper horizons from excessive mixing. Coal combustion-derived FA is often highly reactive in soil, leading to changes in its soil properties both physical and chemical,

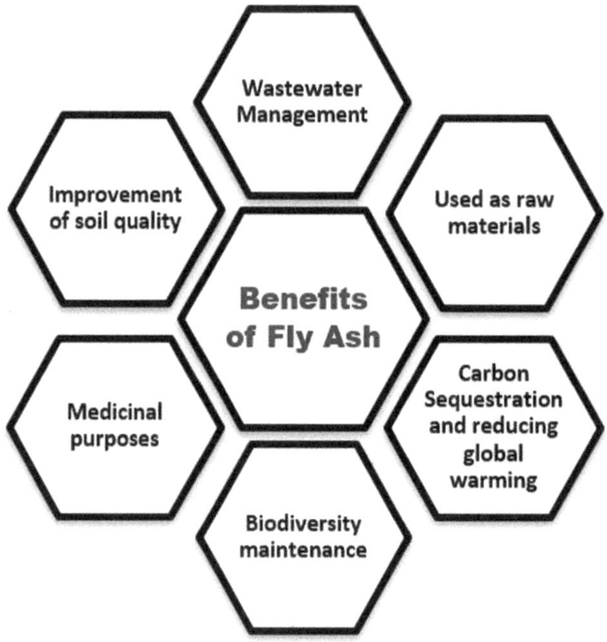

FIGURE 13.1 Benefits of fly ash.

potentially affecting the habitat's trophic status. These alterations can positively influence plant growth and development.

In recent experiments conducted in Indian forest plantations, researchers examined the plant growth and development by impact of FA like *Tecona* sp., *Acacia* sp., and *Caesalpinia* sp. The majority of the species studied showed a growth increase of 20%–30% when compared to those grown without FA. The most optimal outcomes were observed when ash was mixed with soil at a 1:1 ratio (Kumar & Jha, 2014).

FA possesses the potential to enhance soil health, acting as a soil modifier while improving its water retention and fertility (Basu et al., 2009; Wong & Wong, 1990). It contains a range of essential macro and micronutrients, thereby fostering plant growth and increasing crop yields by 10%–40% (Kishor et al., 2010). Specifically, cereal crops have shown a yield increase of 10%–15%, while pulses, oilseeds, and vegetables have exhibited increases of 20%–25% and up to 40%, respectively. Thus, it plays a crucial role in enhancing food security, a primary objective in agricultural production. Additionally, integrating FA into degraded soils aids in their reclamation and expands the available land for food production (Ukwattage et al., 2013). Research indicates that adding FA to cultivated land at 70 tons/ha rate can transform soil texture from clayey to loamy. Furthermore, incorporating FA into clay soil at varying percentages by weight (0%, 5%, 10%, and 15%) enhances properties such as workability, porosity, moisture retention capacity (Dhindsa et al., 2016), and root penetration. Additionally, FA application aids in maintaining soil pH balance (Gorai & Ash, 2018).

13.5.2 Wastewater Treatment

FA, an abundant and cost-efficient substance, is currently being explored for its potential to act as an adsorbent for eliminating a wide range of organic pollutants. Pesticides (malathion, lindane, carbofuran), dyes (methylene blue, crystal violet, malachite green), and phenolic compounds (nitro, chloro, amino) present in wastewater could be effectively removed from the water stream by the use of ash (Ahmaruzzaman, 2009).

Wang et al. investigated several samples containing varying levels of unburned carbon in FA to understand their ability to adsorb dyes such as crystal violet and methylene blue from aqueous solutions (Wang & Wu, 2006). Their findings revealed that unburned carbon is the primary contributor to the adsorption of dyes, while the capability of mineral matter in FA to adsorb is minimal. Higher levels of residual carbon in FA correspond to greater adsorption capacity. For FA devoid of carbon, the adsorption capacities for crystal violet and methylene blue are only 1.0×10^{-6} and 2.0×10^{-6} mol/g, respectively. In contrast, FA enriched with carbon exhibits significantly higher adsorption capacities, measuring 1.0×10^{-4} and 1.2×10^{-4} mol/g for crystal violet and methylene blue, respectively.

Yurum et al. utilized FA as an effective method for extracting boron from water, alongside zeolite and demineralized lignite, each exhibiting varying capacities. FA demonstrated a remarkable 94% boron removal rate. Through batch experiments, they investigated the removal capacities, acquired adsorption isotherms, and analyzed both kinetic and thermodynamic parameters. The solution's pH notably influences the efficiency of boron removal by all adsorbents, with optimal adsorption occurring at a pH level of 10 (Yüksel & Yürüm, 2009).

The release of phosphorus into surface waters poses a significant pollution threat. Developing technologies for phosphorus removal presents an opportunity to mitigate environmental risks and promote recycling. A study by Sinha et al., 2011 investigated how FA can act as a phosphate removal material and its response to variations in temperature (28°C–50°C), retention time (5–30 minutes), and adsorbent quantity (5–10 g/100 mL) on phosphorus removal. The experiments utilized phosphate solutions with concentrations ranging from 20 to 200 mg/L phosphorus. The findings showed that the FA efficiently eliminated all the 20 mg/L phosphorus when 10 g of the adsorbent was used, and the retention time was 30 minutes at a temperature of 28°C. Nonetheless, as the phosphorus concentration increased, the removal capacity decreased, reaching 86.51% removal at a concentration of 200 mg/L phosphorus. The amount of adsorbent had a substantial effect on phosphorus removal, with mean removal rates of 94.81% at 5 g and 97.5% at 10 g. The maximum value of Langmuir adsorption capacity is 40.98 mg/kg for 5 g of FA after reaching equilibrium for 30 minutes at 28°C.

13.5.3 Utilization of FA as Raw Material

13.5.3.1 Concrete Applications

The incorporation of FA into concrete presents a multifaceted approach to addressing environmental and structural concerns in construction (Ilic et al., 2003). By substituting a portion of cement with FA, not only is there a reduction in CO_2 emissions and energy consumption, but also a promotion of sustainable resource management. The spherical shape of FA particles imparts enhanced workability to concrete, reducing viscosity (Ikotun et al., 2017) and frictional losses during pumping while also mitigating water demand. Pozzolanic reactions between FA and cement result in calcium-silicate-hydrate (CSH) gel (Kumar & Prasad, 2019) formation enhancing long-term strength and durability. The precise amount of FA is crucial, as it fills concrete pores and speeds up pozzolanic reactions, particularly when used alongside additional cementitious materials such as GGBS (ground granulated blast furnace slag) (Turk et al., 2013) or SF (silica fume). Moreover, FA-concrete exhibits improved resistance to various aggressive agents, including chloride, sulfate, and acid attacks, due to refined pore structures and reduced permeability. The incorporation of FA also enhances resistance to alkali-silica reaction (ASR) (Zheng et al., 2016) and high temperatures, while offering potential benefits in mitigating hydration heat. Nonetheless, the heterogeneity of FA necessitates further research to fully comprehend its effects on concrete properties, emphasizing the importance of tailored approaches based on FA type, fineness, and chemical composition. These findings underscore the promising role of FA in advancing sustainable concrete technology while highlighting avenues for ongoing exploration and refinement (Xu & Shi, 2018).

Shaikh et al. investigated the durability and CS (compressive strength) properties of HVFAC (High-Volume Fly Ash Concrete) incorporating UFFA (ultra-fine FA) (Shaikh & Supit, 2015). The study revealed that concrete containing 8% UFFA had the greatest CS throughout all the tested time periods. The incorporation of 8% UFFA resulted in an improvement in the CS during the early stages of HVFA (High-Volume Fly Ash) cement. Particularly, a significant enhancement of nearly 200% in CS was noted in HVFAC with 52% FA and 8% UFFA after three days of curing. Furthermore, the addition of 8% UFFA also contributed to enhanced long-term CS. Conventional concrete and HVFAC with 32% FA experienced approximately 55% and 10% enhancements in CS, respectively, after 90 days, when compared to mixes without UFFA. The study attributes these improvements to the characteristics of UFFA, such as its diminutive particle size and elevated amorphous content, which contribute to a faster pozzolanic reaction. This reaction effectively occupies the pores in the concrete, enhancing its CS. Additionally, the high surface area and the small particle size of UFFA, along with its amorphous silica content, enable a pozzolanic reaction that extends over a longer duration compared to PC (Portland cement) (Saha, 2018).

According to Iqbal et al. in self-compacting concrete (SCC), increasing the FA content from 100 to 125 kg/m³ led to a 13.5% increase in CS (Iqbal et al., 2016). However, further increasing the FA dosage to 150 kg/m³ resulted in a 4.5% decrease in CS. This improvement in strength was particularly notable at seven and 28 days, suggesting that it was crucial to determine the ideal dosage of FA to occupy the pores within the concrete adequately.

Biomass ash also contributed to the improvement in concrete production. Juenger and Siddique shows that incorporating cementitious biomass ash improves the performance of concrete (Juenger & Siddique, 2015). Srinivasreddy et al. (2013) shows the importance of adding rice husk ash (RHS) to concrete for improved workability. Concrete mixed with RHS has proven to be better than mixed with the SF.

13.5.3.2 Synthesis of Geopolymeric Material

FA, a byproduct of the petrochemical and electrical sectors, was looked into as a potential foundational component of a novel geopolymeric material (Andini et al., 2008; Swanepoel & Strydom, 2002). It is used in the synthesis of geopolymers due to its advantageous chemical composition, availability, and environmental benefits. It is rich in alumina (Al_2O_3) and silica (SiO_2), which are essential for the geopolymerization process. These components react with alkaline activators to form a three-dimensional polymeric network, creating a strong, durable material. Its widespread availability and low cost make it an attractive raw material for geopolymer synthesis, reducing reliance on virgin materials. Using FA in geopolymers helps divert this industrial waste from landfills reducing environmental pollution. Geopolymers made from FA require lower energy input compared to traditional cement production, leading to significantly reduced CO_2 emissions.

13.5.4 Carbon Sequestration and Reducing Global Warming

Employing the FA as a soil ameliorant presents a promising avenue for reducing carbon dioxide (CO_2) emissions and mitigating global warming (Robertson et al., 2000). Agriculture, a significant contributor to greenhouse gas emissions, offers opportunities for mitigating its impact by adopting alternative practices. One such practice involves substituting traditional agricultural lime (aglime) (Wong, 1995) with FA. Traditionally, the application of aglime was assumed to release all its carbon content as CO_2 into the atmosphere, contributing to significant emissions. However, research indicates that utilizing FA instead of aglime can lead to a reduction in the emission of CO_2 gas which has a positive impact on climate regulation (Van Deventer et al., 2010). It promotes resource efficiency and supports a circular economy, which contributes to overall sustainability and climate goals. However, careful management is needed to avoid potential environmental risks, such as contamination and particulate emissions, which could offset some of the climate benefits.

The IPCC (Intergovernmental Panel on Climate Change) mentions that applying aglime releases CO_2 into the atmosphere. The US EPA in 2001, assessed that 9 Tg (Teragram) of CO_2 emissions resulted from approximately 20 Tg of applied aglime (Wong, 1995). Some researchers have explored using FA instead of aglime to reduce CO_2 emissions. Experimental findings indicate that one ton of FA can capture or store up to 26 kg of CO_2, highlighting its potential for CO_2 mitigation (Montes-Hernandez et al., 2009). Utilizing in place of lime as a soil amendment could decrease net CO_2 emissions, thereby mitigating global warming. Additionally, incorporating FA into concrete production as a cement substitute can reduce CO_2 emissions associated with cement manufacturing (Ferreira et al., 2003). Estimates suggest that the inclusion of one ton of FA in the production of concrete could avert the CO_2 emission in order of two tons typically generated during cement manufacturing, contributing to greenhouse gas reduction and mitigating global warming. Therefore, there are several advantages to using FA in both agricultural and construction sectors, including utilizing a cost-effective raw material, conserving natural resources such as land, water, coal, and lime, reducing waste, and minimizing global warming effects.

13.5.5 Medicinal Purposes

The use of nanostructured aluminosilicate derived from FA waste in industrial and biomedical applications has been widely studied. Aluminosilicates from FA have shown great potential for use in drug delivery systems, wound treatment, bone formation, and intestinal therapeutics. However, there are only a few studies that have explored its application in the medical field. One prominent medical application of aluminosilicates is their use as nanocarriers for small drug molecules and charged ions, ensuring effective drug delivery due to their unique porous structure and excellent biocompatibility. Aluminosilicates are also widely used in floating drug delivery, sustained drug delivery, and microencapsulation with drugs (Ramanathan et al., 2020). Additionally, a study examined zeolite X synthesized from CFA for its anti-proliferative effects on human breast cancer Michigan Cancer Foundation-7 (MCF-7) cells through in vitro studies and assessed the probable molecular mechanism (Subhapriya & Gomathipriya, 2018).

13.5.6 Biodiversity Maintenance

FA is a secondary home for many flora and fauna. Tropek et al. hypothesized that the aeolian sand shares a common physical property with FA and is hence suitable for the habitat of various psammophilous species (Tropek et al., 2013). FA deposits is said to be a secondary habitant for some of the arthropods (Tropek et al., 2016). Kolar et al. shows that FA can improve the freshwater habitant like benthos, zooplankton, phytoplankton, and many more (Kolar et al., 2023).

13.6 CHALLENGES AND LIMITATIONS

Due to its small size and lightweight nature, there is a concern that FA could become airborne, leading to environmental and air pollution (Pandey et al., 2014; Ruhl et al., 2011). The iron and aluminum oxides found atop the FA particles tend to attract harmful trace elements such as Sb, Cd, Be, As, Pb, Se, V, and Hg, which are primarily concentrated on the outer layer of the FA. Increasing the number of FA dump sites causes the local and its surrounding environment to be prone to health issues (Silva & Da Boit, 2011; Smith et al., 2006). Prolonged inhalation of the FA may result in various health issues, including permanent respiratory disorders, exacerbation of pre-existing conditions such as asthma and bronchitis, and potentially even lung cancer.

Utilizing plantations for ecological restoration on FA dumpsites is a global strategy, albeit hindered by slow processing rates. The specific requirements of plants for optimal growth include particular physicochemical and biological conditions, which are often compromised by the harsh environment of FA (Pandey & Rizvi, 2009). This adversity diminishes both plant growth rates and the success of phytoremediation efforts. FA adversely affects soil properties such as water holding (Marimon-Junior et al., 2020) and heat retention capacity (Rao et al., 1998), further impeding plant growth. Additionally, the pH of FA influences the release and availability of metal(loid)s, with acidic conditions observed to decrease metal(loid) availability in soil. The absence of organic matter in FA, coupled with the inhospitable conditions for plant-specific bacteria, poses challenges for vegetation survival and growth on FA-amended sites.

13.7 FUTURE DIRECTIONS AND RESEARCH NEEDS

FA holds significant potential for environmental applications, including air and water pollution treatment, but faces economic barriers to industrial-scale utilization. Modification of FA, such as through conversion into zeolites or geopolymer synthesis, enhances its adsorption capacity and broadens its industrial applications. Additionally, unburned carbon in FA can be separated and utilized for premium carbon materials like activated carbon, offering cost advantages. Despite these opportunities, research into FA utilization remains limited. Further studies are needed to explore its potential in soil enrichment, mineral content enhancement, and other innovative applications. This review catalyzes future research, highlighting the diverse economic prospects and encouraging exploration into the untapped potential of FA.

13.8 CONCLUSION

In this chapter, we delve into the potential of utilizing FA for ecosystem development. Initially regarded as waste from coal combustion, FA has recently gained significant attention for its ecological benefits. With the global shift toward greener practices, FA emerges as a promising candidate for ecosystem development, offering cost-effective solutions across various environmental applications. Its rich composition of organic and inorganic compounds makes it a versatile resource for addressing ecological challenges, ranging from soil stabilization to wastewater treatment. Not only does FA help eliminate toxicity from the environment, but it also contributes to carbon sequestration and promotes sustainable development practices. Advanced characterization techniques such as SEM, XRF, and XRD are instrumental in understanding FA properties and assessing its suitability for ecological applications. As ongoing research uncovers more ecological advantages of FA, it stands out as a valuable asset in fostering environmental sustainability and resilience.

REFERENCES

Abid, S. R., Nahhab, A. H., Al-aayedi, H. K. H., & Nuhair, A. M. (2018). Expansion and strength properties of concrete containing contaminated recycled concrete aggregate. *Case Studies in Construction Materials*, *9*, e00201. https://doi.org/10.1016/j.cscm.2018.e00201

Ahmaruzzaman, M. (2009). Role of fly ash in the removal of organic pollutants from wastewater. *Energy & Fuels*, *23*(3), 1494–1511. https://doi.org/10.1021/ef8002697

Alterary, S. S., & Marei, N. H. (2021). Fly ash properties, characterization, and applications: A review. *Journal of King Saud University-Science*, *33*(6), 101536.

Andini, S., Cioffi, R., Colangelo, F., Grieco, T., Montagnaro, F., & Santoro, L. (2008). Coal fly ash as raw material for the manufacture of geopolymer-based products. *Waste Management*, *28*(2), 416–423. https://doi.org/10.1016/j.wasman.2007.02.001

Awoyera, P., & Adesina, A. (2019). A critical review on application of alkali activated slag as a sustainable composite binder. *Case Studies in Construction Materials, 11*, e00268. https://doi.org/10.1016/j.cscm.2019.e00268

Basu, M., Pande, M., Bhadoria, P. B. S., & Mahapatra, S. C. (2009). Potential fly-ash utilization in agriculture: A global review. *Progress in Natural Science, 19*(10), 1173–1186. https://doi.org/10.1016/j.pnsc.2008.12.006

Dhindsa, H. S., Sharma, R. D., & Kumar, R. (2016). Role of fly ash in improving soil physical properties and yield of wheat (*Triticum aestivum*). *Agricultural Science Digest – A Research Journal, 36*(2). https://doi.org/10.18805/asd.v36i2.10626

Ferreira, C., Ribeiro, A., & Ottosen, L. (2003). Possible applications for municipal solid waste fly ash. *Journal of Hazardous Materials, 96*(2–3), 201–216. https://doi.org/10.1016/S0304-3894(02)00201-7

Gorai, S., & Ash, F. (2018). Utilization of Fly ash for sustainable environment management. *J. Mater. Environ. Sci, 9*(2), 385-393.

Ikotun, B. D., Fanourakis, G. C., & Bhardwaj, S. M. (2017). The effect of fly ash, β-cyclodextrin and fly ash-β-cyclodextrin composites on concrete workability and strength. *Cement and Concrete Composites, 78*, 1-12.

Ilic, M., Cheeseman, C., Sollars, C., & Knight, J. (2003). Mineralogy and microstructure of sintered lignite coal fly ash. *Fuel, 82*(3), 331–336. https://doi.org/10.1016/S0016-2361(02)00272-7

Iqbal, S., Ali, A., Holschemacher, K., Ribakov, Y., & Bier, T. A. (2016). Effect of fly ash on properties of self-compacting high strength lightweight concrete. *Periodica Polytechnica Civil Engineering*. https://doi.org/10.3311/PPci.8171

Iribarne, J., Iribarne, A., Blondin, J., & Anthony, E. J. (2001). *Hydration of combustion ashes Đ a chemical and physical study, Fuel, 80*(6), 773–784.

Iyer, R. S., & Scott, J. A. (2001). *Power station fly ash—A review of value-added utilization outside of the construction industry. Resources, conservation and recycling, 31*(3), 217–228.

Juenger, M. C. G., & Siddique, R. (2015). Recent advances in understanding the role of supplementary cementitious materials in concrete. *Cement and Concrete Research, 78*, 71–80. https://doi.org/10.1016/j.cemconres.2015.03.018

Kikuchi, R. (1999). Application of coal ash to environmental improvement: Transformation into zeolite, potassium fertilizer, and FGD absorbent. *Resources, Conservation and Recycling, 27*(4), 333–346. https://doi.org/10.1016/S0921-3449(99)00030-0

Kishor, P., Ghosh, A. K., & Kumar, D. (2010). Use of fly ash in agriculture: A way to improve soil fertility and its productivity. *Asian Journal of Agricultural Research, 4*(1), 1–14.

Klose, S., & Makeschin, F. (2004). Chemical properties of forest soils along a fly-ash deposition gradient in eastern Germany. *European Journal of Forest Research, 123*(1), 3–11. https://doi.org/10.1007/s10342-004-0018-6

Kolar, V., Chmelová, E., Bílková, M., Borovec, J., Carreira, B. M., Černý, M., Ditrich, T., Horká, P., Hrivniak, Ľ., Hrubý, F., Jan, J., Landeira-Dabarca, A., Lepšová-Skácelová, O., Musilová, Z., Otáhalová, Š., Poláková, M., Polášková, V., Sacherová, V., Špaček, J.,…, & Tropek, R. (2023). Muddying the unexplored post-industrial waters: Biodiversity and conservation potential of freshwater habitats in fly ash sedimentation lagoons. *Science of The Total Environment, 900*, 165803. https://doi.org/10.1016/j.scitotenv.2023.165803

Kolias, S., Kasselouri-Rigopoulou, V., & Karahalios, A. (2005). Stabilisation of clayey soils with high calcium fly ash and cement. *Cement and Concrete Composites, 27*(2), 301–313. https://doi.org/10.1016/j.cemconcomp.2004.02.019

Koukouzas, N., Zeng, R., Perdikatsis, V., Xu, W., & Kakaras, E. (2006). Mineralogy and geochemistry of Greek and Chinese coal fly ash. *Fuel, 85*(16), 2301–2309. https://doi.org/10.1016/j.fuel.2006.02.019

Kruse, A., Funke, A., & Titirici, M. M. (2013). Hydrothermal conversion of biomass to fuels and energetic materials. *Current Opinion in Chemical Biology*, *17*(3), 515–521.

Kumar, V., & Jha, G. K. (2014). Use of fly ash in agriculture: Indian scenario. *International Workshop on Agricultural Coal Ash Uses*, *27*, 29.

Kumar, V. P., & Prasad, D. R. (2019). Influence of supplementary cementitious materials on strength and durability characteristics of concrete. *Advances in Concrete Construction*, *7*(2), 75.

Marimon-Junior, B. H., Hay, J. D. V., Oliveras, I., Jancoski, H., Umetsu, R. K., Feldpausch, T. R., Galbraith, D. R., Gloor, E. U., Phillips, O. L., & Marimon, B. S. (2020). Soil water-holding capacity and monodominance in Southern Amazon tropical forests. *Plant and Soil*, *450*(1–2), 65–79. https://doi.org/10.1007/s11104-019-04257-w

Maroto-Valer, M. M., Taulbee, D. N., & Hower, J. C. (2001). Characterization of differing forms of unburned carbon present in fly ash separated by density gradient centrifugation. *Fuel*, *80*(6), 795–800.

Martins, S., Capitão, S., Santos, L. A., & Rodrigues, C. (2024). Valorisation of pulp and paper industry wastes—Incorporation in bituminous mixtures for road construction. *Sustainability*, *16*(3), 1254. https://doi.org/10.3390/su16031254

Montes-Hernandez, G., Pérez-López, R., Renard, F., Nieto, J. M., & Charlet, L. (2009). Mineral sequestration of CO2 by aqueous carbonation of coal combustion fly-ash. *Journal of Hazardous Materials*, *161*(2–3), 1347–1354. https://doi.org/10.1016/j.jhazmat.2008.04.104

Nath, P., & Sarker, P. (2011). Effect of fly ash on the durability properties of high strength concrete. *Procedia Engineering*, *14*, 1149–1156. https://doi.org/10.1016/j.proeng.2011.07.144

Olatoyan, O. J., Kareem, M. A., Adebanjo, A. U., Olawale, S. O. A., & Alao, K. T. (2023). Potential use of biomass ash as a sustainable alternative for fly ash in concrete production: A review. *Hybrid Advances*, *4*, 100076. https://doi.org/10.1016/j.hybadv.2023.100076

Pandey, K. B., & Rizvi, S. I. (2009). Plant polyphenols as dietary antioxidants in human health and disease. *Oxidative Medicine and Cellular Longevity*, *2*(5), 270–278. https://doi.org/10.4161/oxim.2.5.9498

Pandey, P. K., Kass, P. H., Soupir, M. L., Biswas, S., & Singh, V. P. (2014). Contamination of water resources by pathogenic bacteria. *AMB Express*, *4*(1), 51. https://doi.org/10.1186/s13568-014-0051-x

Pandey, V. C., Bajpai, O., & Singh, N. (2016). Plant regeneration potential in fly ash ecosystem. *Urban Forestry & Urban Greening*, *15*, 40–44. https://doi.org/10.1016/j.ufug.2015.11.007

Querol, X., Moreno, N., Umaña, J. C., Alastuey, A., Hernández, E., López-Soler, A., & Plana, F. (2002). Synthesis of zeolites from coal fly ash: An overview. *International Journal of Coal Geology*, *50*(1–4), 413–423. https://doi.org/10.1016/S0166-5162(02)00124-6

Ram, L. C., & Masto, R. E. (2010). An appraisal of the potential use of fly ash for reclaiming coal mine spoil. *Journal of Environmental Management*, *91*(3), 603–617. https://doi.org/10.1016/j.jenvman.2009.10.004

Ramanathan, S., Gopinath, S. C. B., Arshad, M. K. M., & Poopalan, P. (2020). Nanostructured aluminosilicate from fly ash: Potential approach in waste utilization for industrial and medical applications. *Journal of Cleaner Production*, *253*, 119923. https://doi.org/10.1016/j.jclepro.2019.119923

Robertson, G. P., Paul, E. A., & Harwood, R. R. (2000). Greenhouse gases in intensive agriculture: Contributions of individual gases to the radiative forcing of the atmosphere. *Science*, *289*(5486), 1922–1925. https://doi.org/10.1126/science.289.5486.1922

Roy, W. R., Thiery, R. G., Schuller, R. M., & Suloway, J. J. (1981). Coal fly ash: A review of the literature and proposed classification system with emphasis on environmental impacts. *Environmental Geology No. 096*. https://www.ideals.illinois.edu/items/80202/bitstreams/210637/data.pdf

Ruhl, M., Bonis, N. R., Reichart, G.-J., Damsté, J. S. S., & Kürschner, W. M. (2011). Atmospheric carbon injection linked to end-triassic mass extinction. *Science, 333*(6041), 430–434. https://doi.org/10.1126/science.1204255

Sair, S., Oushabi, A., Kammouni, A., Tanane, O., Abboud, Y., & El Bouari, A. (n.d.). *Case studies in construction materials*. Retrieved 13 April 2024, from https://www.researchgate.net/profile/Omar-Tanane/publication/322978981_Mechanical_and_thermal_conductivity_properties_of_hemp_fiber_reinforced_polyurethane_composites/links/6423532a92cfd54f84357796/Mechanical-and-thermal-conductivity-properties-of-hemp-fiber-reinforced-polyurethane-composites.pdf

Saha, A. K. (2018). Effect of class F fly ash on the durability properties of concrete. *Sustainable environment research, 28*(1), 25–31.

Shaikh, F. U. A., & Supit, S. W. M. (2015). Compressive strength and durability properties of high volume fly ash (HVFA) concretes containing ultrafine fly ash (UFFA). *Construction and Building Materials, 82*, 192–205. https://doi.org/10.1016/j.conbuildmat.2015.02.068

Silva, L. F. O., & Da Boit, K. M. (2011). Nanominerals and nanoparticles in feed coal and bottom ash: Implications for human health effects. *Environmental Monitoring and Assessment, 174*(1–4), 187–197. https://doi.org/10.1007/s10661-010-1449-9

Sinha, A. K., Singh, S., Masto, R. E., Verma, R., Selvi, V. A., & Ram, L. C. (2011). Removal of phosphorus from aqueous solution using lignite fly ash. *Bhartiya Vaigyanic Evam Audhyogik Anusandhan Patrika (Hindi), 19*(2), 183–187.

Smith, K. R., Veranth, J. M., Kodavanti, U. P., Aust, A. E., & Pinkerton, K. E. (2006). Acute pulmonary and systemic effects of inhaled coal fly ash in rats: Comparison to ambient environmental particles. *Toxicological Sciences, 93*(2), 390–399. https://doi.org/10.1093/toxsci/kfl062

Srinivasreddy, A. B., McCarthy, T. J., & Lume, E. (2013). *Effect of rice husk ash on workability and strength of concrete*, in 26th Biennial Concrete Institute of Australia's National Conference, pp. 1–10.

Subhapriya, S., & Gomathipriya, P. (2018). Zeolite X from coal fly ash inhibits proliferation of human breast cancer cell lines (MCF-7) via induction of S phase arrest and apoptosis. *Molecular Biology Reports, 45*(6), 2063–2074. https://doi.org/10.1007/s11033-018-4363-9

Swanepoel, J. C., & Strydom, C. A. (2002). Utilisation of fly ash in a geopolymeric material. *Applied Geochemistry, 17*(8), 1143–1148. https://doi.org/10.1016/S0883-2927(02)00005-7

Takada, T., Hashimoto, I., Tsutsumi, K., Shibata, Y., Yamamuro, S., Kamada, T., Inoue, K., Tsuzura, K., & Yoshida, K. (1995). Utilization of coal ash from fluidized-bed combustion boilers as road base material. *Resources, Conservation and Recycling, 14*(2), 69–77. https://doi.org/10.1016/S0921-3449(95)80001-8

Tekin, R., Bac, N., Warzywoda, J., & Sacco, A. (2015). Effect of reaction mixture composition and silica source on size distribution of zeolite X crystals. *Journal of Crystal Growth, 411*, 45–48. https://doi.org/10.1016/j.jcrysgro.2014.11.017

Tropek, R., Cerna, I., Straka, J., Cizek, O., & Konvicka, M. (2013). Is coal combustion the last chance for vanishing insects of inland drift sand dunes in Europe? *Biological Conservation, 162*, 60–64. https://doi.org/10.1016/j.biocon.2013.03.027

Tropek, R., Cerna, I., Straka, J., Kocarek, P., Malenovsky, I., Tichanek, F., & Sebek, P. (2016). In search for a compromise between biodiversity conservation and human health protection in restoration of fly ash deposits: Effect of anti-dust treatments on five groups of arthropods. *Environmental Science and Pollution Research, 23*(14), 13653–13660. https://doi.org/10.1007/s11356-015-4382-1

Turk, K., Karatas, M., & Gonen, T. (2013). Effect of Fly Ash and Silica Fume on compressive strength, sorptivity and carbonation of SCC. *KSCE Journal of Civil Engineering, 17*(1), 202–209. https://doi.org/10.1007/s12205-013-1680-3

Ukwattage, N. L., Ranjith, P. G., & Bouazza, M. (2013). The use of coal combustion fly ash as a soil amendment in agricultural lands (with comments on its potential to improve food security and sequester carbon). *Fuel*, *109*, 400–408. https://doi.org/10.1016/j.fuel.2013.02.016

Van Deventer, J. S. J., Provis, J. L., Duxson, P., & Brice, D. G. (2010). Chemical research and climate change as drivers in the commercial adoption of alkali activated materials. *Waste and Biomass Valorization*, *1*(1), 145–155. https://doi.org/10.1007/s12649-010-9015-9

Wang, S., & Wu, H. (2006). Environmental-benign utilisation of fly ash as low-cost adsorbents. *Journal of Hazardous Materials*, *136*(3), 482–501.

Weber, J., Strączyńska, S., Kocowicz, A., Gilewska, M., Bogacz, A., Gwiżdż, M., & Debicka, M. (2015). Properties of soil materials derived from fly ash 11 years after revegetation of post-mining excavation. *CATENA*, *133*, 250–254. https://doi.org/10.1016/j.catena.2015.05.016

Wong, J. W. C. (1995). The production of artificial soil mix from coal fly ash and sewage sludge. *Environmental Technology*, *16*(8), 741–751. https://doi.org/10.1080/09593331608616313

Wong, J. W. C., & Wong, M. H. (1990). Effects of fly ash on yields and elemental composition of two vegetables, *Brassica parachinensis* and *B. chinensis*. *Agriculture, Ecosystems & Environment*, *30*(3–4), 251–264. https://doi.org/10.1016/0167-8809(90)90109-Q

Xu, G., & Shi, X. (2018). Characteristics and applications of fly ash as a sustainable construction material: A state-of-the-art review. *Resources, Conservation and Recycling*, *136*, 95–109. https://doi.org/10.1016/j.resconrec.2018.04.010

Yang, J., Zhao, Y., Zyryanov, V., Zhang, J., & Zheng, C. (2014). Physical–chemical characteristics and elements enrichment of magnetospheres from coal fly ashes. *Fuel*, *135*, 15–26. https://doi.org/10.1016/j.fuel.2014.06.033

Yao, Z. T., Ji, X. S., Sarker, P. K., Tang, J. H., Ge, L. Q., Xia, M. S., & Xi, Y. Q. (2015). A comprehensive review on the applications of coal fly ash. *Earth-Science Reviews*, *141*, 105–121. https://doi.org/10.1016/j.earscirev.2014.11.016

Yu, C.-L., Deng, Q., Jian, S., Li, J., Dzantor, E. K., & Hui, D. (2019). Effects of fly ash application on plant biomass and element accumulations: A meta-analysis. *Environmental Pollution*, *250*, 137–142. https://doi.org/10.1016/j.envpol.2019.04.013

Yüksel, S., & Yürüm, Y. (2009). Removal of boron from aqueous solutions by adsorption using fly ash, zeolite, and demineralized lignite. *Separation Science and Technology*, *45*(1), 105–115. https://doi.org/10.1080/01496390903256042

Zheng, K., Adriaensens, P., De Schutter, G., Ye, G., & Taerwe, L. (2016). Studies on the alkali–silica reaction rim in a simplified calcium–alkali–silicate system. *Materials*, *9*(8), 670. https://doi.org/10.3390/ma9080670

14 Policy Review, International Collaboration, and Potential Associated with Ash Management

Status Quo and Future Goals

Manish Yadav, Satya Prakash Sahu,
Charu Jhamaria, Nitin Kumar Singh and Deepika

14.1 INTRODUCTION

The management of coal fly ash (CFA) has emerged as a major environmental concern especially in those countries that continue to rely on coal as their major source of energy (Yao et al., 2015; Vig et al., 2023). As countries of the world shift toward more sustainable practices, the CFA management has featured in national policies and global collaborations for dealing with its environmental effects as well as promoting the safe environment (Zhang, 2014; UNEP, 2017). At present, around 1.2 billion tons of CFA is produced every year and the main contributors are those countries that rely heavily on coal such as China, the United States, and India (Dwivedi and Jain, 2014; Yao et al., 2015; Yadav et al., 2022). However, the level of its utilization is not the same in these countries due to differences in the policy frameworks for the usage of fly ash, the technological advancement in ash utilization, and the market demand for products derived from fly ash (Grabias-Blicharz and Franus, 2023). In most of the developed countries, the management of the CFA is mainly done by landfilling, which is a great concern to the environment as this method leads to groundwater and air pollution (Izquierdo and Querol, 2012; Kumar et al., 2021). On the other hand, fewer countries have adopted sustainable practices, for instance, the reuse of the fly ash in construction materials that not only have a lesser negative impact on the environment but also add value to the economies of these countries (Jones and Roberts, 2019). These considerations have led to the increasing demand for the development and implementation of rules and regulations as well as international cooperation in the management of CFA in different parts of the world.

DOI: 10.1201/9781003498124-14

The management of CFA is a basically a policy-related issue and depending on the country, the policies can be elaborate and different from one country to another (Yunusa et. al., 2006). In India, the Ministry of Environment, Forest and Climate Change (MoEF&CC) has released a policy that construction activities in the vicinity of coal-fired power plants should use CFA. This policy is useful in minimizing the fly ash disposal in mine void and encourage its application in building materials (Ojha et al., 2022). Likewise, the Chinese "Circular Economy Promotion Law" focuses on the recycling of CFA and other industrial waste products for reducing the waste and optimization of resources (Fan and Fang, 2020). Similarly, Japan has put measures that encourage the utilization of CFA in agriculture and construction activities, which serves as a good example of how fly ash management can be incorporated into the circular economy models (Marinina et al., 2021). However, despite the abovementioned policies and/or regulations, the extent of CFA utilization in these countries remained very low in comparison to its generation rates (Ahmaruzzaman, 2010). It is important to mention here that in United States and members of the European Union (EU), due to their strict environmental laws and well-developed markets for fly ash-based products, the utilization rate has reached up to 70% (Gharfalkar et al., 2015; ACAA, 2021). While, the Resource Conservation and Recovery Act (RCRA) of United States allows for the regulation of coal ash disposal where it is classified as non-hazardous waste and at the same time encouraging reuse. Whereas, the EU has adopted the Waste Framework Directive that requires minimization of wastes and recycling and reuse of industrial residues/by-products such as fly ash (Heidrich et al., 2013). In the recent past, even global bodies like the United Nations Environment Programme (UNEP) and the International Energy Agency (IEA) have also come forward to realize the need for regional collaboration in the management of industrial residues including CFA (UNEP, 2018; IEA, 2020). The EU has also played a significant role in promoting the circular economy model whereby CFA can be recycled and reused across the member countries (Van Ewijk, 2018; De Sa and Korinek, 2021). These organizations have helped in sharing of experiences and formulation of policies that member countries can use in enhancing their CFA management systems.

As elaborated in the above background, it is evident that there is a strong need for more stringent policies and international cooperation in dealing the problems associated with CFA management. Also, the coordination of current and future regulations of CFA management is also essential for those countries where transboundary pollution is an issue, including Southeast Asia. Achieving this goal will require the development of new markets for ash-derived products and the implementation of policies that incentivize the recycling and reuse of CFA. Overall, the establishment of national and international networks and partnerships is crucial in ensuring the benefits of sustainable CFA management on a global scale.

14.2 ENVIRONMENTAL AND HEALTH IMPLICATIONS OF CFA

The two main methods for disposing of CFA are settling ponds and landfills. In settling ponds, slurry made from CFA and water is transported through ponds; the water flows off to the second pond while the heavier particles settle at the first one. This process is repeated until the first pond is filled with fly ash. The water can either be returned to the receiving streams or it can be reused. In any case, most of its toxic components make it a dangerous pollutant to the environment. Fly ash is composed of many oxides, including those of silicon, iron, magnesium, aluminum, and calcium. Apart from these, a host of additional trace elements, including radioactive and rare earth elements, have also been found; these are the ones that cause the toxicity. The presence of rare earth elements in fly ash has been linked to low to moderate ecological risks; however, the associated heavy metals present a more severe threat, particularly to children, with hazard indices exceeding safety limits (Slavković-Beškoski et al., 2024). The three main areas that fly ash affects are air, water, and soil (Chen et al., 2024; Kamara et al, 2023; Gopinathan et al., 2022; Hossain et al., 2015). Fly ash particles are often released into the atmosphere, contributing to air pollution. Fine particulate matter ($PM_{2.5}$ and PM_{10}) from fly ash can travel long distances, affecting air quality over large areas (Adar and Kaufman, 2007). Hazardous gases like SO_2, NO_2, etc., produced by the coal combustion units might directly impact the living system. After prolonged inhalation, these particles penetrate deeply into the body and settle in the lungs, which eventually causes cardiovascular disorders, cerebrovascular diseases, and even lung cancer (Gopinathan et al., 2022; Ghio et al., 2012; Mergler et al., 2007; Nriagu and Pacyna, 1988). Improper disposal of fly ash can lead to the leaching of heavy metals such as arsenic, lead, and mercury into the soil and groundwater, posing risks to agriculture and drinking water sources (Adriano et al., 1980). The contamination of water bodies with fly ash can harm aquatic life due to the toxicity of heavy metals and the alteration of pH levels (Kim et al., 2003). While some studies suggest that utilizing CFA in construction materials can immobilize these heavy metals, the overall environmental implications remain concerning, thereby necessitating improved management and disposal strategies to mitigate health risk. Earlier studies on environmental and health implications of coal-fired fly ash with some relevant key findings are listed in Table 14.1.

The CFA is considered as an environmental hazardous substance because of the presence of heavy metals that pollute the soil and water sources if disposed improperly. These include arsenic, mercury, and lead which are dangerous because they are toxic and persistent in nature and can easily enter into the food chain, thus posing a threat to the ecosystems and the biodiversity. Various notable health effects have been highlighted in literature which are directly associated with the exposure to CFA to public health. The respiratory illnesses caused by the fly ash include respiratory and cardiovascular diseases. The long-term exposure of particular matter containing CFA has also been linked to instances of cancer and other related severe diseases in people. Further, the presence of toxic metals in CFA also increased the health risks in severe manner. Therefore, managing and regulating CFA is of utmost importance in order to reduce its effects on the environment and human health. It has been observed that current practices of CFA are much different across the world and there is a lack of standard regulations and enforcement. This can be further strengthened via adopting efficient technologies for fly ash disposal and utilization, and proper monitoring and assessment of risks involved in its handling and disposal. Despite the ongoing research in this direction, there are still some gaps that have not been fully addressed. Further research should be directed toward the effects of CFA on people's health in the long run, the efficiency of different prevention measures, and the ways

TABLE 14.1
Earlier studies on environmental and health implications of coal-fired fly ash

Sl. no.	Author (year)	Key findings	Study area
1	Gopinathan et al. (2022)	Geochemistry, structural characteristics, and environmental impact of *CFA were studied. The presence of Si-based minerals and unburnt carbon was confirmed by FE-SEM analysis.* *Elemental analysis through ICP-OES confirmed the presence of major oxides of Al, Ca, Fe, Mg, Na, P, S, Si, and Ti in the samples.* *Structural analysis through XRD exposed the presence of minerals like quartz, hematite, anatase, muscovite, and rutile.* *Morphological analysis through FESEM confirmed the presence of unburnt carbon.*	Talcher Coalfield, India
2	Verma et al. (2021)	Considerable amounts of heavy metals such as Ni, Pb, Fe, Mn, Cr, and Al were found at selected sites near the ash pond. According to heavy metal evaluation index, groundwater within the range of 5 km showed a heavy load of heavy metal pollution. Moderate level of metal pollution was observed in the range of 7 km, and at a distance of 15 km heavy metal pollution was found to be very low.	Near the ash pond of Parichha Thermal Power Plant (Jhansi), India
3	Hossain et al. (2015)	Chemical composition of the fly ash sample was determined by XRF analysis, and the major components of fly ash were alumina (Al_2O_3: 40%) and silica (SiO_2: 50%).	Barapukuria, Dinajpur, Bangladesh
4	Singh et al. (2014)	Groundwater samples from different locations within the 2 km radius of two thermal power plants were collected and analyzed for heavy metals (Cd, Cr, Cu, Fe, Ni, and Pb). This study indicated contamination of bore well and pond water within the study area with higher concentration levels of these heavy metals.	Delhi, India

(Continued)

TABLE 14.1
(Continued)

Sl. no.	Author (year)	Key findings	Study area
5	Frankel et al. (2023)	The impact of coal ash runoff on Seminole ramshorn snails, providing insights into potential environmental consequences of coal-fired fly ash on freshwater ecosystems.	Maryland, United States
6	Han et al. (2021)	A modified risk assessment approach was utilized to analyze the possible environmental hazards associated with trace elements found in fly ash and gypsum originating from coal-fired power stations in China. The results of the experiments showed that between 92.9% and 98.2% of the total outputs were trace elements found in fly ash and gypsum.	China
7	Khan and Umar (2019)	In order to evaluate the environmental impact of hazardous CFA on the ecological system, physicochemical and hydrogeochemical analyses of soil and groundwater were conducted. The results showed the existence of harmful arsenic mineral phases in addition to the typical minerals found in CFA. Additionally, this indicated that coal fly ash had a significant impact on agricultural soil.	New Delhi, India

of developing an efficient and long-term strategy for managing the health impacts of CFA. Further, the studies about the other applications of CFA that can minimize its negative impact on the environment must be conducted.

Therefore, it can be concluded that CFA is a material that poses a number of environmental and health risks that need further discussion. Solving these problems requires a multi-sectoral approach that draws support from policymakers, research institutions, industry stakeholders, and the society. It becomes vital for different regulatory bodies to employ and develop better regulatory measures, efficient technologies for CFA management and continuous research in order to protect the health of human beings and the environment from the negative impacts of CFA.

14.3 REGULATIONS FOR FLY ASH UTILIZATION

It is estimated that CFA, a by-product of coal combustion, is produced on a colossal scale around the world with an estimated quantity of about 1.2 billion tons on an annual basis. Although the CFA can be used in the manufacturing of construction materials such as cement concrete, wall board, bricks and the like, it is estimated that ~64% of it is utilized globally. For instance, in developing countries like Botswana, 672,000 tons of fly ash is generated annually, and out of this 70% remains unutilized, mainly because of the absence of sophisticated technologies and complicated legal structures. The rules that guide the management and use of CFA differ across countries and this poses a big challenge especially when trying to maximize its use.

Malaysia's "Environmental Quality (Scheduled Wastes) Regulations 2005" has classified fly ash with hazardous metals such as aluminum, arsenic, mercury, and lead as SW 104 which has to be managed as scheduled waste (Government of Malaysia [GOM], 2005). This regulation also makes it possible to handle CFA with a lot of caution so that it does not pollute the environment. Further, SW 406 raises concerns for the disposal of clinker, slag, and ash resulting from the incineration of the scheduled wastes mentioning that such wastes should be handled to eliminate risks to human health and environment. Likewise, Thailand has the "Industrial Waste Disposal B. E. 2548 (2005)" which outlines the ways of handling industrial wastes, including fly ash as per Ministry of Industry (Thailand). This regulation identifies hazardous and non-hazardous fly ash; as each having its own treatment and disposal methods. As CFA contains hazardous elements, its storage and disposal require extra caution, for instance, it should be stored for a limited period, disposal methods must include effective pollution control measures, and the transportation of hazardous wastes must follow certain laid down procedures (Darmansyah et al., 2023).

In China, fly ash is classified as hazardous waste under the "Regulation on Safe Management of Hazardous Chemicals No. 591," primarily due to its high concentrations of dissolved salts and heavy metals, such as cadmium, lead, and zinc. This classification necessitates rigorous management practices to protect human health and the environment. Despite this classification, China has made significant strides in fly ash utilization, achieving a reported 70% reuse rate.

The EU is also progressive in this direction, as it focuses on the management of fly ash, including biomass ash with residues from additives used in the combustion process. The EU regulation enforces that such wastes must be either reused or recycled though it bans on disposal in the landfills. This regulation is in line with other environmental goals of the EU, which promotes the use of sustainable practices and reduction of waste.

The Environmental Protection Agency (EPA) of the United States reacted to the 2008 disaster at Tennessee Valley Authority power plant by putting forward new regulations for Coal Combustion Residuals (CCRs) under RCRA. These regulations, proposed on June 21, 2010, offer two alternative approaches: one, would redesignate CCRs as special waste under subtitle C of RCRA and apply more rigorous disposal controls and the other would keep CCRs under subtitle D of RCRA but would establish national minimum standards for disposal. Both options contain provisions for the stability of surface impoundments in order to avoid further major releases (U.S. Environmental Protection Agency [USEPA], 2024).

India's Fly Ash Notification 2022 has clearly shown a good intent for environmental conservation as 100% of the ash produced from coal or lignite-based thermal power plants (TPPs) is instructed to be mandatorily used. This notification also mandates that ash utilization rates of TPPs should be 80% per annum and must reach 100% in a three-year cycle. It also lays down a step-by-step process for using the ash generated before the notification with the aim of achieving 100% utilization within the next ten years. For the implementation of these mandates, the notification includes the "polluter pays" provision which makes the TPPs liable to pay penalties for non-compliance and promote the sustainable utilization of fly ash in several sectors like road construction, embankments, and construction materials.

Overall, it is observed that different regulations across the various nations are an indication of the issues that arise when dealing with fly ash, the prospects of enhancing its use while at the same time having to consider the impacts on the environment and the people.

14.3.1 Fly Ash Notification (2022), MoEF&CC, India

To address the challenges of land usage for ash disposal in ponds and the pollution caused by ash, the Ministry of Environment, Forest and Climate Change has issued several notifications regarding ash utilization. The first notification was released on September 14, 1999, and was subsequently amended in 2003, 2009, and 2016, with notifications dated August 27, 2003, November 3, 2009, and January 25, 2016, respectively. In 2021, the Ministry issued a new notification on December 31, 2021, which superseded all previous notifications. Additionally, an amendment to this notification was issued on December 30, 2022.

The notification mandates that coal or lignite-based TPPs ensure 100% utilization of fly ash and bottom ash generated by them. This is to be done in an environmentally friendly manner as per the specified guidelines. The fly ash and bottom ash can be utilized in various eco-friendly ways such as:

- Manufacturing of fly ash-based products like bricks, blocks, tiles, etc.
- Use in cement manufacturing, ready-mix concrete, and geopolymer-based construction materials.
- Road construction, including flyovers and embankments.

- Filling of low-lying areas and abandoned mines.
- Construction of dams.
- Agriculture, based on soil testing.
- Export to other countries for similar eco-friendly purposes.

The responsibility of power plants has been fixed which states that TPPs are responsible for ensuring that the fly ash is utilized as per the guidelines. They must achieve 100% utilization within three years, with a minimum of 80% utilization in any given year. In this notification, the utilization of legacy waste has also been discussed, which says that existing stockpiles of fly ash, referred to as "legacy ash," must be fully utilized within ten years from the date of the notification. The use of legacy ash is in addition to the annual utilization targets set for the ongoing generation of fly ash. Penalty provision has been provided stating that penalties are to be imposed on TPPs that fail to meet the utilization targets, based on the Polluter Pays Principle (PPP). A committee chaired by the Central Pollution Control Board (CPCB) will be formed to review, monitor, and recommend the inclusion of new eco-friendly methods for fly ash utilization. This notification aims to promote sustainable development by ensuring the responsible management and utilization of fly ash and bottom ash, minimizing environmental impact, and conserving natural resources.

14.3.2 Fly Ash Utilization in India

Central Electricity Authority is a government body which is involved in the monitoring of fly-ash generation and its utilization in India. As per their report, In the fiscal year 2021–2022, data were received from 200 thermal power stations, which collectively represent an installed capacity of 213,620.50 MW. These power stations consumed a total of 759.02 million tons of coal, resulting in the generation of 270.82 million tons of fly ash as shown in Figure 14.1. Of the fly ash produced, 259.86 million tons were successfully utilized, achieving an impressive utilization rate of 95.95% (Central Electricity Authority [CEA], 2022).

The average ash content in the coal used across these stations was recorded as 35.68%. This data highlight the significant efforts made in the effective management and utilization of fly ash, reflecting a commitment to minimizing environmental impact while maintaining high operational efficiency. Figure 14.2 shows that the largest portion, 68.83 million tons (25.41%), was used in the cement industry, reflecting its significant demand for fly ash as a raw material. Roads and flyovers also utilized a substantial amount, accounting for 45.99 million tons (16.98%). The reclamation of low-lying areas saw the utilization of 35.57 million tons (13.13%), while bricks and tiles manufacturing used 31.62 million tons (11.68%). Other notable uses included mine filling with 18.38 million tons (6.79%) and ash dyke raising with 14.82 million tons (5.47%). A smaller percentage was directed toward agriculture, concrete, and the hydro power sector, collectively accounting for less than 1% of the total. However, 10.96 million tons (4.05%) of fly ash remained unutilized, indicating room for improvement in its efficient management (CEA, 2022).

While India has made promising progress in fly ash utilization, the challenge of achieving 100% utilization remains significant, especially as power demand

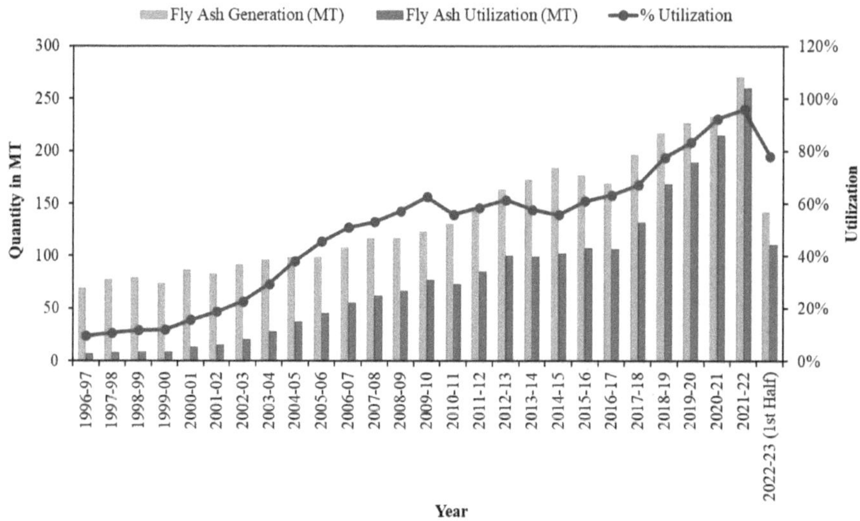

FIGURE 14.1 Fly-ash generation and utilization chart in India.

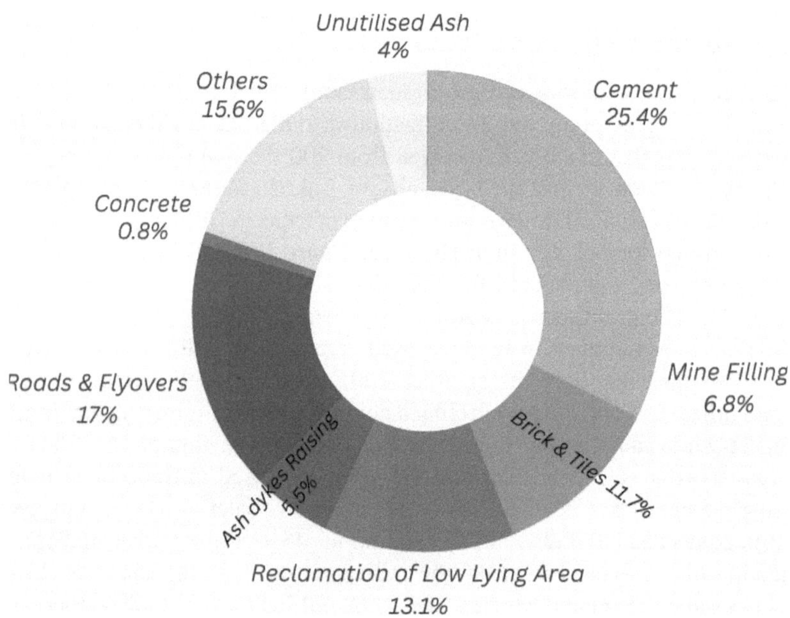

FIGURE 14.2 Ash utilization in different sectors.

continues to rise. Meeting ambitious targets will become increasingly difficult without innovative solutions. Among the various options, mine filling stands out as a viable method for utilizing large quantities of fly ash. However, achieving full utilization will require continued transformation and the adoption of new approaches.

14.4 RECLAMATION OF COAL MINE VOID WITH FLY ASH

As of date filling mine voids with fly ash in an environmentally friendly manner is the most practical solution for bulk utilization. This not only saves a large amount of land for ash disposal, but it also decreases the associated environmental impacts of restoring the mined-out area for productive use. Fly ash in bulk can also be used to store underground mines instead of sand and to fill abandoned open-cast mine voids. This can lead to a greater use of generated fly ash. Fly ash can be

- Used as a replacement for sand.
- Uses 50% less water.
- Flows easily and fills the void/cavity.
- It possesses water-holding capacity, making it suitable for afforestation.

It is investigated and determined that fly ash is generally inert in nature. Moreover, coal utilized in Indian TPPs contains a high ash percentage. As a result, heavy metal enrichment is lower than in fly ash produced by TPPs in developed countries like the United States, Australia, China, etc.

In India, the Ministry of Coal has made substantial steps to ensure proper disposal of fly ash by allocating mine voids for this purpose. In 2023, a Central Level Working Group (CLWG) was formed, chaired by the Additional Secretary, Ministry of Coal. The Central Electricity Authority (CEA) receives an application from interested TPPs for the allocation of mining voids, which is then discussed at the CLWG meeting. In this proactive approach, 19 mines have been allotted to 13 TPPs. This allocation tackles environmental problems related to fly ash disposal and encourages sustainable practices in the coal mining industry. Furthermore, fly ash has been repurposed so far at different coal mines like Gorbi coal mine pit-, South Balanda OCP (Open Cast Project), etc.

CPCB (2019) provided a guideline for the disposal of fly ash for reclamation of the mines. As per the guidelines, prior to beginning ash disposal activities in the mining void, the power plant authorities must do the following research as given below to ensure that there should be no change, damage, or deterioration in water quality and hydrology in and around the planned area:

- Geotechnical study of the pipeline corridor and mine void area
- Pre and post-filling mine water quality including leachability of metals
- Topographical survey of pipeline corridor and mine void area
- Techno-Economic feasibility study for disposal of ash into the quarry
- Ash characterization and leachate study
- Feasibility of transportation of ash to mine void

Regular environmental monitoring will be carried out both during the ash disposal process and after the mine void has been reclaimed. Table 14.2 shows the detailed monitoring program. Minimum three piezometers (upstream, downstream, and control), as illustrated in Figure 14.3, must be created to analyze the water quality surrounding the mine that has been reclaimed with fly ash. In the event of degradation of environmental quality, the same shall be immediately reported to the concerned

TABLE 14.2
Monitoring program during the disposal of ash

Samples	Parameters to be analyzed	Frequency
Ash samples	Chemical parameters in %: SiO_2, Al_2O_3, Fe_2O_3, K_2O, TiO_2, CaO, MgO, Na_2O, P_2O_5, SO_3. Trace elements (mg/kg, Toxicity Characteristic Leaching Procedure (TCLP test): As, Ba, Cd, Co, Cr, Cu, F, Fe, Hg, Mn, Ni, Pb, and Zn. Radioactivity (Bq/kg) includes 238U, 236Ra, 232TH, 228Ra, 230Pb, 40K, and 137Cs.	Once before initiation of filling
Ash leachate analysis	Trace elements (mg/kg, using TCLP test): As, Ba, Cd, Co, Cr, Cu, F, Fe, Hg, Mn, Ni, Pb, Zn	Once a year
Piezometer water samples	Chemical parameters (mg/l, except, pH and EC): pH, EC, TDS, total alkalinity, Ca, Mg, Na, K, Cl, SO_4, NO_3, PO_4. Trace elements (mg/l): As, Ba, Cd, Co, Cr, Cu, F, Fe, Hg, Mn, Ni, Pb, Zn	Monthly
Mine water sample	Same as above	Monthly
Groundwater	Same as above	Twice a year – Pre-monsoon and post-monsoon
Surface water samples	Same as above	Twice a year – Pre-monsoon and post-monsoon
Soil samples	Texture, type, pH, and cation exchange capacity. Trace elements (mg/l): As, Ba, Cd, Co, Cr, Cu, F, Fe, Hg, Mn, Ni, Pb, Zn	Once a year
Survey of flora and fauna	Listing of flora (herbs, shrubs, and trees) and fauna (soil invertebrates and other animals) based on field observations and review of information available. Analysis of trace elements in plants (herbs, shrubs, and trees), the invertebrates· Analysis of trace elements in aquatic fauna from the mine void filled with fly ash bio-accumulation and bio-magnification tests	Once in two years

Source: CPCB (2019).

State Pollution Control Board (SPCB) and appropriate preventive/corrective action will be taken.

14.4.1 Case Study South Balanda Open Cast Mine of Mahanadi Coalfields Limited

In India, the Central Mine Planning and Designing Institute (CMPDI) conducted a feasibility study for ash disposal in an abandoned mine pit, followed by hydrological

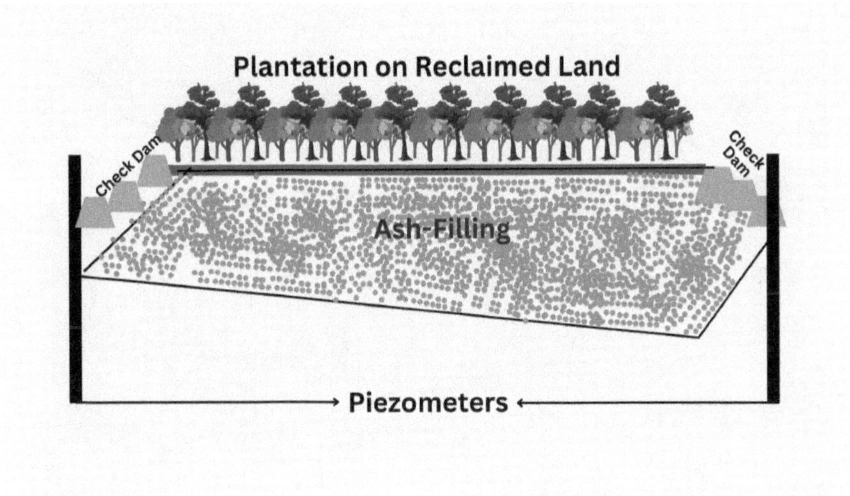

FIGURE 14.3 A typical diagram for development of piezometers in an ash-filled mine.

and environmental impact assessments based on above criteria, specifically for the South Balanda Open Cast Project in Odisha. The findings indicated that disposing of ash in the mine voids posed minimal environmental risk. Based on these results, National Thermal Power Plant's, Talcher Thermal Power Station (TTPS) received approval to discharge ash slurry into the South Balanda mine voids. The State Pollution Control Board, in coordination with Mahanadi Coalfields Limited (MCL) and the power generation units, consented to this disposal method. Currently, ash slurry is transported via a ten-km pipeline to the abandoned mine pit. The entire void, with a capacity of 14 million cubic meters, has been successfully filled, conserving approximately 500 ha of fertile land that would have otherwise been required for constructing an ash pond by TTPS (NTPC). The reclaimed mine with fly ash has been shown in Figure 14.4.

14.5 CONCLUSIONS

The CFA is a complex material containing several oxides of elements such as silicon, iron, magnesium, aluminum, calcium, and a few other trace elements that are toxic in nature such as radioactive and rare earth elements. Heavy metals such as arsenic, mercury, and lead are particularly dangerous in CFA as they can build up in the ecosystem and in food chains, thus exerting adverse effects on ecosystems and biodiversity. The huge quantity of CFA produced worldwide makes it necessary for stricter measures to be put in place and for the world to come up with measures to address the issue. In this view, many countries of the world have started to respond to these challenges in different ways. The Chinese law on circular economy specifically provides for the recycling of industrial waste residues such as fly ash. Japan has put in place measures that have enhanced the use of CFA in its circular economy model and its utilization in agriculture and construction has been encouraged. In the United

FIGURE 14.4 Balanda OCP after the backfilling of ash slurry.

States, the RCRA is the governing law that guides the disposal of coal ash. Malaysia's Environmental Quality (Scheduled Wastes) Regulations 2005 classifies CFA containing hazardous metals like aluminum, arsenic, mercury, and lead as scheduled wastes and thus suggests proper treatment. The Thai Industrial Waste Disposal B. E. 2548 (2005) provides specific measures of handling industrial waste inclusive of fly ash. India's Fly Ash Notification 2022 enforces the utilization of CFA produced from coal or lignite-fired TPPs in full measure with a large proportion used in cement industry, road construction, among other uses including the Balanda Open Cast Mine.

REFERENCES

ACAA (2021) Production and Use of Coal Combustion Products in the US, Annual Survey Results 2014–2020. American Coal Ash Association. https://acaa-usa.org/publications/production-use-reports/

Adar, S. D., & Kaufman, J. D. (2007). Cardiovascular disease and air pollutants: Evaluating and improving epidemiological data implicating traffic-related air pollution as a risk factor. *Environmental Health Perspectives, 115*(6), 898–904.

Adriano, D. C., Page, A. L., Elseewi, A. A., Chang, A. C., & Straughan, I. (1980). Utilization and disposal of fly ash and other coal residues in terrestrial ecosystems: A review. *Journal of Environmental Quality, 9*(3), 333–344.

Ahmaruzzaman, M. (2010). A review on the utilization of fly ash. *Progress in Energy and Combustion Science, 36*(3), 327–363.

Central Electricity Authority (CEA). (2022). *Report on fly ash generation at coal/lignite-based thermal power stations and its utilization in the country for the year 2021-22.* New Delhi: Central Electricity Authority, Civil Design Division.

Central Pollution Control Board (CPCB). (2019). *Guidelines for disposal/utilisation of fly ash for reclamation of low lying areas and in stowing of abandoned mines/quarries.* New Delhi: Central Pollution Control Board.

Chen, Y., Fan, Y., Huang, Y., Liao, X., Xu, W., & Zhang, T. (2024). A comprehensive review of toxicity of coal fly ash and its leachate in the ecosystem. *Ecotoxicology and Environ mental Safety, 269*, 115905.

Darmansyah, D., You, S. J., & Wang, Y. F. (2023). Advancements of coal fly ash and its prospective implications for sustainable materials in Southeast Asian countries: A review. *Renewable and Sustainable Energy Reviews, 188*, 113895.

de Sa, P. and J. Korinek (2021-03-01), "Resource efficiency, the circular economy, sustainable materials management and trade in metals and minerals", OECD Trade Policy Papers, No. 245, OECD Publishing, Paris. http://dx.doi.org/10.1787/69abc1bd-en

Dwivedi, A., & Jain, M. K. (2014). Fly ash–waste management and overview: A review. *Recent Research in Science and Technology, 6*(1), 30–35.

Fan, Y., & Fang, C. (2020). Circular economy development in China-current situation, evaluation and policy implications. *Environmental Impact Assessment Review, 84*, 106441.

Frankel, T. E., Crowell, C., Giancarlo, L., Hydorn, D., & Odhiambo, B. K. (2023). Investigating the potential impacts of coal ash runoff on the freshwater Seminole ramshorn snail (*Planorbella duryi*) under laboratory conditions. *Chemosphere, 310*, 136815.

Gharfalkar, M., Court, R., Campbell, C., Ali, Z., & Hillier, G. (2015). Analysis of waste hierarchy in the European waste directive 2008/98/EC. *Waste Management, 39*, 305–313.

Ghio, A. J., Smith, C. B., & Madden, M. C. (2012). Lung injury after exposure to fly ash particles. *Environmental Health Perspectives, 120*(2), 215–220.

Gopinathan, P., Santosh, M. S., Dileepkumar, V. G., Subramani, T., Reddy, R., Masto, R. E., & Maity, S. (2022). Geochemical, mineralogical and toxicological characteristics of coal fly ash and its environmental impacts. *Chemosphere, 307*, 135710.

Government of Malaysia (GOM). (2005). *Environmental quality (scheduled wastes) regulations 2005 [P.U.(A) 294/2005].* Federal Subsidiary Legislation under the Environmental Quality Act 1974 (Act 127). Percetakan Nasional Malaysia Berhad.

Government of Thailand, Ministry of Industry. (2005). *Notification of Ministry of Industry re: Industrial waste disposal B.E. 2548 (2005).* Government Gazette, Volume 123, Special Part 11d.

Grabias-Blicharz, E., & Franus, W. (2023). A critical review on mechanochemical processing of fly ash and fly ash-derived materials. *Science of the Total Environment, 860*, 160529.

Han, D., Xu, L., Wu, Q., Wang, S., Duan, L., Wen, M.,…, & Liu, K. (2021). Potential environmental risk of trace elements in fly ash and gypsum from ultra–low emission coal–fired power plants in China. *Science of the Total Environment, 798*, 149116.

Heidrich, C., Feuerborn, H. J., & Weir, A. (2013, April). Coal combustion products: A global perspective. In *World of coal ash conference* (Vol. 22, p. 25).

Hossain, M. N., Paul, S. K., & Hasan, M. M. (2015). Environmental impacts of coal mine and thermal power plant to the surroundings of Barapukuria, Dinajpur, Bangladesh. *Environmental Monitoring and Assessment, 187*, 1–11.

International Energy Agency (IEA). (2020). *Energy technology perspectives 2020*. International Energy Agency, Paris, France.

Izquierdo, M., & Querol, X. (2012). Leaching behaviour of elements from coal combustion fly ash: An overview. *International Journal of Coal Geology, 94*, 54–66.

Jones, A., & Roberts, M. (2019). Utilization of fly ash in construction materials. *Journal of Construction Engineering, 45*(4), 567–578.

Kamara, S., Foday Jr, E. H., & Wang, W. (2023). A review on the utilization and environmental concerns of coal fly ash. *American Journal of Chemistry and Pharmacy, 2*(2), 53–65.

Khan, I., & Umar, R. (2019). Environmental risk assessment of coal fly ash on soil and groundwater quality, Aligarh, India. *Groundwater for Sustainable Development, 8*, 346–357.

Kim, A. G., Kazonich, G., & Dahlberg, M. (2003). Relative solubility of cations in class F fly ash. *Environmental Science & Technology, 37*(19), 4507–4511.

Kumar, R., Singh, N. K., & Gupta, A. (2021). Advances in fly ash management and utilization. *Journal of Environmental Management, 277*, 111–123.

Marinina, O., Nevskaya, M., Jonek-Kowalska, I., Wolniak, R., & Marinin, M. (2021). Recycling of coal fly ash as an example of an efficient circular economy: A stakeholder approach. *Energies, 14*(12), 3597.

Mergler, D., Anderson, H. A., Chan, L. H., Mahaffey, K. R., Murray, M., & Sakamoto, M. (2007). Methylmercury exposure and health effects in humans: A worldwide concern. *AMBIO: A Journal of the Human Environment, 36*(1), 3–11.

Moon, S. T., & Turner, A. B. (2009). Coal ash law and regulation in the United States: An overview. *Southeastern Environmental Law Journal, 18*, 173.

Nriagu, J. O., & Pacyna, J. M. (1988). Quantitative assessment of worldwide contamination of air, water and soils by trace metals. *Nature, 333*(6169), 134–139.

Ojha, P. N., Singh, B., Kaura, P., & Satyakam, R. (2022). Ash utilization strategy in India—A way forward. In *Advances in sustainable materials and resilient infrastructure* (pp. 145–157). Singapore: Springer Singapore.

Singh, R. K., Gupta, N. C., & Guha, B. K. (2014). Assessment of ground water contamination for heavy metals in the proximity of ash ponds. *Elixir Pollution, 75*, 28016–28019.

Slavković-Beškoski, L., Ignjatović, L., Ćujić, M., Vesković, J., Trivunac, K., Stojaković, J.,…, & Onjia, A. (2024). Ecological and health risks attributed to rare earth elements in coal fly ash. *Toxics, 12*(1), 71.

United Nations Environment Programme (2017) Towards a Pollution-Free Planet Background Report. United Nations Environment Programme, Nairobi, Kenya.

U.S. Environmental Protection Agency. *EPA history: Resource Conservation and Recovery Act*. EPA. Retrieved [08-28, 2024], from https://www.epa.gov/history/epa-history-resource-conservation-and-recovery-act

Van Ewijk, S; (2018) *Resource Efficiency and the Circular Economy: Concepts, Economic Benefits, Barriers, and Policies*. UCL Institute for Sustainable Resources: London, UK.

Verma, C., Hussain, A., Madan, S., & Kumar, V. (2021). Assessment of heavy metal pollution in groundwater with respect to distance from ash pond by using heavy metal evaluation index. *Applied Water Science, 11*, 1–6.

Vig, N., Ravindra, K., & Mor, S. (2023). Environmental impacts of Indian coal thermal power plants and associated human health risk to the nearby residential communities: A potential review. *Chemosphere*, *341*, 140103.

Yadav, V. K., Gacem, A., Choudhary, N., Rai, A., Kumar, P., Yadav, K. K.,…, & Islam, S. (2022). Status of coal-based thermal power plants, coal fly ash production, utilization in India and their emerging applications. *Minerals*, *12*(12), 1503.

Yao, Z. T., Ji, X. S., Sarker, P. K., Tang, J. H., Ge, L. Q., Xia, M. S., & Xi, Y. Q. (2015). A comprehensive review on the applications of coal fly ash. *Earth-Science Reviews*, *141*, 105–121.

Yunusa, I. A. M., Eamus, D., DeSilva, D. L., Murray, B. R., Burchett, M. D., Skilbeck, G. C., & Heidrich, C. (2006). Fly-ash: an exploitable resource for management of Australian agricultural soils. *Fuel*, *85*(16), 2337–2344.

Zhang, X. (2014). Management of coal combustion wastes. *IEA Clean Coal Centre*, 2–68.

15 Life Cycle Assessment and Cost Analysis of Fly Ash

Sandeep Singh, Gaurav Kabra and Debopriyo Roy

15.1 INTRODUCTION

Uninterrupted energy flow is one of the basic requirements of any industry (manufacturing and service sectors). The production of energy to fulfill any country's industrial and domestic requirements is a crucial challenge for any government. Power generation from coal-based plants is popular among other options, especially in underdeveloped and developing countries. Coal-based power generation results in massive coal fly ash (CFA) production as a waste (Curpen et al., 2023). The generation of CFA happens at around 1,200°C–1,700C and consists of various organic and inorganic substances (Blissett & Rowson, 2012). These substances pose adverse environmental impacts due to the presence of some potentially hazardous elements (Saha & Roychowdhury, 2023).

Biomass-based captive power plants using rice husk and sugarcane bagasse are another critical source of FA production (Kerdsuwan & Laohalidanond, 2022). Disposing of FA generated from various sources is a significant concern for all stakeholders (Rathnayake et al., 2018). Literature highlights that FA is causing severe threats to the environment as well as living organisms (Fernando et al., 2021). Ensuring sustainable practices in coal-based and biomass-based power generation and other activities demands the proper utilization or disposal of FA with minimal environmental impact (Deokar & Pathak, 2023).

15.1.1 Fly Ash and Associated Applications

With the increased awareness among the stakeholders, the utilization of FA with minimal impact on the environment has started in various sectors of the construction and manufacturing industry (Curpen et al., 2023). Broadly, three types of FAs are generated from industrial processes, namely CFA, rice husk ash (RHA), and bagasse ash (BA) (Huang et al., 2017). While CFA is produced in coal-based power plants, RHA and BA are produced in agricultural activities, namely rice milling and sugarcane milling (Kerdsuwan & Laohalidanond, 2022). CFA has great capability in terms of binding and also the load-bearing characteristics that make it applicable to many industrial needs (Curpen et al., 2023). CFA is mainly applied in the construction

 DOI: 10.1201/9781003498124-15

sector particularly in concrete blocks as a partial replacement to Portland cement due to its strength and durability (Wang et al., 2017a, 2017b).

High silica content in RHA makes it useful in many industries (Zulqar Nain & Kasilingam, 2023). It serves as a pozzolanic material in cement and concrete and enhances the mechanical properties and durability of the structures (Zulqar Nain & Kasilingam, 2023). It is widely used in ceramic and refractory products due to its high-temperature resistance properties (Jamora et al., 2023). It is instrumental in enhancing soil health and fertility, improving crop yield and sustainability (Itam et al., 2022). BA is the residue that remains after burning the sugarcane bagasse. Using BA in concrete and bricks helps enhance the strength and reduce the overall cost of building materials (Itam et al., 2022). It acts as a soil conditioner, providing essential nutrients and improving the soil structure, promoting better plant growth and increased agriculture productivity. All these applications help industries reduce carbon emissions and attain sustainable development. However, finding innovative uses for these ashes will help industries manage their wastes by repurposing material, which would otherwise be discarded.

15.1.2 Environmental Considerations of Ashes

The production of industrial waste (or by-products), such as CFA, RHA, and BA, brings both challenges and opportunities from an environmental perspective (Dunmade, 2012). These wastes, if not treated, can adversely impact the environment. However, appropriate handling and repurposing of these wastes can help protect the environment and promote sustainability (Huang et al., 2017). These ashes consist of fine particulates, which may cause airborne diseases, respiratory problems, and other health issues in humans and animals. Uncontrolled burning of coal and biomass releases particulates and greenhouse gases (GHGs) such as CO_2 and CH_4 into the atmosphere, causing air pollution (Teixeira et al., 2016). When these ashes are disposed of in landfills or ash ponds, there is a risk of releasing toxic heavy metals (arsenic, mercury, and lead) and other compounds (silica) into water bodies (Dunmade, 2012). These elements and compounds adversely affect the water quality and aquatic ecosystem. It also poses a risk to the communities using these water sources for their daily needs. The ill-treatment of these ashes can also cause soil contamination and reduce their fertility (Teixeira et al., 2019). Their heavy accumulation in the soil can alter the pH and cause nutrient imbalance, affecting plant growth and soil productivity. The heavy metals in them can act as carcinogens and reach the human body through food (Saha & Roychowdhury, 2023).

Apart from all the negative impacts discussed, various positives are associated with the ashes (CFA, RHA, and BA) if appropriately utilized. Developing a framework to channel the flow of ashes from their sources to the construction industries will help reduce the use of Portland cement and contain the emission of CO_2 to a large extent (Rathnayake et al., 2018). Compared to other industrial wastes, ashes are suitable soil stabilizers and improve the structural strength of the soil with minimal negative impact (Adiansyah, 2023). It helps reclaim degraded land with enhanced structural integrity. Using RHA and BA as ingredients of manure will reduce farmers' reliance on industrial fertilizers and reduce the emission of various GHGs and other pollutants (Fernando et al., 2021). It will also help maintain the soil's nutritional properties, improve productivity, and promote sustainable agriculture practices. Yet, there is a need for framework and evaluation tools that could assist in determining the effects of ashes on the environment and their socio-economic consequences (Rebitzer et al., 2004). It is highly imperative to set up such tools and frameworks in academia to ensure that the general people benefit from such interventions (Dunmade et al., 2019). One such tool to evaluate life cycle and encourage the sustainable utilization of ashes is the life cycle analysis (Tsiropoulos et al., 2015).

15.2 LIFE CYCLE ASSESSMENT AS A SUSTAINABILITY TOOL

LCA is an environment assessment and evaluation tool which helps in comparing impacts of a product throughout its life cycle, from manufacturing and distribution to end-of-life disposal (Terlouw et al., 2021). LCA is preferred over traditional environmental assessment methods, which typically consider only one stage of the product life cycle (either production or disposal). This approach ensures that the environmental benefits realized at one stage do not become detrimental impacts at another stage (Gabisa et al., 2019). Thus, due to delivering specific information about the environmental effects of varying stages of product life cycles, LCA facilitates appropriate choices for action (Dunmade et al., 2019). In a life cycle, it will not be difficult to know which stage or phase has a high effect on the environment. These hotspots are the areas that can easily be targeted to enhance environmental performance by the stakeholders. LCA has the advantage of enabling relative ease in comparison and evaluation of the environmental effects of different designs of an individual product (Jamora et al., 2023). Thus, it offers scientific conclusions on how to correctly determine the norms and procedures that should be adopted to minimize the impact of the product on the environment (Zulqar Nain & Kasilingam, 2023).

The findings of some research work on the environmental and economic effects of the reuse of FA and other industrial waste materials are presented in Table 15.1. Huang et al. (2017) applied municipal solid waste incinerator (MSWI) FA as an alkali reagent which undergoes the Waelz process at an electric arc furnace (EAF) ash recycling plant. It was followed by LCA and cost-benefit analysis of the application with other conventional methods like landfill disposal after stabilization/solidification, reuse in cement kiln, and use in brick aggregates. The study provides evidence that this approach is cheaper and has a lesser impact on the environment than these methods. Wang et al. (2017a, 2017b) used LCA in performing the life cycle environmental impact assessment of MFCF, magnetized fly-ash compound fertilizer. MFCF recommends streamlining for transport cleaning and balanced fertilizing in order to lessen environmental impacts. Jangde et al. (2024) explore the biotic and abiotic pathways for carbonization of FA to get nutrients plus status of compost. Some of the advantages of this process include the reduction of waste, reuse, and the recovery of resources, as well as the improvement of the soil.

In another study, Zhang et al. (2023) carried out LCA to compare the environmental impacts of the complete chain of brown coal bricks manufacturing with that of the Portland cement bricks. The outcomes of LCA reveal that the major environmental burdens of brown coal bricks are associated with raw material

TABLE 15.1
Studies on LCA

Author	Aims of study	Focus	Major findings	Limitations
Jangde et al. (2024)	Conversion of FA into compost within the circular economy framework.	Biological and chemical processes involved in converting FA into nutrient-rich compost.	FA can be turned into valuable compost, thereby reducing waste and enhancing soil fertility.	Effectiveness depends on FA composition and composting methods.
Wang et al. (2017a, 2017b)	Environmental impact of MFCF through life cycle.	Production and use of MFCF.	MFCF production leads to non-renewable resource depletion and eutrophication.	Based on hypothetical data and may not reflect actual conditions.
Huang et al. (2017)	Reuse of municipal solid waste incineration (MSWI) FA in the Waelz process using LCA and cost-benefit analysis.	Applications in various processes.	MSWI FA as an alkali reagent has low environmental impact and economic benefits.	Data specific to Taiwan and may not be applicable elsewhere.
McAvoy et al. (2021)	Integration of LCA and system dynamics (SD) for impact assessment.	Temporal dynamics in impact assessment.	Combining LCA and SD improves understanding of dynamic systems.	Results may not differ significantly in relatively static systems.
Zhang et al. (2023)	Environmental impacts of geopolymer concrete bricks from brown coal FA.	Comparison with Portland cement bricks.	Geopolymer bricks from brown CFA show lower impacts compared to Portland cement bricks.	
Li et al. (2019)	Review of LCA and LCCA studies on recycling solid wastes in highway pavement.	LCA and LCCA of recycled solid waste materials in highway pavement.	Recycling reduces energy use, emissions, and costs, but transport and leaching concerns exist.	Results may vary based on conditions and assumptions.
Habibi et al. (2023)	Reviews environmental impacts of concrete mixtures from an LCA perspective.	Impacts including emissions and toxicity.	Recycled aggregates and supplementary cementitious materials (SCMs) are not well-studied for durability.	Lack of durability studies and use stage assessments.
Navaratnam et al. (2023)	Environmental benefits of waste-based cementitious materials for wall plaster.	Performance of waste-based materials.	FA mortar has lower environmental impacts and can reduce cement use.	Production impacts of industrial waste materials.

consumption and utilization. However, the assessment reveals that there are large reductions in: ozone depletion, water depletion, and metal depletion for these bricks. These results are important as they show that the environmental advantages of brown coal bricks in the construction industry can challenge the current practices. In their review, Li et al. (2019) focused on LCA and Life Cycle Cost Analysis (LCCA) studies of recycling solid waste in highway pavement. According to the previous studies, the use of recycled materials such as asphalt pavement, steel slag, and CFA in highway construction is effective in minimizing energy consumption, greenhouse emission as well as the costs. Among the pioneering reviews on applying system thinking methods with LCA, McAvoy et al. (2021) explain that different forms of LCA can benefit from this approach. It specifically examines the two primary approaches to combining system dynamics models with LCA: incorporating life cycle inventory and impact assessment factors into a system dynamic model and integrating system dynamics model results into a life cycle assessment (LCA).

15.3 LCA FRAMEWORK

LCA framework is shown in Figure 15.1. There are four steps in developing the LCA framework: (1) defining the goal and scope, (2) conducting a life cycle inventory analysis, (3) carrying out a life cycle impact assessment, and (4) interpreting the findings (Dunmade et al., 2019).

15.3.1 Goal Definition

Our goal is to understand and compare the environmental impacts of CFA, RHA, and BA in the entire life cycle. This means starting from how they are extracted and processed, to how they're transported, used, and eventually disposed of or recycled. We want to measure and compare the environmental footprints of each type of ash from the moment they're created to their final disposal or reuse. Another objective is to see how the use of these ashes could be beneficial instead of traditional materials, providing valuable insights for managers, practitioners, and policymakers (Dunmade, 2012; Fernando et al., 2021; Gabisa et al., 2019).

15.3.2 Scope Definition

The scope outlines the system boundaries, functional units, and assumptions for the LCA (Dunmade, 2012; Fernando et al., 2021; Gabisa et al., 2019).

15.3.3 Functional Unit

The functional unit in LCA is the unit of analysis or measurement used to quantify a product's or process's environmental impact. It usually is expressed in mass (kilogram or tons), but many a time, mass-based units are unable to capture the nutritional values or quality of products (McAuliffe et al., 2023). Hence, alternate functional units need to be proposed, like the nutrition quality index (McAuliffe et al., 2023).

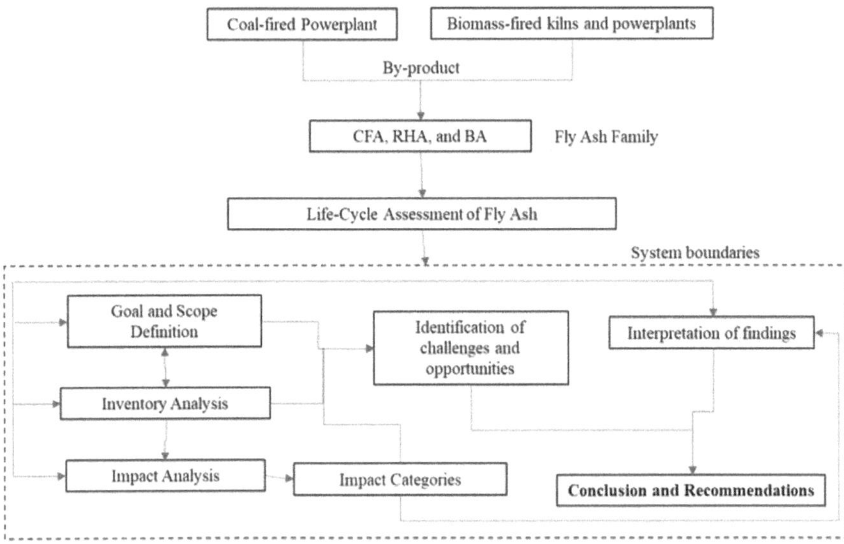

FIGURE 15.1 An illustration of conventional LCA process steps.

In the case of LCA of CFA, RHA, and BA, a mass-based functional unit is represented by one ton of ashes utilized or disposed of. The study assumes that the ashes are generated in typical industrial and agricultural setups. It also assumes that the standard practices are used in the transportation, processing, end-use, and disposal of these ashes.

15.3.4 System Boundaries

The system boundaries in LCA provide the scope of analysis within a defined process, activities, and life cycle stages of the product or process under consideration (Das et al., 2022). Setting system boundaries for LCA is essential as it enhances the clarity and understanding of the environmental impact associated with the product's entire life cycle, such as CFA, RHA, and BA, in the present case (Demirel et al., 2019). These boundaries allow for more accurate and holistic environmental assessment by covering all phases of ashes (starting from generation to end life). The system boundaries are discussed below.

15.3.4.1 Generation of Ash

The initial stage of ashes involves the combustion of raw materials including coals, rice husks, and sugarcane bagasse. The fine particles remain with the gases when these materials are burned up in coal and biomass-fired boilers to generate electricity and steam (Zhang et al., 2024). Such particles are trapped in pollution control devices such as electrostatic precipitators or fabric filters and generate CFA, RHA, BA consisting of high amounts of silica, alumina, and calcium. The combustion process effectiveness and its temperature affect the quality and quantity of ash from the combustion (Janga et al., 2024).

15.3.4.2 Collection and Processing of Ash

The second one is the collecting and processing stage, which is the most sensitive phase of the ash's life cycle (Demirel et al., 2019). This phase is very important because if ashes are not collected and processed they may pollute the environment (Habibi et al., 2023). This process also determines the quality and usefulness of generated ashes. Collection of ash can be done from the flue gas stream in the form of CFA, or directly from the combustion chamber in the form of RHA and BA using filtration equipment's such as electrostatic precipitators/bag houses (Li et al., 2019). The effectiveness of these filtration systems affects the environmental loading of this phase of the LCA (Gabisa et al., 2019). The collected ashes require grinding, sieving treatment, chemical treatment, and finally drying. They also bring improvements in the physical/chemical characterization of ashes for suitability in several uses (Deokar & Pathak, 2023). However, as mentioned earlier, processing may need some more energy and could therefore result in emissions in the atmosphere (Gabisa et al., 2019). In addition, efforts should be made to minimize resource usage and effectively manage the waste by-products which are always generated during this process in a way that it will have no adverse impact in the environment (Demirel et al., 2019).

15.3.4.3 Transportation of Ashes

This phase connects where the ashes are created to where they are used (Jamora et al., 2023). The phase is of paramount importance because it interacts with external factors like weather changes and natural disasters, which are fundamental to LCA (Jangde et al., 2024). It can be noted that the chosen mode of transportation plays a critical role in affecting the environment (Orozco et al., 2024). Other factors like fuel rate, emissions as well as chances of spillage during transit also play an essential role in determining environmental impact (Deokar & Pathak, 2023). However, there are several other factors that are very significant for conducting the assessment of this situation. Among them are the extent of separation between the aforementioned points and the manner of handling the ash during transportation (Das et al., 2022).

15.3.4.4 Utilization of Ash

The phase of the life cycle of ash brings a positive effect on the environment. Some of the uses of ash are in construction industry, agriculture, ceramics, refractory materials, and in plastics and rubber industry as a filler. All of these applications contribute to the reduction of pollution since they involve the use of waste material instead of fresh materials (Rathnayake et al., 2018).

15.3.4.5 End-of-Life Disposal or Recycling

This is the final stage within the system boundaries, which involves disposing or recycling unused ashes (Li et al., 2019). Suppose the ashes cannot be used in any applications. In that case, it is disposed of in landfills or recycled for use in new applications or recovery of valuable materials (e.g. extraction of silica from RHA) (Dunmade, 2012). Recycling extends the life cycle of the ash and reduces the need for virgin materials, which in turn reduces the environmental impact. The environmental impact of disposal depends on the following factors: the toxicity of the ash, the design of the landfill, and the possibility of remediation in the future. The environmental benefit of recycling depends on the efficiency of the recycling process, the quality of recycled products, and the market demand for recycled materials (Li et al., 2019). However, disposal and recycling have their own consistent problems that need to be addressed properly.

15.4 ENVIRONMENTAL IMPACT CATEGORIES

The LCA framework is classified according to the categories of environmental impacts (Adiansyah, 2023). Depending on the location of the impact in the product life cycle, the indicators are distinguished as midpoint and the endpoint impacts (Orozco et al., 2024). Depending on the severity of the impact, these are referred to as toxic and/or non-toxic indicators. While endpoint indicators offer a general view of the effects of environmental interferences, midpoint indicators help in evaluating the situation in protected areas (Thorne et al., 2024). On the other hand, toxic indicators take into account problematic impacts of the product or process in the spatial and temporal context of the assessment, in contrast to non-toxic indicators, which address other environmental hotspots (Orozco et al., 2024). In this context, it should be noted that the use of midpoint and endpoint indicators in LCA varies depending on the goal of the assessment being conducted. As in the case of CFA, RHA, and BA, these two categories for indicators can be used for the assessment. In the next subsections, the classification of environmental impacts for the CFA, RHA, and BA in terms of sustainable development according to these subcategories is described.

15.5 MIDPOINT VS ENDPOINT INDICATORS

Some of the midpoint indicators are global warming potential (GWP), acidification potential, eutrophication potential, human toxicity potential, ecotoxicity potential, resource depletion, energy, and land use which describe the impact of CFA, RHA, and BA on the environment (Demirel et al., 2019). The emission of GHGs is relatively higher with the CFA in comparison with the RHA and BA. Due to high content of NO_2 and SO_x, CFA has a high acidification potential compared to RHA and BA. Since coal (the source of CFA) has a high nutrient content compared to the rice husk and sugarcane bagasse, the eutrophic potential of CFA is higher than that of the RHA and BA (Thorne et al., 2024). Since coal is mined from the earth while sugarcane and rice husk are agriculture by-products, the toxic metal (i.e. arsenic, mercury, and lead) content is high in coal compared to the other two, and hence the human toxic potential is high for CFA compared to the RHA and BA (have silica as the main component). Coal is a non-renewable source while rice husk and bagasse are agriculture by-products, the resource depletion potential of CFA is very high compared to RHA and BA (Petlickaitė et al., 2024). The production of CFA is highly energy-intensive and requires a huge land for power generation; the same is the case with the bagasse, and hence the energy and land use potential is high with CFA and BA while low with the RHA (Zulqar Nain & Kasilingam, 2023).

Endpoint indicators present the aggregated effect of midpoint impact on human health, ecosystem quality, and resource availability. These indicators provide a broader perspective on the environmental consequences of using CFA, RHA, and

BA. It incorporates the cumulative effect of exposure to toxic elements, respiratory issues from particulate matter, and potential long-term health risks like cancer. Using CFA in construction and other applications can mitigate some health risks if it replaces more harmful materials. RHA and BA generally have lower impacts on human health, but respiratory issues from dust inhalation and potential exposure to trace toxic substances are still concerns. The impact of CFA on the ecosystem is high compared to RHA and BA due to the presence of heavy metals in CFA, which can contaminate the water bodies and soil, increase acidification, and disrupt local flora and fauna. The improper disposal or excessive use of RHA and BA can still disrupt the soil ecosystem and water quality. Some of the drawbacks associated with CFA are as follows: non-renewable resources such as coal have been depleted, and the excessive use of this product may cause resource shortage in the future. There is no such concern with relation to the production of RHA and BA as both of them are obtained from renewable resources.

15.6 TOXIC VS NON-TOXIC INDICATOR

Environmental impacts can also be classified with reference to the toxicity level (Guinee, 2002). These indicators assist in expressing the degree of hazards that these materials pose to human health, ecosystems, and environment (Pennington et al., 2002). For LCA studies, there are toxic and non-toxic impact categories that give a more extensive view of the environmental impacts of CFA, RHA, and BA. Toxic indicators such as human toxicity, ecotoxicity, and carcinogenicity assess the potential harm that materials can cause to humans, ecosystems, and the likelihood of inducing cancer, respectively (Sala et al., 2022). They are hazardous to health and the environment and can alert the dangers of CFA especially because of the inclusion of heavy metals and other toxic compositions (Zhang et al., 2023).

More desirable, non-toxic markers including GWP, resource consumption, energy use, and land use offer broader understanding of environmental impacts of this ash. They give information about the sustainable aspects of utilization so that the uses and drawbacks related with the life cycle of ash can be comprehended. In many cases, RHA and BA obtained from renewable biomass are considered less hazardous to the global environment than CFA due to their less toxic effects. Through evaluation of both toxic and non-toxic markers, both the stakeholders can come to the right decision concerning the utilization and disposal of CFA, RHA, and BA by balancing the need to avoid adverse environmental and health impacts while practicing sustainable management (Pennington et al., 2002).

15.7 LCA ANALYSIS OF FLY ASH USAGE FOR DIFFERENT APPLICATIONS

CFA, RHA, and BA are industrial by-products obtained by coal, rice husks, and sugarcane bagasse combustion. These materials have gained attention for their potential reuse in various applications. Through LCA, we can evaluate the environmental impact of these materials across different uses. By analyzing factors like resource efficiency, energy consumption, emissions, and waste management, LCA offers a comprehensive understanding of the sustainability of these materials in diverse applications.

15.7.1 Construction Material

The construction industry is a major consumer of resources and a significant contributor to environmental degradation. Thus, the practitioners have increasing interest in utilizing industrial by-products like CFA, RHA, and BA as alternative construction materials (Zhang et al., 2023). While all three materials can reduce the environmental footprint of construction, their sustainability depends on sourcing, processing, transportation, and end-of-life considerations (Blissett & Rowson, 2012). CFA-enhanced concrete reduces the overall carbon footprint and reduces energy demand in its production. It also helps divert the CFA from landfills (Demirel et al., 2019). Using CFA as construction material in concrete alleviates the environmental burden associated with its disposal (Navaratnam et al., 2023). LCA studies indicate a significant reduction (up to 30%) in CO_2 emissions from concrete production (Blissett & Rowson, 2012). However, LCA also suggests the management of emissions during the use phase or demolition of the structure to mitigate the risk of leaching of toxic elements in soil and water bodies (Petlickaitė et al., 2024).

Using RHA as a partial substitute for cement reduces the demand for virgin raw materials such as limestone and clay (Zulqar Nain & Kasilingam, 2023). When used as a replacement for cement, it can reduce CO_2 emissions associated with concrete production. RHA has proved to improve the durability of concrete, particularly in aggressive environments. The use of RHA in construction also presents some environmental challenges (Fernando et al., 2021). The production of RHA of suitable quality for construction applications is an energy-intensive process (Guinee, 2002). The properties of RHA can vary significantly depending on the process and type of husk used, which can affect the performance of concrete (Demirel et al., 2019). Therefore, the LCA of RHA needs to consider the energy consumption and emissions associated with processing, as well as the need for additional processing or quality control measures (Das et al., 2022).

The use of BA in construction helps to promote the circular economy where an agricultural by-product has been converted into a resource of high value (Zulqar Nain & Kasilingam, 2023). It minimizes the reliance on conventional resources like cement in this sector (Das et al., 2022). BA has pozzolanic properties. Therefore, it can react with calcium hydroxide in concrete to form additional cementitious compounds that improve the strength and durability of concrete, resulting in more durable structures and reduced environmental impacts over time (Orozco et al., 2024). However, the environmental concerns associated with BA are the same as those of RHA (Fernando et al., 2021).

15.7.2 As Embankment Material

One of the advantages of CFA as an embankment material is that it replaces natural soils and thus the degradation of natural soils (Habibi et al., 2023). This is helpful to prevent habitat destruction, erosion, and biodiversity loss (Demirel et al., 2019). It also provides a good solution for the disposal of the by-product. This reduces the waste that is sent to landfills, as this runs counter to traditional circular economy models. As CFA contains a high percentage of carbon, the introduction of conventional fill material reduces the overall carbon content by eliminating the extraction, processing, and transportation of the conventional materials (Navaratnam et al., 2023). Nevertheless, the ability of CFA to be used as an embankment material creates problems such as leaching of trace toxic chemicals into the soil and water. The LCA of CFA must therefore take into account the risk factor of contamination, especially in regions with high rainfall or high groundwater levels (Fernando et al., 2021). For these reasons, adequate containment measures as well as lifetime assessment and aftercare must be implemented. When RHA and BA are used as backfill materials, they contribute to the preservation of the environment and can therefore be used as a raw material for disposal with additional benefits for sustainable agricultural practices and waste management systems. LCA studies indicate that incorporating RHA and BA into embankments can significantly lower the project's carbon footprint, particularly when RHA is sourced locally (Thorne et al., 2024). However, the LCA must also account for the energy consumption and emissions related to the production and additional processing required to ensure consistent material properties suitable for embankment use (Jamora et al., 2023).

15.7.3 Landfilling Material

The incorporation of CFA, RHA, and BA as industrial wastes in landfilling is increasingly being adopted in environmental management and in waste disposal strategies (Li et al., 2019). These materials present possibilities of developing progressive waste management strategies. LCA gives a systematic approach to the assessment of environmental impacts of using CFA, RHA, and BA as landfill materials (Zulqar Nain & Kasilingam, 2023). It provides a rational understanding of the sustainability of the material under study and helps in deciding the optimal utilization of the material under consideration (Demirel et al., 2019). CFA increases landfill compaction and stabilization, which saves space for waste disposal, which is a critical issue in areas where landfill space is scarce and expensive to acquire. However, LCA must consider the risks of leachate formation, especially in landfills located in areas with high rainfall or fluctuating groundwater levels (Fernando et al., 2021). All three ashes offer resource efficiency benefits by reducing the need for virgin materials in landfilling (Das et al., 2022). However, RHA and BA stand out for their sustainability advantages, as they promote the use of agricultural residues and help in reducing waste (Zulqar Nain & Kasilingam, 2023).

15.7.4 Adsorbent

Employment of the industrial and agricultural by-products in the removal of water and air pollutants have received immense interest in the industry (Guinee, 2002). It important to mention here that out of all the by-products, CFA, RHA, and BA have enormous potential to be utilized in the adsorption process as they provide a cost-efficient solution than the conventional adsorbent such as activated carbon (Zulqar Nain & Kasilingam, 2023). These by-products also consume less energy to work as adsorbents as compared to the activated carbon. RHA is renewable and has a high silica content that is important for adsorption purposes (Demirel et al., 2019). In adsorption processes, BA can effectively minimize the emission of GHGs during the purification of industrial emissions or in wastewater treatment (Blissett & Rowson, 2012). However, the LCA of CFA reveals that log deposited to landfills has relatively higher toxicity and leachability risks attributed to the disposal of heavy metals. However, RHA and BA obtained from biomass are comparatively non-toxic and hence safe for the environment and human beings when applied in the right manner (Zulqar Nain & Kasilingam, 2023). However, the LCA should consider the risks of leaching and contamination with all three materials, especially in water treatment.

15.7.5 Other Applications

Apart from the aforementioned uses, CFA, RHA, and BA have other uses in industries such as in ceramics as well as refractories (Rathnayake et al., 2018). These by-products have specific characteristics which make it possible to employ them in ceramic and refractory materials. The incorporation of such ashes in ceramics and refractories aids in the valorization of these industrial by-products (Demirel et al., 2019) thus enabling a partial or total substitution of raw materials like clay and feldspar. The application of these ashes in ceramics also leads to lowering of firing temperature which is commonly used in the manufacturing process (Khalil et al., 2025). However, as we've seen, there are advantages in using ash in ceramics and refractory materials, but the practice is not without complications. There are some gaps for LCA to consider; first, environmental and health risks and more specifically the risk related to contaminated heavy metals in the products coming from CFA. This also influences the quality and performance characteristics of ceramics refractories and other uses of these ashes, which may require specific quality checks and tests.

15.8 CONCLUSIONS AND THE WAY FORWARD

The LCA of CFA, RHA, and BA in different industrial applications reveals the importance of using these products in addition to showing the various strengths and weaknesses of their utilization. Although all three materials provide the potential for decreasing the environmental impacts of industrial processes, their sustainability depends on several factors, including purchasing, processing, transportation, and recycling or disposal. Optimized management of these by-products can thus translate to improved conservation of resources and energy and reduced wastage. However, issues like risks associated with having high levels of heavy metals, variations in the properties of waste materials, and the risk of emissions are apparent. Overall, with the help of LCA, stakeholders can effectively avoid or manage risks related to these environmental benefits and therefore rationally and effectively use CFA, RHA, and BA in various industrial conditions. Additionally, it supports the circular economy and contributes to achieving long-term environmental and socio-economic goals.

REFERENCES

Adiansyah, J. S. (2023). Environmental performance of three coal ash management strategies in a copper mine site using endpoint impact category. *Chemical Engineering Transactions*, *106*, 289–294. https://doi.org/10.3303/CET23106049

Blissett, R. S., & Rowson, N. A. (2012). A review of the multi-component utilisation of coal fly ash. *Fuel*, *97*, 1–23. https://doi.org/10.1016/j.fuel.2012.03.024

Curpen, S., Teutsch, N., Kovler, K., & Spatari, S. (2023). Evaluating life cycle environmental impacts of coal fly ash utilization in embankment versus sand and landfilling. *Journal of Cleaner Production*, *385*, 135402. https://doi.org/10.1016/j.jclepro.2022.135402

Das, P., Cheela, V. R. S., Mistri, A., Chakraborty, S., Dubey, B., & Barai, S. V. (2022). Performance assessment and life cycle analysis of concrete containing ferrochrome slag and fly ash as replacement materials – A circular approach. *Construction and Building Materials*, *347*, 128609. https://doi.org/10.1016/j.conbuildmat.2022.128609

Demirel, S., Öz, H. Ö., Güneş, M., Çiner, F., & Adın, S. (2019). Life-cycle assessment (LCA) aspects and strength characteristics of self-compacting mortars (SCMs) incorporating fly ash and waste glass PET. *The International Journal of Life Cycle Assessment*, *24*(6), 1139–1153. https://doi.org/10.1007/s11367-018-1562-5

Deokar, S. K., & Pathak, P. D. (2023). Utilization of rice and sugarcane ashes in wastewater treatment: A case study for pesticide removal from aqueous solution. In P. D. Pathak, & S. A. Mandavgane (Eds.), *Biorefinery: A sustainable approach for the production of biomaterials, biochemicals and biofuels* (pp. 233–246). Springer Nature. https://doi.org/10.1007/978-981-19-7481-6_8

Dunmade, I. (2012). Recycle or dispose off? Lifecycle environmental sustainability assessment of paint recycling process. *Resources and Environment*, *2*(6), 291–296.

Dunmade, I., Madushele, N., Adedeji, P. A., & Akinlabi, E. T. (2019). A streamlined life cycle assessment of a coal-fired power plant: The South African case study. *Environmental Science and Pollution Research*, *26*(18), 18484–18492. https://doi.org/10.1007/s11356-019-05227-6

Fernando, S., Gunasekara, C., Law, D. W., Nasvi, M. C. M., Setunge, S., & Dissanayake, R. (2021). Life cycle assessment and cost analysis of fly ash-rice husk ash blended alkali-activated concrete. *Journal of Environmental Management*, 295, 113140. https://doi.org/10.1016/j.jenvman.2021.113140

Gabisa, E. W., Bessou, C., & Gheewala, S. H. (2019). Life cycle environmental performance and energy balance of ethanol production based on sugarcane molasses in Ethiopia. *Journal of Cleaner Production*, *234*, 43–53. https://doi.org/10.1016/j.jclepro.2019.06.199

Guinee, J. B. (2002). Handbook on life cycle assessment operational guide to the ISO standards. *The International Journal of Life Cycle Assessment*, *7*(5), 311–313. https://doi.org/10.1007/BF02978897

Habibi, A., Tavakoli, H., Esmaeili, A., & Golzary, A. (2023). Comparative life cycle assessment (LCA) of concrete mixtures: A critical review. *European Journal of Environmental and Civil Engineering*, *27*(3), 1285–1303. https://doi.org/10.1080/19648189.2022.2078885

Huang, T. Y., Chiueh, P. T., & Lo, S. L. (2017). Life-cycle environmental and cost impacts of reusing fly ash. *Resources, Conservation and Recycling*, *123*, 255–260. https://doi.org/10.1016/j.resconrec.2016.07.001

Itam, Z., Muhammad Mubin Shaikh Ahmad Fadzil, S., Syamsir, A., Zainoodin, M., & Beddu, S. (2022). Application of sugarcane ash as an additional cementitious material in concrete. *Materials Today: Proceedings*, *66*, 2824–2829. https://doi.org/10.1016/j.matpr.2022.06.523

Jamora, J. B., Go, A. W., Gudia, S. E. L., Giduquio, M. B., & Loretero, M. E. (2023). Evaluating the use of rice residue ash in cement-based industries in the Philippines – Greenhouse gas reduction, transportation, and cost assessment. *Journal of Cleaner Production*, *398*, 136623. https://doi.org/10.1016/j.jclepro.2023.136623

Janga, S., Raut, A. N., & Murmu, A. L. (2024). Assessment of thermal and mechanical pro perties of fly ash based geopolymer blocks with a sustainability perspective using multi-criteria decision-making approach. *Journal of Building Engineering*, *88*, 109261. https://doi.org/10.1016/j.jobe.2024.109261

Jangde, V., Choubey, S., & Markam, D. J. (2024). Sustainable circular economy: Transforming fly ash into valuable compost. *Journal of Environmental Nanotechnology*, *13*(2), 470–476. https://doi.org/10.13074/jent.2024.06.242606

Kerdsuwan, S., & Laohalidanond, K. (2022). Biomass conversion to power in southeast asian countries: Current situation and perspectives. *Green Energy and Technology*, 523–553. Scopus. https://doi.org/10.1007/978-981-16-2648-7_22

Khalil, A. K. A., Bouaziz, I., Jaber, L., Abushawish, A., Almanassra, I. W., Ali Abdelkareem, M., & Ali Atieh, M. (2025). Fly ash as zero cost material for water treatment applications: A state of the art review. *Separation and Purification Technology*, *354*, 129104. https://doi.org/10.1016/j.seppur.2024.129104

Li, J., Xiao, F., Zhang, L., & Amirkhanian, S. N. (2019). Life cycle assessment and life cycle cost analysis of recycled solid waste materials in highway pavement: A review. *Journal of Cleaner Production*, *233*, 1182–1206. https://doi.org/10.1016/j.jclepro.2019.06.061

McAuliffe, G. A., Takahashi, T., Beal, T., Huppertz, T., Leroy, F., Buttriss, J., Collins, A. L., Drewnowski, A., McLaren, S. J., Ortenzi, F., van der Pols, J. C., van Vliet, S., & Lee, M. R. F. (2023). Protein quality as a complementary functional unit in life cycle assessment (LCA). *The International Journal of Life Cycle Assessment*, *28*(2), 146–155. https://doi.org/10.1007/s11367-022-02123-z

McAvoy, S., Grant, T., Smith, C., & Bontinck, P. (2021). Combining Life Cycle Assessment and System Dynamics to improve impact assessment: A systematic review. *Journal of Cleaner Production*, *315*, 128060. https://doi.org/10.1016/j.jclepro.2021.128060

Navaratnam, S., Tushar, Q., Jahan, I., & Zhang, G. (2023). Environmental sustainability of industrial waste-based cementitious materials: A review, experimental investigation and life-cycle assessment. *Sustainability*, *15*(3), 1873. https://doi.org/10.3390/su15031873

Orozco, C., Babel, S., Tangtermsirikul, S., & Sugiyama, T. (2024). Comparison of environmental impacts of fly ash and slag as cement replacement materials for mass concrete and the impact of transportation. *Sustainable Materials and Technologies*, *39*, e00796. https://doi.org/10.1016/j.susmat.2023.e00796

Pennington, D., Crettaz, P., Tauxe, A., Rhomberg, L., Brand, K., & Jolliet, O. (2002). Assessing human health response in life cycle assessment using ED10s and DALYs: Part 2—Noncancer effects. *Risk Analysis*, *22*(5), 947–963. https://doi.org/10.1111/1539-6924.00263

Petlickaitė, R., Jasinskas, A., Venslauskas, K., Navickas, K., Praspaliauskas, M., & Lemanas, E. (2024). Evaluation of multi-crop biofuel pellet properties and the life cycle assessment. *Agriculture*, *14*(7), Article 7. https://doi.org/10.3390/agriculture14071162

Rathnayake, M., Julnipitawong, P., Tangtermsirikul, S., & Toochinda, P. (2018). Utilization of coal fly ash and bottom ash as solid sorbents for sulfur dioxide reduction from coal fired power plant: Life cycle assessment and applications. *Journal of Cleaner Production*, *202*, 934–945. https://doi.org/10.1016/j.jclepro.2018.08.204

Rebitzer, G., Ekvall, T., Frischknecht, R., Hunkeler, D., Norris, G., Rydberg, T., Schmidt, W.-P., Suh, S., Weidema, B. P., & Pennington, D. W. (2004). Life cycle assessment: Part 1: Framework, goal and scope definition, inventory analysis, and applications. *Environment International*, *30*(5), 701–720. https://doi.org/10.1016/j.envint.2003.11.005

Saha, D., & Roychowdhury, T. (2023). Characterisation of coal and its combustion ash: Recognition of environmental impact and remediation. *Environmental Science and Pollution Research*, *30*(13), 37310–37320. https://doi.org/10.1007/s11356-022-24864-y

Sala, S., Biganzoli, F., Mengual, E. S., & Saouter, E. (2022). Toxicity impacts in the environmental footprint method: Calculation principles. *The International Journal of Life Cycle Assessment*, *27*(4), 587–602. https://doi.org/10.1007/s11367-022-02033-0

Teixeira, E. R., Mateus, R., Camões, A., & Branco, F. G. (2019). Quality and durability properties and life-cycle assessment of high volume biomass fly ash mortar. *Construction and Building Materials*, *197*, 195–207. https://doi.org/10.1016/j.conbuildmat.2018.11.173

Teixeira, E. R., Mateus, R., Camões, A. F., Bragança, L., & Branco, F. G. (2016). Comparative environmental life-cycle analysis of concretes using biomass and coal fly ashes as partial cement replacement material. *Journal of Cleaner Production*, *112*, 2221–2230. https://doi.org/10.1016/j.jclepro.2015.09.124

Terlouw, T., Treyer, K., Bauer, C., & Mazzotti, M. (2021). Life cycle assessment of direct air carbon capture and storage with low-carbon energy sources. *Environmental Science & Technology*, *55*(16), 11397–11411. https://doi.org/10.1021/acs.est.1c03263

Thorne, J., Bompa, D. V., Funari, M. F., & Garcia-Troncoso, N. (2024). Environmental impact evaluation of low-carbon concrete incorporating fly ash and limestone. *Cleaner Materials*, *12*, 100242. https://doi.org/10.1016/j.clema.2024.100242

Tsiropoulos, I., Faaij, A. P. C., Lundquist, L., Schenker, U., Briois, J. F., & Patel, M. K. (2015). Life cycle impact assessment of bio-based plastics from sugarcane ethanol. *Journal of Cleaner Production*, *90*, 114–127. https://doi.org/10.1016/j.jclepro.2014.11.071

Wang, J., Wang, Y., Sun, Y., Tingley, D. D., & Zhang, Y. (2017a). Life cycle sustainability assessment of fly ash concrete structures. *Renewable and Sustainable Energy Reviews*, *80*, 1162–1174. https://doi.org/10.1016/j.rser.2017.05.232

Wang, P., Wang, J., Qin, Q., & Wang, H. (2017b). Life cycle assessment of magnetized fly-ash compound fertilizer production: A case study in China. *Renewable and Sustainable Energy Reviews*, *73*, 706–713. https://doi.org/10.1016/j.rser.2017.02.005

Zhang, J., Fernando, S., Law, D. W., Gunasekara, C., Setunge, S., Sandanayake, M., & Zhang, G. (2023). Life cycle assessment for geopolymer concrete bricks using brown coal fly ash. *Sustainability*, *15*(9), 7718. https://doi.org/10.3390/su15097718

Zhang, T., Wen, Z., Tan, Y., Shi, X., Sun, Y., & Ekins, P. (2024). Advancing circular economy of pallets: A comprehensive evaluation framework. *Resources, Conservation and Recycling*, *211*, 107874. https://doi.org/10.1016/j.resconrec.2024.107874

Zulqar Nain, M., & Kasilingam, S. (2023). Influence of rice husk ash and bagasse ash on durability of concrete. *Materials Today: Proceedings*, *93*, 71–78. https://doi.org/10.1016/j.matpr.2023.06.464

16 Circular Economy Capabilities Toward Fly-Ash Management

Siddhartha Pandey, Tuhin Sengupta, Gaurav Kabra, Swati Verma and Tarun Kumar

16.1 INTRODUCTION

The increasing environmental challenges posed by industrial activities, especially coal-based power generation, have spurred the need for sustainable waste management solutions. One of the most significant by-products of this industry is fly ash, a residue generated from the combustion of coal in thermal power plants. Fly ash management has become a critical issue due to the large quantities produced globally and its potential environmental hazards if improperly disposed of. However, as industries and governments worldwide shift toward circular economy principles, fly ash is being reimagined as a valuable resource rather than waste. This paradigm shift has opened new avenues for sustainable infrastructure, where fly ash is repurposed in construction, water management, and soil reclamation. Traditional waste management approaches, such as landfilling, have proven inadequate for dealing with the massive quantities of fly ash generated annually. In countries like India, China, and the United States, which are among the world's largest producers of fly ash, the environmental and health risks associated with fly ash disposal are immense. These risks include air and water pollution from leached heavy metals, soil contamination, and respiratory diseases from airborne particulate matter (Nguyen et al., 2023a). Thus, there is an urgent need to explore alternative methods for managing fly ash, focusing on transforming it from an environmental liability into an asset for sustainable infrastructure. Fly ash offers several advantages, including its pozzolanic properties, which make it suitable for use in concrete, road construction, bricks, and other building materials. Its reuse not only reduces the environmental impact of traditional construction materials like cement but also contributes to waste reduction by diverting fly ash from landfills. This dual benefit of environmental protection and resource conservation has positioned fly ash as a key component in the development of green infrastructure.

However, despite its potential, fly ash utilization faces several barriers, including regulatory challenges, technological limitations, and market acceptance (Seidler and Malloy, 2020). The need to address these barriers through a comprehensive framework of regulatory support, technological innovation, and market development is a key motivation for this chapter. Additionally, this chapter

DOI: 10.1201/9781003498124-16

aims to contribute to the growing body of literature on the circular economy by highlighting the role of public-private partnerships (PPPs), sustainable transportation, and innovative applications of fly ash in green infrastructure. While there has been significant progress in fly ash management and its integration into various sectors, critical gaps remain in the widespread adoption of sustainable practices for fly ash utilization. One of the primary gaps is the lack of a unified global regulatory framework that governs fly ash disposal and reuse. Different countries have varying standards for fly ash quality, storage, and transportation, creating inconsistencies in its application across industries and geographies (Ghodeswar and Oliver, 2022). For instance, while the European Union (EU) has stringent guidelines for fly ash reuse in construction, many developing countries still rely heavily on landfilling due to a lack of infrastructure and policy enforcement. Another significant gap lies in the technological capabilities required to fully unlock the potential of fly ash. While technologies exist to process fly ash for use in construction materials, soil amendment, and water filtration, the variability in the composition of fly ash due to differences in coal types, combustion processes, and emission controls poses challenges for standardization (Wang et al., 2020). Additionally, the high carbon content in certain fly ash types can limit their use in specific applications, necessitating further advancements in processing technologies. Furthermore, the market demand for fly ash-based products remains limited in many regions, partly due to a lack of awareness and resistance from traditional industries. Without sufficient market pull, the economic viability of large-scale fly ash utilization remains in question. This gap highlights the need for comprehensive strategies to develop new markets for fly ash products and raise awareness about their environmental and economic benefits.

16.2 FRAMEWORK FOR FLY ASH MANAGEMENT IN THE CIRCULAR ECONOMY

Circular economy offers an effective framework for tackling environmental challenges by encouraging resource efficiency, waste minimization, and sustainability. In this context, the sustainable management of fly ash, a by-product of coal combustion in power plants, is essential. Fly ash, when improperly disposed of, poses significant environmental and health risks, such as air and water pollution. However, when managed effectively within a circular economy framework, fly ash can be transformed into a valuable resource, particularly for industries like construction, agriculture, and environmental remediation (Ondrasek et al., 2021). The framework developed in the diagram illustrates a strategic approach for fly ash management within the circular economy. It highlights six key strategies that work in synergy to create sustainable systems for the use, regulation, transportation, and market development of fly ash. These strategies include factors as per the Figure 16.1:

Each of these strategies addresses critical aspects of fly ash management and presents opportunities for sustainable development. Below is a detailed explanation of each component and its significance in the overall framework.

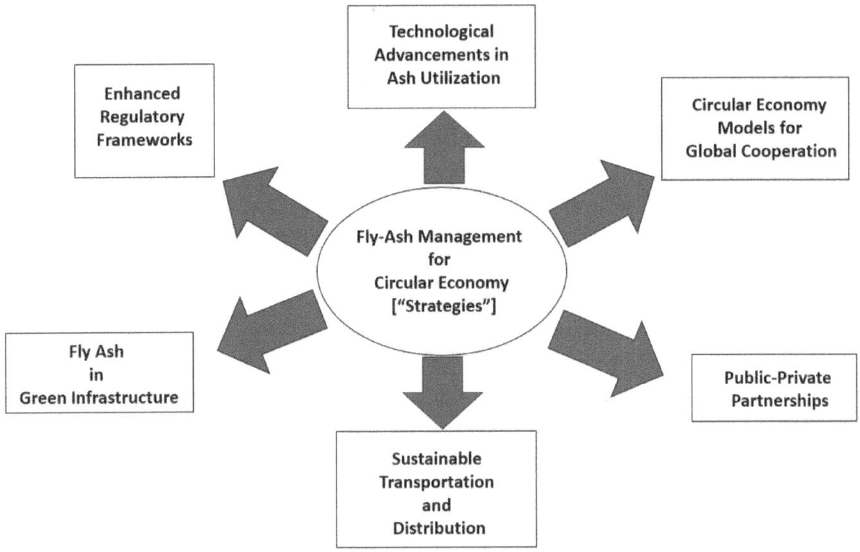

FIGURE 16.1 Strategic framework on fly-ash management for circular economy.

16.2.1 Enhanced Regulatory Frameworks for Ash Management in a Circular Economy

As environmental concerns related to industrial by-products such as coal fly ash (CFA) continue to grow, enhanced regulatory frameworks have emerged as critical components of sustainable ash management (Darmansyah et al., 2023). These frameworks serve as the backbone of policies that guide the responsible disposal, utilization, and management of fly ash, ensuring that its environmental footprint is minimized while promoting its role in a circular economy. The increasing production of CFA, especially in countries dependent on coal-based power generation like India, China, and the United States, has spurred the development of policies to encourage the reuse of ash in industrial applications and construction.

16.2.1.1 The Importance of Enhanced Regulatory Frameworks

Regulatory frameworks are essential in ensuring the effective management of CFA and other types of industrial waste. The increased demand for electricity and energy, particularly from coal-based power plants, has resulted in significant ash production. In India, for example, the production of CFA increased from 40 million tons in the early 1990s to over 200 million tons in 2018, with expectations to reach 275 million tons per year by 2032 (Ram and Mohanty, 2022). This rapid growth in ash production necessitates stringent regulations that mandate its safe disposal and utilization to prevent environmental degradation. At the heart of these regulatory frameworks is the need to protect air, water, and soil from contamination by heavy metals and other toxic substances found in CFA. Fly ash contains oxides of silicon, iron, magnesium, calcium, and trace elements that pose significant risks to ecosystems and public health if improperly managed. In addition to addressing environmental risks, regulations also aim to promote the circular economy by encouraging the reuse of fly ash in applications such as road construction, cement production, and mine backfilling (Chen et al., 2024).

16.2.1.2 International Approaches to Ash Management

Countries around the world have taken different approaches to regulating fly ash disposal and utilization. In the United States, the Environmental Protection Agency (EPA) has classified coal ash as a non-hazardous waste under the Resource Conservation and Recovery Act (RCRA), allowing for its reuse in certain industrial applications (Locke et al., 2020). The EU, on the other hand, has adopted the Waste Framework Directive, which promotes the minimization of industrial waste and encourages recycling and reuse. The circular economy model promoted by the EU plays a pivotal role in the management of CFA, ensuring that the ash is repurposed in a manner that reduces environmental impact while fostering economic benefits. China, one of the largest producers of CFA, has implemented the "Circular Economy Promotion Law," which focuses on the recycling of industrial by-products like fly ash to reduce waste and optimize resources (Labianca et al., 2024). This law not only incentivizes the recycling of CFA but also mandates its use in construction materials and agriculture. Other countries, such as Japan and Malaysia, have enacted similar regulations to manage the environmental impact of ash while encouraging its reuse in various sectors. In India, the Ministry of Environment, Forest and Climate Change (MoEF&CC) has been at the forefront of fly ash regulation, issuing several notifications aimed at increasing the utilization of fly ash (Rajasugunasekar and Devi, 2022). The most recent **Fly Ash Notification of 2022** mandates that all coal or lignite-based thermal power plants ensure 100% utilization of the ash produced within three years, with penalties for non-compliance based on the "Polluter Pays Principle" (PPP). These regulations have been critical in driving the country's efforts toward achieving sustainable ash management.

16.2.1.3 Key Challenges in Regulatory Implementation

Despite the progress made through regulatory frameworks, there remain significant challenges in implementing these policies effectively. One of the major obstacles is the lack of infrastructure and technology for the efficient handling, transportation, and processing of fly ash. In developing countries, the absence of advanced technologies for ash utilization often leads to landfilling or improper disposal, resulting in air and water pollution (Darmansyah et al., 2023). Moreover, even in countries with stringent regulations, enforcement remains a critical issue. Power plants often fail to meet the mandated utilization targets due to limited market demand for fly ash-based products or logistical barriers. Another challenge is the lack of standardization in global regulatory approaches. While developed countries have made significant strides in ash management, many developing nations still struggle to adopt comprehensive policies. The disparity in regulatory frameworks across countries creates challenges for international cooperation in fly ash management, limiting the potential for cross-border collaborations (Awino and Apitz, 2024).

16.2.2 Technological Advancements in Ash Utilization for a Circular Economy

The rise of coal and biomass-based power generation has led to the accumulation of vast amounts of fly ash. This by-product of energy production, composed of fine particles released during combustion, presents both an environmental challenge and an opportunity. Fly ash, when improperly managed, can cause soil, air, and water pollution due to its heavy metal content and fine particulate matter. However, technological advancements are offering promising solutions to transform this waste into a valuable resource, aligning with circular economy principles.

16.2.2.1 The Role of Technological Advancements in Ash Utilization

Technological advancements have been critical in overcoming the environmental challenges posed by fly ash. As power plants and industries seek to reduce their environmental footprints, innovative technologies have paved the way for the sustainable management and reuse of fly ash. These advancements allow for its repurposing in sectors such as construction, agriculture, and manufacturing, reducing the need for virgin materials while minimizing the waste sent to landfills (Oyejobi et al., 2024). Fly ash is predominantly composed of silica, alumina, and calcium, which gives it properties suitable for use in construction materials, such as cement, concrete, and bricks. However, it also contains trace amounts of hazardous elements like arsenic, mercury, and lead, which present health risks if not properly managed. Technological innovations focus on processing fly ash to minimize its toxicity while maximizing its utility in industrial applications. The integration of these technologies is essential for achieving the goals of a circular economy by turning waste into a resource.

16.2.2.2 Innovations in Fly Ash-Based Construction Materials

One of the most significant advancements in ash utilization technology is the development of fly ash-based construction materials. Fly ash has long been used as a partial replacement for Portland cement in concrete production. This practice not only diverts fly ash from landfills but also reduces the carbon footprint of concrete, as cement production is one of the largest industrial sources of CO_2 emissions (Orozco et al., 2024). By replacing a portion of the cement with fly ash, the overall emissions associated with construction are reduced. Recent innovations in fly ash technology have led to the development of geopolymers, a type of eco-friendly building material that is formed by activating fly ash with alkali solutions. Geopolymers exhibit high compressive strength, low shrinkage, and excellent thermal and chemical resistance, making them ideal for use in infrastructure projects (Naik et al., 2023). These materials have a lower carbon footprint compared to traditional concrete and are gaining attention as a sustainable alternative in the construction industry. Furthermore, advanced processing techniques now allow for the production of fly ash bricks, which have several environmental advantages. Fly ash bricks are lighter than traditional clay bricks, require less energy to produce, and reduce the demand for virgin clay. The production of fly ash bricks also minimizes greenhouse gas emissions, as it eliminates the need for the high-temperature firing process used in traditional brick manufacturing. These advancements in construction materials demonstrate how fly ash can be transformed into high-performance, eco-friendly products that contribute to sustainable development and reduce environmental impacts.

16.2.2.3 Fly Ash in Agriculture and Soil Improvement

Technological advancements have also expanded the use of fly ash in the agricultural sector. Fly ash, particularly from biomass sources like rice husk and bagasse, is rich in essential nutrients such as silica, calcium, magnesium, and potassium. These properties make it a valuable soil amendment that can improve soil structure, enhance fertility, and increase crop yields (Varshney et al., 2022). Research has shown that fly ash can improve the water retention capacity of soil, reduce soil compaction, and increase aeration, all of which are crucial for plant growth. Additionally, the use of fly ash as a soil conditioner reduces the need for chemical fertilizers, which are often associated with harmful environmental impacts like water eutrophication and greenhouse gas emissions. This application of fly ash in agriculture is a prime example of how technology can turn waste into a resource that supports sustainable agricultural practices. However, the use of fly ash in agriculture requires careful management to prevent the leaching of heavy metals into the soil and groundwater. Technological solutions such as chemical stabilization and encapsulation have been developed to mitigate the risks associated with heavy metals in fly ash. These processes involve treating the fly ash with chemicals that bind the toxic elements, preventing them from leaching into the environment (Wang et al., 2020). Through these innovations, fly ash can be safely used in agriculture, contributing to both soil health and the circular economy.

16.2.2.4 Fly Ash in Wastewater Treatment and Environmental Remediation

Another promising area of technological advancement in ash utilization is its application in wastewater treatment and environmental remediation. Fly ash has been found to possess excellent adsorptive properties due to its high surface area and porous structure, making it effective in removing contaminants from water and air (Nguyen et al., 2023a, 2023b). This has led to the development of fly ash-based adsorbents that can be used in various purification processes. For instance, fly ash has been successfully used to remove heavy metals such as lead, cadmium, and chromium from industrial wastewater. The porous nature of fly ash allows it to trap and immobilize these contaminants, preventing them from entering natural water bodies. In addition to heavy metals, fly ash can also be used to remove organic pollutants like dyes and pesticides, further demonstrating its versatility as an environmental remediation tool. In air pollution control, fly ash has been used in flue gas desulfurization systems to remove sulfur dioxide (SO_2) from the emissions of coal-fired power plants. By acting as a sorbent for SO_2, fly ash helps reduce acid rain and air pollution, while simultaneously being repurposed for beneficial uses (He et al., 2024). These advancements in environmental remediation highlight the potential of fly ash to address pollution issues while contributing to the circular economy. By repurposing fly ash in wastewater treatment and air pollution control, industries can reduce their environmental impact and create a more sustainable waste management system.

16.2.2.5 Advanced Processing Technologies for Fly Ash

The potential of fly ash utilization is further enhanced by the development of advanced processing technologies. These technologies aim to improve the quality of fly ash by removing impurities and enhancing its physical and chemical properties. Mechanical activation, for example, involves grinding fly ash to reduce its particle size, increasing its reactivity and making it more suitable for use in high-performance materials like geopolymers (Naik et al., 2023). Another important advancement is the use of thermal processing to remove unburnt carbon from fly ash. Fly ash with high carbon content is often unsuitable for certain applications, such as in cement production, because it can affect the setting time and strength of the final product. By subjecting fly ash to thermal treatment, the carbon can be oxidized and removed, resulting in a purer form of fly ash that is more versatile and valuable for industrial use. In addition to mechanical and thermal processing, chemical treatments have been developed to modify the surface properties of fly ash, making it more effective in applications like adsorption and catalysis. These treatments involve coating the surface of fly ash particles with chemicals that enhance their reactivity, enabling them to be used in a wider range of applications. By improving the quality and versatility of fly ash through advanced processing technologies, industries can expand the range of products and applications in which fly ash can be used. This not only reduces the need for virgin materials but also contributes to the overall sustainability of industrial processes.

16.2.3 Circular Economy Models for Global Cooperation in Ash Management

The concept of the circular economy has gained widespread recognition as a framework for addressing the environmental and resource challenges of modern economies. Unlike the traditional linear economic model, which is based on a "take, make, dispose" approach, circular economy emphasizes resource efficiency, recycling, and waste minimization. By keeping materials and products in use for as long as possible, the circular economy aims to close the loop on waste, transforming by-products into resources. In the context of industrial waste like fly ash, circular economy models are particularly relevant. Fly ash, a by-product of coal and biomass-based power generation, not only poses significant environmental challenges but also offers opportunities for sustainable reuse.

16.2.3.1 The Circular Economy and Its Relevance to Ash Management

The circular economy is centered around principles such as designing out waste, keeping materials in use, and regenerating natural systems. These principles are especially applicable to ash management, where large volumes of waste are generated through industrial processes. The production of CFA, rice husk ash (RHA), and bagasse ash (BA) has increased significantly due to the growing demand for energy and industrial materials (Fořt et al., 2021). Traditionally, these ashes were disposed of in landfills, leading to pollution and resource loss. However, through circular economy models, fly ash can be repurposed into valuable products, reducing environmental impact while contributing to resource efficiency. Circular economy models for ash management focus on reusing and recycling fly ash in various industrial applications, such as construction, agriculture, and manufacturing. By integrating these models into global supply chains, industries can reduce their reliance on virgin materials, lower carbon emissions, and minimize the environmental footprint of their operations. At the same time, international cooperation is essential for scaling these efforts and ensuring that best practices and technologies are shared across borders.

16.2.3.2 The Role of Global Cooperation in Circular Economy Models

Global cooperation is critical for the successful implementation of circular economy models in ash management. Many countries face similar challenges related to industrial waste, particularly fly ash, which is produced in large quantities by coal-fired power plants and biomass-based energy generation. Many countries have developed policies and regulations to manage fly ash; the level of utilization and technological advancement vary widely. Global cooperation can help harmonize standards, share technological innovations, and promote best practices in fly ash reuse. International organizations such as the United Nations Environment Programme (UNEP) and the International Energy Agency (IEA) have recognized the importance of global cooperation in promoting circular economy principles. These organizations have launched initiatives to support countries in adopting sustainable practices for ash management, including recycling, reuse, and pollution prevention. By facilitating knowledge sharing and technology transfer, global cooperation can help countries accelerate the adoption of circular economy models and improve their ash management systems.

16.2.3.3 Circular Economy Models for Ash Utilization

A range of circular economy models can be applied to ash management, each offering opportunities for global cooperation. These models include resource recovery, industrial symbiosis, and closed-loop systems, all of which emphasize the continuous use of materials and the minimization of waste. By applying these models to ash management, industries can create sustainable solutions that benefit both the environment and the economy.

16.2.3.3.1 Resource Recovery and Reuse

One of the core principles of the circular economy is the recovery and reuse of materials that would otherwise be discarded as waste. In the context of fly ash, resource recovery involves extracting valuable components from the ash and repurposing them in industrial applications. For example, CFA is rich in silica, alumina, and calcium, which can be used as raw materials in the construction industry to produce cement, concrete, and bricks (Kelechi et al., 2022). The reuse of fly ash in construction not only reduces the demand for virgin materials but also lowers the carbon footprint of building materials by replacing traditional energy-intensive components like Portland cement. International cooperation can play a vital role in scaling up resource recovery efforts. For instance, countries that have successfully implemented fly ash recycling programs, such as China and the EU, can share their knowledge and technologies with countries that are still developing their ash management systems. This collaboration can help improve the efficiency and effectiveness of resource recovery initiatives, ensuring that more countries benefit from the environmental and economic advantages of fly ash reuse.

16.2.3.3.2 Industrial Symbiosis

Another important circular economy model is industrial symbiosis, which involves the sharing of resources between different industries to optimize the use of materials and reduce waste. In an industrial symbiosis system, the by-products of one industry are used as raw materials for another, creating a closed-loop system in which waste is minimized. Fly ash is an ideal candidate for industrial symbiosis, as it can be used in a variety of industries, including construction, agriculture, and manufacturing (Khan et al., 2023). For example, power plants that produce fly ash can collaborate with cement manufacturers to supply them with the ash needed for concrete production. Similarly, fly ash can be used as a soil amendment in agriculture, improving soil structure and fertility while reducing the need for chemical fertilizers. Industrial symbiosis can also extend to other sectors, such as ceramics and plastics, where fly ash can be used as a filler material. By fostering cooperation between industries, industrial symbiosis helps create more efficient and sustainable supply chains. Global cooperation is essential for promoting industrial symbiosis on a larger scale. Governments can facilitate partnerships between industries by providing incentives for companies to collaborate on waste management and resource recovery. International agreements and trade policies can also support the exchange of by-products like fly ash between countries, ensuring that industries can access the materials they need for sustainable production.

16.2.3.3.3 Closed-Loop Systems

Closed-loop systems are a key component of the circular economy, in which products are designed to be reused, remanufactured, or recycled at the end of their life cycle. In a closed-loop system for fly ash management, the ash is continuously repurposed in industrial applications, ensuring that it remains in use rather than being disposed of in landfills. This model not only reduces waste but also promotes resource efficiency by keeping valuable materials in circulation. Fly ash can be integrated into closed-loop systems in several ways. For example, in the construction industry, fly ash-based concrete can be reused in new construction projects after a building is demolished. Similarly, fly ash used as a soil amendment can be recovered and reapplied in agricultural systems, maintaining soil health and productivity over time (Đolić et al., 2021). The development of closed-loop systems for fly ash requires coordination between industries, governments, and researchers to ensure that the necessary technologies and infrastructure are in place. Global cooperation is crucial for developing closed-loop systems, as it enables countries to share best practices and technologies for ash recovery and reuse. International organizations can play a leading role in facilitating these efforts by providing funding and technical assistance to countries that are working to implement closed-loop systems. Additionally, international standards for fly ash quality and reuse can help ensure that industries have access to the materials they need to close the loop on waste.

16.2.3.4 Challenges to Global Cooperation in Circular Economy Models

While global cooperation offers significant opportunities for advancing circular economy models in ash management, several challenges must be addressed to ensure successful implementation. One of the primary challenges is the disparity in regulatory frameworks and enforcement mechanisms across different countries. While some countries have stringent regulations for fly ash management, others lack the necessary policies to promote recycling and reuse. This inconsistency can create barriers to international cooperation, as industries may face different requirements for ash utilization depending on the country in which they operate. Another challenge is the lack of infrastructure and technology for processing and repurposing fly ash in many countries. Developing countries may struggle to invest in the technologies needed to convert fly ash into valuable products, limiting their ability to participate in circular economy models (Kusuma et al., 2022). To overcome this challenge, global cooperation must include efforts to provide financial and technical support to countries that need assistance in developing their ash management systems. Finally, market demand for fly ash-based products can vary significantly between countries. In some regions, there may be limited awareness or interest in using fly ash in construction, agriculture, or other industries, making it difficult to scale up recycling efforts. Global cooperation can help address this challenge by promoting the benefits of fly ash reuse and encouraging the development of new markets for ash-based products.

16.2.4 Public-Private Partnerships for Sustainable Ash Management in a Circular Economy

PPPs are emerging as an effective approach to address complex environmental challenges, particularly in the realm of industrial waste management. By bringing together the resources, expertise, and innovation of both the public and private sectors, PPPs offer a collaborative framework to promote sustainable practices and implement circular economy models. In the context of ash management, PPPs play a vital role in ensuring that fly ash, a by-product of coal and biomass-based power plants, is effectively managed, recycled, and repurposed.

16.2.4.1 The Importance of Public-Private Partnerships in Ash Management

The increasing production of CFA and other industrial by-products like RHA and BA has created a pressing need for innovative solutions that go beyond traditional waste disposal methods. Governments and industries face significant challenges in managing the environmental impact of ash, which can contaminate air, water, and soil if not properly handled (Khajuria et al., 2022). PPPs provide a mechanism for addressing these challenges by leveraging the strengths of both sectors to create sustainable ash management solutions. In a PPP, the public sector (government agencies) collaborates with the private sector (industries, corporations, and non-governmental organizations) to develop and implement projects that serve the public interest. These partnerships can take various forms, ranging from joint ventures and contractual agreements to long-term collaborations on specific initiatives. For ash management, PPPs can focus on areas such as resource recovery, recycling, infrastructure development, and environmental protection, all of which are key components of a circular economy.

16.2.4.2 How Public-Private Partnerships Work in Ash Management

PPPs in ash management typically involve the government providing regulatory oversight, policy frameworks, and financial incentives, while the private sector contributes technical expertise, innovation, and investment. Together, these two sectors work to create solutions that not only reduce the environmental impact of ash but also transform it into a valuable resource for industrial and commercial use. The collaborative nature of PPPs allows for the development of integrated systems that align with circular economy principles, where waste materials are continuously repurposed and reintegrated into supply chains. An example of how PPPs can function in ash management is through the development of ash recycling facilities. In such partnerships, the government can provide land, regulatory approvals, and financial subsidies, while private companies invest in the construction and operation of facilities that process fly ash into usable products such as cement, bricks, and construction materials. By sharing the risks and rewards, both sectors benefit from the collaboration, while society gains from the reduced environmental footprint and increased resource efficiency (Kumar et al., 2021).

Another example is infrastructure projects where fly ash is used as a construction material for roads, embankments, and mine backfilling. In this case, the government might fund the initial project, while private construction firms incorporate fly ash-based materials into the design and execution. This approach not only diverts fly ash from landfills but also creates a market for ash-based products, driving economic growth while reducing the reliance on virgin materials.

16.2.4.3 Benefits of Public-Private Partnerships in Ash Management

PPPs offer several benefits in the context of ash management, particularly when it comes to promoting sustainability and the circular economy (Cui et al., 2020). These benefits include:

a. **Leveraging Resources and Expertise:** PPPs allow governments to tap into the technical expertise and financial resources of the private sector, which is often better equipped to handle large-scale industrial projects. Private companies bring innovative technologies and processes that can improve the efficiency and effectiveness of ash management. For example, private companies specializing in waste management may have access to advanced processing technologies that can transform fly ash into high-performance construction materials.
b. **Risk-Sharing:** By working together, the public and private sectors can share the risks associated with ash management projects. This risk-sharing arrangement makes it easier to invest in new technologies and infrastructure, as the financial burden is distributed between both sectors. For instance, a PPP for building a fly ash processing plant might involve the government covering the initial capital costs, while the private sector takes responsibility for the operation and maintenance of the facility.
c. **Accelerating Innovation:** PPPs foster innovation by creating an environment where both sectors can experiment with new technologies and business models. For ash management, this might involve developing new recycling techniques, finding novel uses for fly ash in manufacturing, or improving the efficiency of ash collection and transportation systems. The private sector's focus on profitability drives the need for continuous improvement, while the public sector ensures that these innovations align with environmental regulations and sustainability goals.

d. **Creating New Markets for Ash-Based Products:** PPPs can play a crucial role in developing new markets for ash-based products, such as fly ash concrete, bricks, and tiles. Governments can promote the use of ash-based materials in public infrastructure projects, providing a steady demand for these products and encouraging private companies to invest in ash-processing technologies. At the same time, the private sector can help develop marketing strategies and distribution channels to expand the use of ash-based products in commercial construction and other industries.
e. **Meeting Environmental Goals:** By promoting the circular economy through ash management, PPPs contribute to national and international environmental goals, such as reducing carbon emissions, conserving natural resources, and preventing pollution. Fly ash recycling reduces the need for virgin materials like limestone and clay, lowering the environmental impact of construction projects. Additionally, by diverting fly ash from landfills, PPPs help prevent air and water pollution, contributing to cleaner, healthier ecosystems.

16.2.4.4 Challenges of Public-Private Partnerships in Ash Management

Despite the potential benefits of PPPs in ash management, several challenges can hinder their success. These challenges include:

a. **Regulatory Complexity:** Ash management involves navigating a complex regulatory landscape, with different rules governing the disposal, transportation, and reuse of fly ash in various regions (Putilova and Zroichikov, 2024). For PPPs to succeed, governments must establish clear and consistent regulations that encourage the sustainable use of ash while protecting public health and the environment. However, inconsistent policies or bureaucratic hurdles can slow down projects, leading to delays and increased costs.
b. **Alignment of Objectives:** In a PPP, the public and private sectors often have different objectives. While the government may prioritize environmental sustainability and public welfare, private companies are primarily driven by profit. Achieving a balance between these goals can be challenging, especially if the profitability of ash-based products is uncertain or if the private sector perceives the investment as too risky. To address this challenge, governments may need to provide financial incentives, such as tax breaks or subsidies, to encourage private investment in ash management.
c. **Long-Term Commitment:** Successful PPPs require long-term commitments from both the public and private sectors. Ash management projects, particularly those focused on infrastructure development and resource recovery, often involve significant upfront investments and long payback periods. Maintaining momentum over the long term can be difficult, especially if political or economic conditions change. Governments must ensure that contracts and agreements are structured to provide stability and continuity, even in the face of changing circumstances.
d. **Technological and Market Uncertainty:** While technological advancements have made it possible to repurpose fly ash in various industries, there is

still uncertainty about the long-term market demand for ash-based products (Tosti et al., 2020). For example, while fly ash concrete and bricks have proven to be environmentally beneficial, their adoption in some regions remains limited due to a lack of awareness or resistance from traditional industries. To overcome this challenge, governments and private companies must work together to educate consumers and create demand for sustainable products.

16.2.5 Sustainable Transportation and Distribution in Ash Management: A Key Component of the Circular Economy

Transportation and distribution are integral parts of the supply chain for any industry, and their environmental impact is substantial. When it comes to ash management, the challenge of transporting and distributing CFA, RHA, and BA in a sustainable manner becomes crucial. Ash is generated as a by-product of coal and biomass-based power plants, and its management involves not only finding suitable applications for reuse but also ensuring that its transportation does not negate the environmental benefits of these applications. Sustainable transportation and distribution systems for ash management are key to realizing the goals of a circular economy by reducing carbon emissions, conserving energy, and optimizing resource use (Sultana et al., 2021).

16.2.5.1 The Importance of Sustainable Transportation in Ash Management

Transportation accounts for a significant portion of the carbon footprint in the industrial sector, especially when moving heavy materials like ash over long distances. As industries strive to adopt circular economy principles, one of the key areas of focus is minimizing the environmental impact of transportation. This is particularly important in ash management, where the distribution of ash from power plants to processing or reuse facilities plays a vital role in determining the overall sustainability of the operation. Sustainable transportation in ash management involves reducing fuel consumption, optimizing routes, and using low-emission vehicles. It also includes strategies for minimizing the distance between ash generation sites and end-users to reduce travel time and energy consumption. Sustainable distribution systems for ash can support the circular economy by ensuring that the environmental benefits of reusing ash are not offset by the carbon emissions associated with its transportation.

16.2.5.2 Challenges in Transporting Fly Ash and Other Ash Types

Transporting fly ash presents several unique challenges due to the material's characteristics and the nature of the industries involved (Rastogi and Paul, 2020). These challenges include:

a. **Distance Between Ash Generation Sites and End-Users:** One of the biggest challenges in ash transportation is the distance between power plants that generate ash and industries that can utilize it. For example, in construc tion, ash may need to be transported to cement factories, brick manufacturers, or road construction sites, which may be located far from the power plant. The longer the distance, the greater the energy consumption and carbon emissions associated with transportation.

b. **Ash Characteristics:** Fly ash is a fine particulate material that can pose health and environmental risks if not handled properly during transportation. If fly ash is transported in open vehicles or inadequately sealed containers, there is a risk of dust emissions, which can contribute to air pollution and pose respiratory health risks. Additionally, the high volume and weight of ash require specialized handling and transportation equipment.
c. **Infrastructure Limitations:** In many regions, the infrastructure for transporting industrial by-products like fly ash is underdeveloped. Poor road conditions, limited access to rail or water transportation, and inadequate loading and unloading facilities can create bottlenecks in the transportation process, leading to delays, increased costs, and higher emissions. Furthermore, in developing countries, the lack of investment in sustainable transport infrastructure exacerbates these challenges.
d. **Cost of Sustainable Transportation:** The cost of implementing sustainable transportation solutions, such as investing in electric or low-emission vehicles, optimizing logistics, and developing infrastructure for multimodal transport, can be high. While these solutions offer long-term environmental and economic benefits, the upfront investment may be a barrier for some companies and governments.

16.2.5.3 Sustainable Transportation Solutions for Ash Management

To address the challenges of ash transportation and distribution, several innovative solutions can be implemented to reduce carbon emissions, improve efficiency, and support the goals of the circular economy (Behera et al., 2021). These solutions include:

a. **Multimodal Transportation:** Multimodal transportation involves using more than one mode of transport such as road, rail, and water to move materials efficiently. For ash management, multimodal transport can significantly reduce carbon emissions by shifting from high-emission road transport to more sustainable options like rail or water. Rail and water transportation have lower per-ton-kilometer emissions compared to road transport, making them ideal for long-distance ash transportation. For example, in India, where large amounts of fly ash are generated by coal-based thermal power plants, using rail networks to transport ash to cement factories or road construction sites can reduce the reliance on trucks and lower emissions. Similarly, transporting ash by barge along rivers can further reduce transportation-related carbon footprints. Governments and private companies can collaborate to invest in the necessary infrastructure, such as loading and unloading facilities for rail and water transport.
b. **Electric and Low-Emission Vehicles:** Another solution for sustainable ash transportation is the adoption of electric vehicles (EVs) and low-emission trucks powered by compressed natural gas (CNG) or hydrogen. While the use of such vehicles is still in its early stages in many industries, they offer a promising way to reduce the environmental impact of ash transportation. EVs produce zero tailpipe emissions, while CNG and hydrogen-powered

vehicles have significantly lower carbon emissions compared to diesel-powered trucks. Private companies involved in ash transportation can invest in electric or hybrid truck fleets, particularly for short-distance transport between power plants and nearby construction sites or factories. Governments can support this transition by providing financial incentives, such as tax breaks or subsidies, for companies that invest in sustainable transportation technologies.

c. **Route Optimization and Logistics Management:** Optimizing transportation routes and logistics is another critical strategy for reducing the environmental impact of ash transportation. Route optimization involves using advanced software to analyze traffic patterns, fuel consumption, and distance to determine the most efficient routes for transporting ash. By minimizing travel time and fuel consumption, companies can reduce both costs and carbon emissions. In addition to route optimization, logistics management systems can help companies coordinate the movement of ash between multiple generation sites and end-users. For example, fly ash from several power plants can be consolidated and transported to a single processing facility, reducing the number of trips and optimizing truck or rail capacity.
d. **Localizing Ash Reuse:** One of the most effective ways to reduce the environmental impact of ash transportation is to localize the reuse of ash by encouraging industries located near power plants to use the ash. By promoting local markets for ash-based products, such as fly ash cement, bricks, and concrete, transportation distances can be minimized, significantly reducing carbon emissions. Governments can play a key role in supporting local markets for ash reuse by incentivizing industries located near power plants to adopt ash-based materials in their production processes. Additionally, policies that require local construction projects to use ash-based products can create demand for fly ash and reduce the need for long-distance transportation.
e. **Bulk Transportation and Containerization:** Bulk transportation, where large volumes of ash are transported in a single trip, can help reduce the number of trips required and lower fuel consumption. This is particularly useful for transporting ash to cement factories, where large quantities are needed. Containerization, where ash is transported in sealed containers, can also minimize the risk of dust emissions during transport. By using bulk transport vehicles or specialized containers designed to handle fly ash, companies can improve the efficiency of transportation while also ensuring that environmental safety standards are met.

16.2.6 Fly Ash in Green Infrastructure: A Pathway to Sustainability

Green infrastructure has become a crucial element in modern urban development, offering solutions to the environmental challenges posed by urbanization, climate change, and resource depletion. At the heart of green infrastructure is the concept of integrating natural processes and sustainable materials into the design and construction of urban systems to create more resilient, eco-friendly, and energy-efficient cities. One material that is increasingly playing a significant role in green infrastructure is fly ash, a by-product of coal combustion in power plants. Fly ash has proven to be a versatile and sustainable alternative in various construction applications, particularly in green infrastructure projects, where reducing carbon footprints, conserving resources, and enhancing environmental quality are priorities (Shukla et al., 2023).

16.2.6.1 The Role of Fly Ash in Green Infrastructure

Fly ash is one of the most abundant industrial by-products in the world, primarily generated by coal-fired power plants. Although historically considered a waste product, fly ash has emerged as a valuable resource for construction due to its pozzolanic properties, which allow it to replace traditional materials like Portland cement in concrete production. This substitution not only reduces the environmental impact of construction but also aligns with the principles of green infrastructure by promoting resource efficiency, waste reduction, and sustainable development. Green infrastructure focuses on building systems that mimic natural processes, such as stormwater management, urban cooling, and carbon sequestration. Fly ash fits into this model by offering an environmentally friendly alternative to conventional building materials, lowering the overall carbon footprint of construction projects. Additionally, its reuse in construction diverts fly ash from landfills, reducing the environmental risks associated with ash disposal, such as air and water pollution.

16.2.6.2 Environmental Benefits of Using Fly Ash in Green Infrastructure

The use of fly ash in green infrastructure offers several key environmental benefits, including:

a. **Reduction in Carbon Emissions:** One of the most significant benefits of using fly ash in green infrastructure is its ability to reduce carbon emissions in construction. Cement production is one of the largest industrial sources of carbon dioxide (CO_2) emissions, accounting for approximately 8% of global emissions. By substituting a portion of cement with fly ash in concrete, emissions associated with the production of cement are reduced. Fly ash requires less energy to produce than cement, and its use in concrete can lower the overall carbon footprint of construction projects.
b. **Resource Conservation:** Fly ash helps conserve natural resources by reducing the demand for raw materials like limestone and clay, which are traditionally used in cement production. By incorporating fly ash into construction materials, industries can decrease the need for mining and quarrying activities that deplete natural resources and disturb the ecosystem. The use of fly ash in green infrastructure is a prime example of how industrial by-products can be repurposed to promote resource efficiency and sustainable material use.
c. **Waste Reduction:** The reuse of fly ash in construction diverts large quantities of ash from landfills, where it would otherwise pose environmental risks. Fly ash contains trace elements such as arsenic, lead, and mercury, which can leach into soil and groundwater if not properly managed. By incorporating fly ash into green infrastructure, these pollutants are

stabilized within the concrete matrix, reducing the potential for environmental contamination.

d. **Improved Durability and Longevity of Structures:** Fly ash enhances the durability and longevity of concrete structures, which are essential for green infrastructure projects that aim to provide long-term environmental benefits. Concrete made with fly ash is more resistant to cracking, sulfate attack, and alkali-silica reaction, which can extend the life of buildings, roads, and other infrastructure components. This increased durability reduces the need for maintenance and reconstruction, further minimizing the environmental impact of construction over time.

16.2.6.3 Economic Benefits of Using Fly Ash in Green Infrastructure

In addition to its environmental advantages, the use of fly ash in green infrastructure offers significant economic benefits, making it an attractive option for governments, industries, and developers (Schneider et al., 2020). These benefits include:

a. **Cost Savings in Material Production:** Fly ash is often available at a lower cost than traditional construction materials like cement and asphalt, making it an economically viable option for large-scale infrastructure projects. The use of fly ash in concrete and other building materials can result in cost savings for developers, particularly in regions where fly ash is readily available. Additionally, fly ash reduces the need for energy-intensive material production processes, further lowering construction costs.
b. **Lower Maintenance and Lifecycle Costs:** Structures made with fly ash-enhanced materials are more durable and resistant to environmental degradation, resulting in lower maintenance and repair costs over time. This is particularly important for green infrastructure projects, where long-term sustainability and cost-efficiency are priorities. The reduced need for maintenance also minimizes the environmental impact of ongoing construction activities, contributing to the overall sustainability of the project.
c. **Job Creation and Economic Development:** The development of fly ash-based industries, such as fly ash brick manufacturing and ash recycling facilities, can create jobs and stimulate economic growth. By investing in green infrastructure projects that use fly ash, governments can promote local industries and support economic development while also addressing environmental challenges. This creates a win-win scenario where both the environment and the economy benefit from the adoption of sustainable practices.

16.2.6.4 Future Potential of Fly Ash in Green Infrastructure

As the demand for sustainable urban development continues to grow, the potential for fly ash in green infrastructure is vast. Ongoing research and development efforts are exploring new ways to incorporate fly ash into innovative materials and technologies that support the goals of the circular economy (Senadheera et al., 2024). From 3D-printed fly ash concrete structures to advanced water filtration systems, the possibilities for using fly ash in green infrastructure are expanding. Additionally, as countries around the world work to meet climate goals and reduce greenhouse gas emissions, the role of fly ash in decarbonizing the construction industry will become even more critical. Governments, industries, and researchers must continue to collaborate on developing policies, technologies, and markets that support the sustainable use of fly ash in green infrastructure projects.

16.3 CONCLUSION

By adopting strategic frameworks that promote enhanced regulatory frameworks, technological advancements, PPPs, and sustainable transportation, industries can significantly reduce the environmental impact associated with fly ash. The research underscores the need for effective regulatory oversight to ensure the safe disposal and reuse of fly ash while encouraging innovation in processing technologies to expand its applications, particularly in green infrastructure projects like sustainable construction and water management systems. PPPs are identified as a critical driver for advancing fly ash utilization, facilitating investments and collaboration between industries and governments to develop recycling infrastructure. Additionally, the role of sustainable transportation is crucial, as optimizing logistics and adopting low-emission vehicles can further reduce the carbon footprint of fly ash management. This chapter also highlights the potential of fly ash in green infrastructure, where it can be used in eco-friendly building materials, permeable pavements, and stormwater management systems, promoting resource efficiency and reducing reliance on virgin materials. By integrating fly ash into urban development, cities can enhance their resilience to climate change while contributing to sustainable resource management.

REFERENCES

Awino, F. B., & Apitz, S. E. (2024). Solid waste management in the context of the waste hierarchy and circular economy frameworks: An international critical review. *Integrated Environmental Assessment and Management*, *20*(1), 9–35.

Behera, U., Das, S. K., Mishra, D. P., Parhi, P. K., & Das, D. (2021). Sustainable transportation, leaching, stabilization, and disposal of fly ash using a mixture of natural surfactant and sodium silicate. *ACS Omega*, *6*(35), 22820–22830.

Chen, Y., Fan, Y., Huang, Y., Liao, X., Xu, W., & Zhang, T. (2024). A comprehensive review of toxicity of coal fly ash and its leachate in the ecosystem. *Ecotoxicology and Environmental Safety*, *269*, 115905.

Cui, C., Liu, Y., Xia, B., Jiang, X., & Skitmore, M. (2020). Overview of public-private partnerships in the waste-to-energy incineration industry in China: Status, opportunities, and challenges. *Energy Strategy Reviews*, *32*, 100584.

Darmansyah, D., You, S. J., & Wang, Y. F. (2023). Advancements of coal fly ash and its prospective implications for sustainable materials in Southeast Asian countries: A review. *Renewable and Sustainable Energy Reviews*, *188*, 113895.

Đolić, M., Karanac, M., Radovanović, D., Umićević, A., Kapidžić, A., Veličković, Z.,…, & Kamberović, Ž. (2021). Closing the loop: As (V) adsorption onto goethite impregnated coal-combustion fly ash as integral building materials. *Journal of Cleaner Production*, *303*, 126924.

Fořt, J., Šál, J., Ševčík, R., Doleželová, M., Keppert, M., Jerman, M.,…, & Černý, R. (2021). Biomass fly ash as an alternative to coal fly ash in blended cements: Functional aspects. *Construction and Building Materials*, *271*, 121544.

Ghodeswar, A., & Oliver, M. E. (2022). Trading one waste for another? Unintended consequences of fly ash reuse in the Indian electric power sector. *Energy Policy*, *165*, 112940.

He, Y., Jia, S., Yi, H., Tang, X., Yu, Q., Gao, F.,…, & Zhao, S. (2024). Utilization of steel slag in air pollution and greenhouse gas emission reduction—Application, mechanism and challenge: A review. *Journal of Environmental Chemical Engineering*, 12(5), 114090.

Kelechi, S. E., Adamu, M., Uche, O. A. U., Okokpujie, I. P., Ibrahim, Y. E., & Obianyo, I. I. (2022). A comprehensive review on coal fly ash and its application in the construction industry. *Cogent Engineering*, *9*(1), 2114201.

Khajuria, A., Atienza, V. A., Chavanich, S., Henning, W., Islam, I., Kral, U.,…, & Li, J. (2022). Accelerating circular economy solutions to achieve the 2030 agenda for sustainable development goals. *Circular Economy*, *1*(1), 100001.

Khan, Z. A., Chowdhury, S. R., Mitra, B., Mozumder, M. S., Elhaj, A. I., Salami, B. A.,…, & Rahman, S. M. (2023). Analysis of industrial symbiosis case studies and its potential in Saudi Arabia. *Journal of Cleaner Production*, *385*, 135536.

Kumar, G., Meena, P., & Difrancesco, R. M. (2021). How do collaborative culture and capability improve sustainability? *Journal of Cleaner Production*, *291*, 125824.

Kusuma, R. T., Hiremath, R. B., Rajesh, P., Kumar, B., & Renukappa, S. (2022). Sustainable transition towards biomass-based cement industry: A review. *Renewable and Sustainable Energy Reviews*, *163*, 112503.

Labianca, C., Farina, I., Colangelo, F., Singh, N., Todaro, F., De Gisi, S.,…, & Tsang, D. C. (2024). Life cycle and economic assessment on different utilization and treatment strategies of combustion and incineration residues. In *Treatment and Utilization of Combustion and Incineration Residues*, (Ed) Lei Wang, Dan Tsang, Jianhua Yan, (pp. 551–565). Elsevier.

Prairie Research Institute, 2020, Coal Ash Response Team final report: Champaign, Prairie Research Institute, 61 p.,https://prairie.illinois.edu/illinois-issues/coal-ash/

Naik, S. S., Pandey, S., Pawar, S. N., Shinde, B. H., & Prakash, C. (2023). Innovative and interactive methodology for development of geopolymer mortar using fly ash of agricultural waste briquettes. *International Journal on Interactive Design and Manufacturing (IJIDeM)*, *18*(8), 5643-5651.

Nguyen, H. H. T., Nguyen, H. T., Ahmed, S. F., Rajamohan, N., Yusuf, M., Sharma, A.,…, & Vo, D. V. N. (2023a). Emerging waste-to-wealth applications of fly ash for environmental remediation: A review. *Environmental Research*, *227*, 115800.

Nguyen, T. H., Pham, Q. V., Nguyen, T. P. M., Vu, V. T., Do, T. H., Hoang, M. T.,…, & Minh, T. B. (2023b). Distribution characteristics and ecological risks of heavy metals in bottom ash, fly ash, and particulate matter released from municipal solid waste incinerators in northern Vietnam. *Environmental Geochemistry and Health*, *45*(5), 2579–2590.

Ondrasek, G., Kovačić, M. B., Carević, I., Štirmer, N., Stipičević, S., Udiković-Kolić, N.,…, & Rengel, Z. (2021). Bioashes and their potential for reuse to sustain ecosystem services and underpin circular economy. *Renewable and Sustainable Energy Reviews*, *151*, 111540.

Orozco, C., Babel, S., Tangtermsirikul, S., & Sugiyama, T. (2024). Comparison of environmental impacts of fly ash and slag as cement replacement materials for mass concrete and the impact of transportation. *Sustainable Materials and Technologies*, *39*, e00796.

Oyejobi, D. O., Firoozi, A. A., Fernandez, D. B., & Avudaiappan, S. (2024). Integrating circular economy principles into concrete technology: Enhancing sustainability through industrial waste utilization. *Results in Engineering*, 24, 102846.

Putilova, I. V., & Zroichikov, N. A. (2024). Review of regulatory and technical documentation in the field of ash and slag handling at TPPs. *Thermal Engineering*, *71*(6), 471–481.

Rajasugunasekar, D., & Devi, R. S. (2022). Legal framework in India to protect the environment. *Ecological Environment: A New Perspective*, 468-486.

Ram, A. K., & Mohanty, S. (2022). State of the art review on physiochemical and engineering characteristics of fly ash and its applications. *International Journal of Coal Science &* Technology, 9(1), 9.

Rastogi, A., & Paul, V. K. (2020). A critical review of the potential for fly ash utilisation in construction-specific applications in India. *Environmental Research, Engineering and M*anagement, 76(2), 65–75.

Schneider, P., Schwerdt, S., Schulz, K., Fiebig, S., & Mirschel, D. (2020). Feasibility of substitute building materials for circular use in urban green infrastructure: Dresden Nexus Conference 2020–Session 4–Circular Economy for Building with Secondary Construction Materials to Minimise Resource use and Land use. *Civil Engineering Design*, *2*(5–6), 159–168.

Seidler, M., & Malloy, K. (2020). A comprehensive survey of coal ash law and commercialization: Its environmental risks, disposal regulation, and beneficial use markets. https://doi.org/10.2172/1869929

Senadheera, S. S., Withana, P. A., Lim, J. Y., You, S., Chang, S. X., Wang, F.,…, & Ok, Y. S. (2024). Carbon negative biochar systems contribute to sustainable urban green infrastructure: A critical review. *Green Chemistry*. ***Green Chem.***, 2024,**26**, 10634-10660

Shukla, B. K., Gupta, A., Gowda, S., & Srivastav, Y. (2023). Constructing a greener future: A comprehensive review on the sustainable use of fly ash in the construction industry and beyond. *Materials Today: Proceedings, 93*, 257-264.

Sultana, S., Ahsan, S., Tanvir, S., Haque, N., Alam, F., Yellishetty, M. (2021). Coal Fly Ash Utilisation and Environmental Impact. In: Jyothi, R.K., Parhi, P.K. (eds) *Clean Coal Technologies*. Springer, Cham. https://doi.org/10.1007/978-3-030-68502-7_15

Tosti, L., van Zomeren, A., Pels, J. R., Damgaard, A., & Comans, R. N. (2020). Life cycle assessment of the reuse of fly ash from biomass combustion as secondary cementitious material in cement products. *Journal of Cleaner Production*, *245*, 118937.

Varshney, A., Dahiya, P., Sharma, A., Pandey, R., & Mohan, S. (2022). Fly ash application in soil for sustainable agriculture: an Indian overview. *Energy, Ecology and E*nvironment, 7(4), 340–357.

Wang, N., Sun, X., Zhao, Q., Yang, Y., & Wang, P. (2020). Leachability and adverse effects of coal fly ash: A review. *Journal of Hazardous Materials*, *396*, 122725.

Index